MEMBRANES AND DISEASE

Membranes and Disease

Editors:

Liana Bolis, M.D.
Professor of General Physiology
School of Pharmacy
University of Milan
Milan, Italy

Joseph F. Hoffman, Ph.D.
Eugene Higgins Professor and
Chairman of Physiology
Department of Physiology
Yale University School of Medicine
New Haven, Connecticut

Alexander Leaf, M.D.
Honorary Chairman
Jackson Professor of Clinical Medicine
Harvard Medical School
and Chief, Medical Services
Massachusetts General Hospital
Boston, Massachusetts

Raven Press ▪ New York

Raven Press, 1140 Avenue of the Americas, New York, New York 10036

Made in the United States of America

International Standard Book Number 0–89004–082–6
Library of Congress Catalog Card Number 75–30235

Preface

This volume contains contributions presented at the biennial International Conference on Biological Membranes, which was held at Crans-sur-Sierre, Switzerland, June 15–21, 1975. The topic of the Conference this year was Membranes and Disease.

During the past 25 years a great deal has been learned about the structure and function of membranes of cells and subcellular organelles. Although their basic molecular structure is still a matter for further research, the importance of their functions to the life of the cell and organism is now well appreciated. In addition to simply providing partitions or separations between the cell and its environment or of compartmentalizing the cell into individual organelles, membranes are the site of many important activities. The flow of substrates past membrane-bound enzymes determines both the rate and directions of the chemical reactions upon which life depends. By their permeability characteristics and with the contribution of specific transport mechanisms, membranes can regulate reactions by providing or withholding access of molecules to reactive sites within the cell. Membranes provide the wrappings for both endocytosis and exocytosis (phagocytosis and secretion), respectively. Membranes further contain in their surface coats the communications systems that keep the individual cells in a tissue or organ functioning in a cooperative manner. There are the receptor sites for specific hormones and drugs. Other recognition sites contain antigens and the identity tags that prevent the body's surveillance system from destroying the cell as an outsider, whereas still other membrane sites recognize neighboring cells and maintain respect for the "civil rights of neighbors," by constraining the innate capacity of the cell to proliferate.

With so many important functions in health it is not surprising that aberrations in one or several factors may also occur in disease states—in fact, be the basis of disease.

Investigators have tended to become preoccupied with the membranes of the particulate cell or organ system in which they are interested, out of which a kind of organ-specific isolation has developed. The Conference was organized on the assumption that techniques and concepts that had developed in this isolationism might well prove beneficial when applied to other cells and tissues. In fact, the communication at the Conference among various groups of investigators working in the different areas was excellent. It is hoped that this book will serve similarly to enhance communication but with a larger audience of clinical scientists engaged in studying diseases—genetic and acquired—as well as physiologists and biochemists studying basic aspects of membrane structure and function, since Nature often reveals the normal through the aberrations created in disease.

Alexander Leaf, Honorary Chairman
Boston, Massachusetts
(October 1975)

Acknowledgment

We wish to express our thanks for support by:

Italian Ministry of Public Education
Ministry of Foreign Affairs
Cassa di Risparmio delle Province Lambarde
The Laboratory LIRCA
Phillips Italy

Academy of Medical Science
Etat du Valais
Commune de Chermignon, Lens, Randogne
Ecole des Roches
Migros
Golf Club de Crans
Société de Banques Suisses
Placette
Pharmacie Rouvinez, Crans
Maison Orsat, Martigny
Maison Favre, Sion

Hoffmann-La Roche, Inc.
Amicon Corporation
Searle Laboratories
Burroughs Wellcome Company
The Upjohn Company
Siemens
Merck & Company, Inc.
National Institute of General Medical Sciences, NIH, DHEW
National Institute of Arthritis and Metabolic Diseases, NIH, DHEW

Contents

Red Blood Cells

Subcellular Organelles and Plasma Membranes

Muscle Membranes

Gastrointestinal Membranes

Kidney Membranes

Nerve Membranes

Contributors

Stanley H. Appel
Division of Neurology
Duke University Medical Center
Durham, North Carolina 27710

G. D. Aurbach
Metabolic Diseases Branch
National Institute of Arthritis, Metabolism, and Digestive Diseases
National Institutes of Health
Bethesda, Maryland 20014

A. G. Awad
Department of Neurochemistry
Clarke Institute of Psychiatry
Toronto, Ontario, Canada

Klaus G. Bensch
Stanford University School of Medicine
Stanford, California 94305

Marilena Bertini
Department of Histology and Embryology
University of Torino
School of Medicine
10126 Torino, Italy

Martin Blank
Department of Physiology
Columbia University College of Physicians and Surgeons
New York, New York 10032

Floyd E. Bloom
Laboratory of Neuropharmacology
Division of Special Mental Health Research, IRP
National Institute of Mental Health
Saint Elizabeths Hospital
Washington, D.C. 20032

H. Harm Bodemann
Medizinische Universitätsklinik
Freiburg im Breisgau, West Germany

S. H. Bryant
Department of Pharmacology and Therapeutics
University of Cincinnati
College of Medicine
Cincinnati, Ohio 45267

C. A. Buck
Wistar Institute of Anatomy and Biology
Philadelphia, Pennsylvania 19104

J. Cerda
University of Florida
Medical School, Division of Gastro-enterology
Gainesville, Florida 32611

Hsia-fei Wang Chang
The Enzyme Research Laboratory
Massachusetts General Hospital
and
Department of Biological Chemistry
Harvard University Medical School
Boston, Massachusetts 02114

A. Church
Hematology Research Laboratory
Division of Hematology
The Saint Vincent Hospital
and
University of Massachusetts Medical School
Worcester, Massachusetts 01610
and
The Worcester Foundation for Experimental Biology
Shrewsbury, Massachusetts 01545

Paolo M. Comoglio
Department of Human Anatomy
University of Torino
Faculty of Medicine
Corso Massimo d'Azeglio, 52
10126 Torino, Italy

D. Conte-Camerino
Istituto di Farmacologia
Università di Bari
Bari, Italy

Carl F. Cori
The Enzyme Research Laboratory
Massachusetts General Hospital
and
Department of Biological Chemistry
Harvard University Medical School
Boston, Massachusetts 02114

R. K. Crane
Department of Physiology
College of Medicine and Dentistry of New Jersey
Rutgers Medical School
Piscataway, New Jersey 08854

Thomas P. Dousa
Mayo Clinic and Foundation
and
Mayo Medical School
Rochester, Minnesota 55901

Rupert Engelhardt
Medizinische Universitätsklinik
78 Freiburg, West Germany

G. Fairbanks
Hematology Research Laboratory
Division of Hematology
The Saint Vincent Hospital
and
University of Massachusetts Medical School
Worcester, Massachusetts 01610
and
The Worcester Foundation for Experimental Biology
Shrewsbury, Massachusetts 01545

J. Ferrero
Département de Pharmacologie
Ecole de Médecine
CH-1211 Geneva 4, Switzerland

Kay L. Fields
MRC Neuroimmunology Project
Department of Zoology
University College
London WC1E 6BT, England

Jorge Flores
Biochemical Pharmacology Unit
Medical Services
Massachusetts General Hospital
Boston, Massachusetts 02114

G. Gardos
National Institute of Haematology and Blood Transfusion
Budapest, Hungary

Rebecca C. Garland
The Enzyme Research Laboratory
Massachusetts General Hospital
and
Department of Biological Chemistry
Harvard University Medical School
Boston, Massachusetts 02114

M. C. Giuliano
Regina Elena Institute for Cancer Research
Rome, Italy

J. Glowinski
Groupe NB (INSERM U.114)
Collège de France
75231 Paris Cedex 5, France

D. D. Godse
Department of Neurochemistry
Clarke Institute of Psychiatry
Toronto, Ontario, Canada

P. Grof
Lithium Clinic
Hamilton Psychiatry Hospital
Hamilton, Ontario, Canada

Hans-Alfred Habicht
Enzyme Laboratory
Medical Polyclinic
Department of Internal Medicine
University of Basel
Basel, Switzerland
and
Institute of Pathology
University of Tübingen
Tübingen, Germany

D. C. Hixson
Department of Biochemistry
University of Texas System Cancer Center
M. D. Anderson Hospital and Tumor Institute
Houston, Texas 77025

S. R. Hollan
National Institute of Haematology and Blood Transfusion
Budapest, Hungary

E. R. Jakoi
Department of Anatomy
Duke University
Durham, North Carolina 27710

Harry W. Jarrett
Kenan Laboratories of Chemistry and Dental Research Center
University of North Carolina
Chapel Hill, North Carolina 27514

P. Jirounek
Département de Pharmacologie
Ecole de Médecine
CH-1211 Geneva 4, Switzerland

G. J. Jones
Departement de Pharmacologie
Ecole de Médecine
CH-1211 Geneva 4, Switzerland

Rolf Kinne
Max-Planck Institut für Biophysik
Frankfurt/Main, Germany

P. L. La Celle
Departments of Medicine and of Radiation Biology and Biophysics
University of Rochester School of Medicine and Dentistry
Rochester, New York 14642

P. Lefresne
Groupe NB (INSERM U. 114)
Collège de France
75231 Paris Cedex 5, France

D. Menard
Department of Physiology
College of Medicine and Dentistry of New Jersey
Rutgers Medical School
Piscataway, New Jersey 08854

R. Murthy
Department of Pharmacology
Faculty of Medicine
University of Toronto
Toronto, Ontario, Canada

G. Neri
Regina Elena Institute for Cancer Research
Rome, Italy

J. Palek
Hematology Research Laboratory
Division of Hematology
The Saint Vincent Hospital
and
University of Massachusetts Medical School
Worcester, Massachusetts 01610
and
The Worcester Foundation for Experimental Biology
Shrewsbury, Massachusetts 01545

John T. Penniston
Kenan Laboratories of Chemistry and Dental Research Center
University of North Carolina
Chapel Hill, North Carolina 27514

Gordon A. Plishker
Kenan Laboratories of Chemistry and Dental Research Center
University of North Carolina
Chapel Hill, North Carolina 27514

H. Preiser
Department of Physiology
College of Medicine and Dentistry of New Jersey
Rutgers Medical School
Piscataway, New Jersey 08854

Thomas Reid
Kenan Laboratories of Chemistry and Dental Research Center
University of North Carolina
Chapel Hill, North Carolina 27514

J. M. Ritchie
Department of Pharmacology
Yale University School of Medicine
New Haven, Connecticut 06510

John D. Roberts
Kenan Laboratories of Chemistry and Dental Research Center
University of North Carolina
Chapel Hill, North Carolina 27514

J. D. Robertson
Department of Anatomy
Duke University
Durham, North Carolina 27710

Leon E. Rosenberg
Department of Human Genetics
Yale University School of Medicine
New Haven, Connecticut 06520

Allen D. Roses
Division of Neurology
Duke University Medical Center
Durham, North Carolina 27710

A. Salamin
Département de Pharmacologie
Ecole de Médecine
CH-1211 Geneva 4, Switzerland

P. A. Santillo
Departments of Medicine and of Radiation Biology and Biophysics
University of Rochester School of Medicine and Dentistry
Rochester, New York 14642

Udo Schmidt
Enzyme Laboratory
Medical Policlinic
Department of Internal Medicine
University of Basel
CH-4056 Basel, Switzerland

Stanley L. Schrier
Stanford University School of Medicine
Stanford, California 94305

Werner Schröter
Department of Pediatrics
University of Göttingen
and
Department of Cell Physiology
Max-Planck Institute of Biophysics
Frankfurt/Main, West Germany

Irving L. Schwartz
Department of Physiology and Biophysics
Mount Sinai Medical and Graduate Schools of the City University of New York
New York, New York 10029

G. Semenza
Laboratorium für Biochemie
Eidgenössische Technische Hochschule
Universitätstrasse 16
CH-8006 Zurich, Switzerland

A. K. Sen
Department of Pharmacology
Faculty of Medicine
University of Toronto
Toronto, Ontario, Canada

Geoffrey W. G. Sharp
Biochemical Pharmacology Unit
Medical Services
Massachusetts General Hospital
Boston, Massachusetts 02114

Stephen B. Shohet
Cancer Research Institute
School of Medicine
University of California
San Francisco, California

Richard L. Sidman
Department of Neuropathology
Harvard Medical School
and
Department of Neuroscience
Children's Hospital Medical Center
Boston, Massachusetts 02115

Philip Siekevitz
The Rockefeller University
New York, New York 10021

H. C. Stancer
Department of Neurochemistry
Clarke Institute of Psychiatry
Toronto, Ontario, Canada

R. W. Straub
Département de Pharmacologie
Ecole de Médecine
CH-1211 Geneva 4, Switzerland

I. Szasz
National Institute of Haematology and Blood Transfusion
Budapest, Hungary

Guido Tarone
Department of Human Anatomy
University of Torino
School of Medicine
Torino, Italy

A. Trouet
International Institute of Cellular and Molecular Pathology
and
Université Catholique de Louvain
B-1200 Brussels, Belgium

P. Tulkens
International Institute of Cellular and Molecular Pathology
and
Université Catholique de Louvain
B-1200 Brussels, Belgium

Klaus Ungefehr
Department of Pediatrics
University of Göttingen
and
Department of Cell Physiology
Max-Planck Institute of Biophysics
Frankfurt/Main, West Germany

Lucy Vaughn
Kenan Laboratories of Chemistry and Dental Research Center
University of North Carolina
Chapel Hill, North Carolina 27514

Diter von Wettstein
Department of Physiology
Carlsberg Laboratory
DK-2500 Copenhagen
and
Institute of Genetics
University of Copenhagen
DK-1353 Copenhagen, Denmark

E. F. Walborg, Jr.
Department of Biochemistry
University of Texas System Cancer Center
M. D. Anderson Hospital and Tumor Institute
Houston, Texas 77025

L. Warren
Wistar Institute of Anatomy and Physiology
Philadelphia, Pennsylvania 19104

R. I. Weed
Departments of Medicine and of Radiation Biology and Biophysics
University of Rochester School of Medicine and Dentistry
Rochester, New York 14642

Milton M. Weiser
Department of Medicine
Harvard Medical School
and
Gastrointestinal Unit
Massachusetts General Hospital
Boston, Massachusetts 02114

James S. Wiley
Hematology-Oncology Section
Department of Medicine
Hospital of the University of Pennsylvania
Philadelphia, Pennsylvania

G. Zampighi
Department of Anatomy
Duke University
Durham, North Carolina 27710

Membranes and Disease, edited by L. Bolis, J. F. Hoffman, and A. Leaf. Raven Press, New York, © 1976.

Pathophysiologic Significance of Abnormalities of Red Cell Shape

P. L. La Celle, R. I. Weed, and P. A. Santillo

Departments of Medicine and of Radiation Biology and Biophysics, University of Rochester School of Medicine and Dentistry, Rochester, New York 14642

In addition to reversible shear stress-induced changes of the equilibrium discocyte shape of the normal erythrocyte, the cell may undergo a variety of reversible or irreversible shape alterations. The discocyte–echinocyte transformation, studied in detail by Jolly (1), Teitel-Bernard (2), Ponder (3), and, more recently, by Weed, La Celle, and Merrill (4) and Bessis and Lessin (5), is a well-known example of reversible shape change. In a number of disease states, particularly hemolytic anemias, one may encounter changes in erythrocyte shape due to such events as fragmentation within the microcirculation, or changes related to the intracellular hemoglobin. Shape changes of the erythrocyte are significant in their potential adverse modification of the rheologic properties of the erythrocyte, and thus the survival of cells, as well as the potential compromise of exchange processes at the capillary level. An understanding of the mechanisms of shape change has importance for the insights into structural organization and dynamics of the membrane protein and lipid components and environmental factors, which affect membrane properties and integrity.

The purpose of this chapter is to describe the reversible discocyte–echinocyte and discocyte–stomatocyte transformations in normal and pathologic cells; to discuss a variety of stimuli which induce such shape changes and potential mechanisms to explain the observed phenomena; and to present data which characterize the membrane material properties of echinocytes and stomatocytes as well as pathologic cells having abnormal shapes.

REVERSIBLE "DISC-SPHERE" SHAPE CHANGES IN NORMAL CELLS

Discocyte–Echinocyte Transformation

Figure 1 illustrates the discocyte–echinocyte transformation, one of two fundamental shape changes which normal human erythrocytes may undergo in reversible fashion. Jolly (1) was first to describe this shape change and to describe its interconvertibility with the discocyte–stomatocyte transforma-

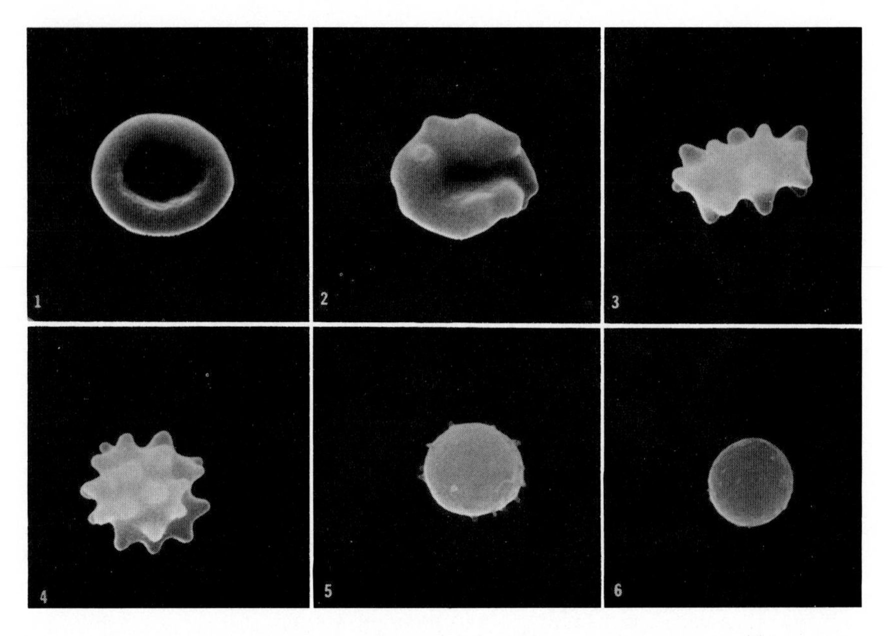

FIG. 1. Discocyte–echinocyte transformation: **1:** Normal discocyte. **2:** Echinocyte I. **3:** Echinocyte II. **4:** Echinocyte III. **5** and **6:** Spheroechinocyte. Scanning electron microscopy (SEM).

tion, and Ponder (3) recognized this transformation to be independent of volume change. When one observes the discocyte–echinocyte transformation, the initial change in the biconcave conformation is the appearance of coarse spicules around the rim, then over the entire cell surface, after which the cell assumes a basically spherical shape. The spicules subsequently become finer, are generally uniformly distributed and may number from approximately 30 to 50 per cell. Ultimately, particularly with prolonged incubation and metabolic depletion, the cell may lose portions of spicules, which are, in fact, losses of entire pieces of membrane, recognized biochemically by loss of membrane lipid and often traces of hemoglobin. When membrane is lost the transformation becomes irreversible and a spherocyte results. The intermediate stages of the transformation can be recognized relatively easily by phase microscopy, but the end-stage spherocyte, produced by the discocyte–echinocyte transformation, may be indistinguishable from that which evolves from the discocyte–stomatocyte shape change.

The echinocyte transformation may be induced by intrinsic and extrinsic factors (6). Deuticke (7) has noted that echinocytogenic agents are nonpolar or anionic amphiphilic compounds; examples include lysolecithin, bile acids, salicylate, dipyridamole, and barbiturates. Normal plasma may become echinocytogenic after prolonged incubation, whereas fresh plasma heated does not become echinocytogenic and heat does not destroy the formed echinocytogenic factor. The plasma factor appears to be lysolecithin (8–10) and the enzyme lecithin-cholesterol acyl transferase. The echinocyte shape change is reversible up to the point of membrane loss, independent of volume, although simple elevation of pH alone will produce this shape change along with the predicted decrease in cell volume. As first demonstrated by Hoffman (11) in 1952 with uranyl salts, recovery of cells left in echinocytogenic medium occurs spontaneously. This reversion to normal does not depend on disappearance of the active agent from the medium, since fresh cells introduced into the same medium undergo the echinocyte transformation and then revert to discocytes. If calcium is added to cells to a concentration of 5×10^{-3} M in buffered medium, there is no effect on red cell shape; however, the presence of calcium, or its removal by a chelator such as EDTA, enhances or reduces, respectively, the effect of the echinocytogenic agent (12).

Nakao et al. (13) observed that the discocyte–echinocyte change occurs when erythrocyte ATP content is reduced by incubation, and Weed et al. (4) noted that a progressive loss of erythrocyte deformability occurred in parallel with reduction of ATP, accumulation of calcium, and echinocyte shape change. Because acute changes in external calcium concentration are without effect, the locus of the Ca-induced shape change appears to be within or at the inner surface of the erythrocyte membrane.

Echinocytes occur *in vivo,* notably in uremia, bleeding peptic ulcer, heart disease, and carcinoma of the stomach, as well as in certain anemias. The term echinocyte (Greek = sea urchin) appears to be an appropriate general descriptive term for all disorders in which the erythrocyte spicules are regularly arranged over the cell surface and are reversible, while the term "acanthocyte" should be reserved for irregularly spiculated cells which do not assume the discocyte shape after washing and suspension in fresh plasma. As it is not possible to distinguish the underlying specific disorders from observation of cell morphology, other terms such as "burr" or "spur cell" are misleading.

Discocyte–Stomatocyte Transformation

The reversible isovolumic stomatocyte transformation is illustrated by the sequence of cell shapes in Fig. 2. In this transformation, the end result is a spherocytic cell as observed by light microscopy and, prior to the stage of membrane loss, is reversible. Initially the cell becomes uniconcave, followed by progressive effacement of the remaining concavity, and ultimately endovesicle formation occurs at the region of curvature, separating the periphery from the remaining shallow concavity until the cell becomes spherical. The dimples seen in this scanning electron micrograph prove to be vesicles when the cell is examined in cross section.

The agents causing stomatocytic shape change, identified in large part by

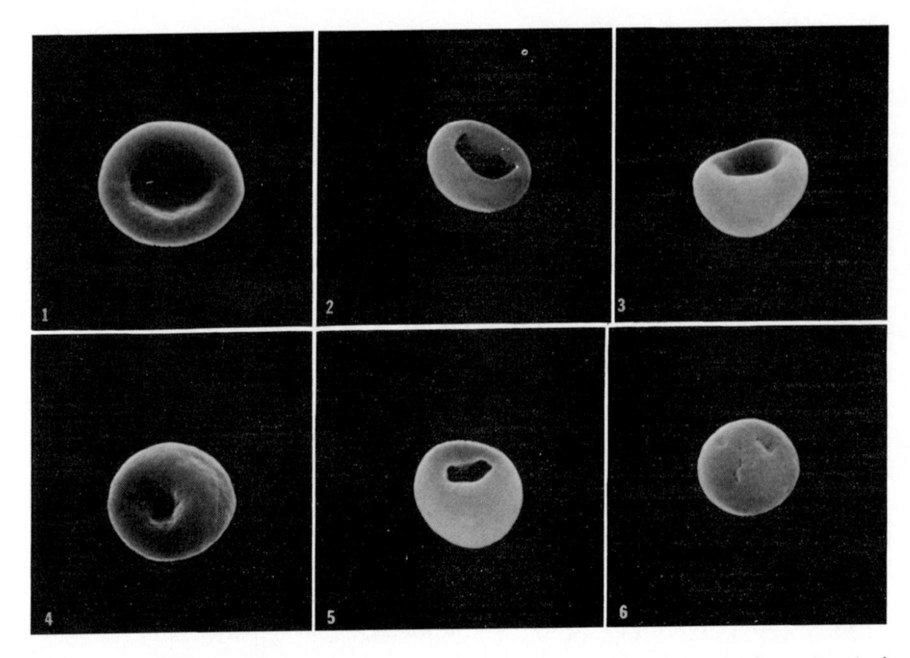

FIG. 2. Discocyte–stomatocyte transformation: **1:** Normal discocyte. **2:** Stomatocyte I. **3:** Stomatocyte II. **4:** Stomatocyte III. **5** and **6:** Spherostomatocyte. (SEM.)

Deuticke (7), are either cationic amphiphilic compounds or nonpenetrating anions. The nonpenetrating anion effect can be overcome by raising the extracellular chloride concentration, which led Deuticke to conclude that in this case the effect was related to an increased transmembrane pH gradient. Lowering the pH of the medium below a value of 6 will produce an identical stomatocytic change, although simple lowering of external pH also produces a change in volume. The stomatocytogenic effect of these agents is enhanced by the presence of external calcium, as is the echinocytogenic effect of elevated pH. It is of note that, at a given concentration of stomatocytogenic or echinocytogenic agent or at specific pH, there is heterogeneity of shape change in the population of erythrocytes, indicating that membrane parameters uniquely determine the effect of a specific internal stimulus.

MECHANISMS OF THE REVERSIBLE SHAPE TRANSFORMATION

No satisfactory model to account for the mechanism of reversible shape transformation has been devised. Deuticke's (7) proposal related the stomatocytogenic effect of nonpenetrating anions to their ability to change the transmembrane pH gradient. Hoffman (14) suggested that the echinocytic shape change depends on a transmembrane concentration gradient of the agent and that, when the agent disappears, the discocyte shape recurs.

Because both echinocytogenic and stomatocytogenic agents are augmented by the presence of calcium (12), the modification of pH or drug-induced shape changes after preincubation of the erythrocytes in a calcium-free medium, and the appearance of the echinocytic shape when cells depleted of ATP accumulate excess calcium led Weed and Chailley (12) to suggest that the echinocytic shape change is correlated with an increase in calcium permeability with stimulation of the calcium extrusion mechanism.

In this model the arrangement of a small number of calcium pump sites over the membrane surface might lead to inhomogeneity of local membrane calcium concentrations and account for the spicules. In this regard, it is of interest to note that the number of spicules on the surface of an echinocyte corresponds closely (30–50) to the number of calcium-pump sites required in the normal erythrocyte membrane ghost, to transport the observed quantity of calcium, per unit time, assuming a pump rate similar to enzyme-mediated processes. Low pH and stomatocytogenic agents may inhibit the calcium pump and cause the characteristic shape change by a calcium-dependent mechanism, which involves reduction of membrane area at constant cell volume. The blocked calcium pump may cause redistribution or local accumulation or altered membrane interaction with calcium to effect the stomatocytic shape change. It is evident that the echinocytogenic and stomatocytogenic agents do not act directly at the identical membrane locus, for both transformations may be superimposed on each other (15).

The amount of calcium involved in these shape changes may be very small, well below levels detectable by atomic absorption spectrophotometry. The small amount of calcium ions involved, however, may account for the rapidity of induction and reversal of shape change, observable with various agents and pH change. The erythrocyte protein, spectrin, located at the inner surface of the intact membrane, possesses several classes of calcium binding sites (16), including a high affinity site having an apparent association constant $K = 3 \times 10^6$ liters/mole. Calcium-induced spectrin aggregation, with resultant change in membrane configuration and area, could account for the shape transformations. However, in solution, spectrin aggregation requires $[Ca^{2+}] \simeq 10^{-3}$ M, whereas in the intact ghost 10^{-7} M calcium causes changes. Under normal circumstances spectrin may be phosphorylated (17), and thus the dephosphospectrin may be the form that aggregates. It has been hypothesized that dephosphospectrin binds strongly to actin or actin-like protein of the erythrocyte membrane and that a dephosphospectrin–actin interaction could augment a calcium-induced spectrin aggregation, thereby multiplying the effect of a small number of calcium ions (18).

Such a calcium-modulated model has a certain analogy in the sarcoplasmic reticulum, which, for the sake of discussion, could be considered an inside-out erythrocyte membrane. Both lysolecithin, noted to be an echinocytogenic agent for the erythrocyte, and chlorpromazine, stomatocytogenic in the erythrocyte, block calcium accumulation in the sarcoplasmic reticulum;

however, the two substances act by very different mechanisms. Lysolecithin causes permeability and resultant calcium leak from the vesicles, while chlorpromazine inhibits the calcium pump (19). Chlorpromazine is known to inhibit the erythrocyte calcium pump (20), and lysolecithin, well known to cause cation leak and hemolysis at higher concentrations, causes calcium leak at very low concentrations.

IRREVERSIBLE SHAPE ALTERATIONS

The described reversible echinocytic and stomatocytic shape changes generally are not accompanied by volume change; however, at high concentrations of chemical agents, the cells become spherical, usually with decrease in volume. The loss of volume indicates that the surface area has decreased, implying either thickening or small folds in the membrane or loss of membrane material at the end stage of these transformations, as described above. The membranes of erythrocytes which have undergone extreme discocyte–echinocyte transformation lose membrane and become spherical; they lose membrane by myelin forms, or spontaneous microspherulation. The comparable progression to irreversible membrane alteration in the stomatocytic transformation involves reduction of membrane surface area by the process of endovesiculation or esotropic vesiculation. At the end stage of each transformation, sufficient membrane area loss has occurred so that reversal is not possible, i.e., insufficient membrane exists to allow a larger membrane surface area:volume ratio.

ECHINOCYTIC AND STOMATOCYTIC TRANSFORMATIONS IN PATHOLOGIC ERYTHROCYTES

Figure 3 illustrates abnormal sickle-cell shape at low oxygen tension. If sickle cells are incubated with an echinocytic agent, then exposed to reduced oxygen tension, the sickled echinocyte or echinodrapanocyte shown in Fig. 4 results. Although one can identify the folds in the membrane and the truncated appearance of the spicules indicating the underlying sickled hemoglobin structure, the predominant shape is that of an echinocyte. Figure 5 depicts a sickled stomatocyte or stomatodrapanocyte produced by reducing the oxygen tension of a sickle hemoglobin-containing erythrocyte first exposed to chlorpromazine. The sickling is evidenced by the obvious surface irregularities, but the dominant conformation is that of the stomatocyte, and under the light microscope it erroneously might be concluded that chlorpromazine had blocked sickling. These illustrations emphasize that the reversible echinocytic and stomatocytic shape changes are fundamental conformational changes of membrane, for in each case, as Weed has emphasized (11), the conformation of the membrane has de-

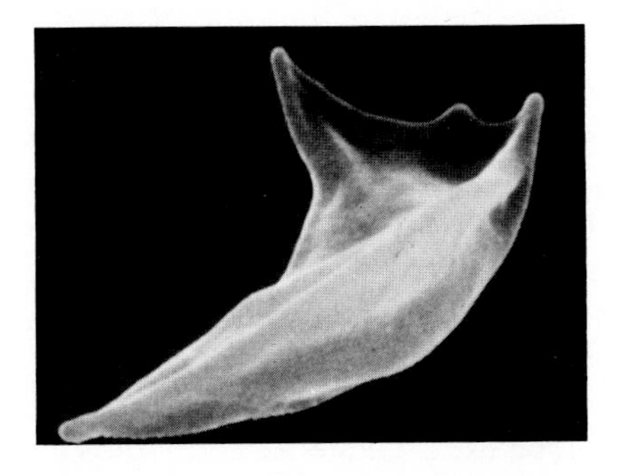

FIG. 3 (upper left). Sickled erythrocyte. This sickle cell, capable of reversible sickling, illustrates the "holly leaf" form and the ridges produced by the conformation of the sickle hemoglobin underlying the membrane. (SEM.)

FIG. 4 (upper right). Sickle echinocyte (echinodrapanocyte). Prior to reduction of oxygen tension to induce sickling, this cell was exposed to an echinocytogenic agent. The echinocytic shape determined the orientation of the sickled hemoglobin. (SEM.)

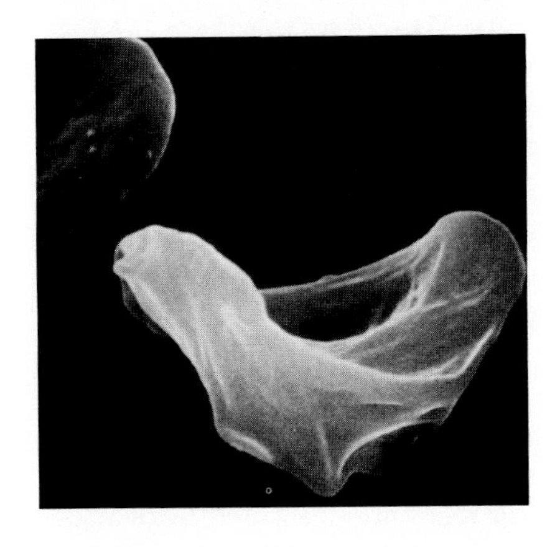

FIG. 5 (lower left). Sickle stomatocyte (stomatodrapanocyte). The stomatocyte shape, caused by exposure to a stomatocytogenic agent prior to induction of sickling, has provided the general orientation for the sickled hemoglobin, clearly evident by the ridges in the membrane. (SEM.)

termined the direction of polymerization of the sickle hemoglobin within the cell.

Pathologic erythrocytes, regardless of their shape, e.g., poikilocytes or acanthocytes, can respond to the same stimuli to develop complex echinocytic shapes in which the echinocytic configuration of spicules is superimposed on the original shape. Echinopoikilocytes, echinoacanthocytes, etc. result (21). Similarly, stomatocytogenic agents produce stomatocyte transformations superimposed on the primary structural pathology of the cell (22).

MEMBRANE STRUCTURE AND MATERIAL PROPERTIES

The lipid bilayer, containing a variety of specialized intrinsic proteins, is an inadequate model to account for the observed elastic properties of the erythrocyte membrane. A lipid membrane has a high bulk modulus, proportionate in area incompressibility, but cannot exhibit the elasticity, i.e., capacity to store energy as a function of deformation, and indeed the lipid membrane is disrupted at low force. From the mechanical standpoint, the erythrocyte membrane behaves as a hyperelastic solid and is anisotropic with a two-dimensional material character that is decoupled from the thickness dimension. It is important to recognize that the membrane may be regarded as a continuum in two dimensions, but has a molecular character in the third; hence, the membrane cannot be regarded simply as a three-dimensional elastic solid. Recent analysis of various models of membrane structure by Bull and Brailsford (22) suggests that a protein chain meshwork, interacting to provide a skeletal network, would provide mechanical properties corresponding to those observed in the erythrocyte. Skalak, Tozeren, Zarda, and Chien (23) and Evans (24) have developed general two-dimensional elastic material models to describe the continuum mechanical behavior, and Evans (25) has applied this material concept to an analysis of deformed erythrocytes. The latter work demonstrates that an elastomer network model will accommodate the observed behavior of the erythrocyte membrane undergoing fluid shear with microfilament (microtether) and teardrop formation; the large deformations and fragmentation demonstrated in micropipette aspiration experiments; and the large strains with elastic recoil typical of cells during capillary flow. A statistical thermodynamic calculation to relate the derived elastic shear modulus to the molecular weight of the subchains of the proposed protein network of the erythrocyte membrane indicates a molecular weight in the order of 10^5, similar to that of spectrin (24,26), which suggests the potential role of spectrin in membrane elasticity.

A material such as the erythrocyte membrane may be adequately described by several constants representing intrinsic properties. Thus, the solid material properties of the elastomer network are described by the elastic constant (μ), corresponding to the elastic shear modulus; η_e, the viscous material constant characterizing viscous dissipation during elastic deformation (η_e depends on rate of deformation); η_p, the membrane surface viscosity in plastic flow; and T_0, the yield tension or tension beyond which the approximately linear stress–strain relationship does not hold.

EVALUATION OF MATERIAL PROPERTIES BY MEMBRANE DEFORMATION

Our experiments to define the solid elastic properties of normal and pathologic erythrocyte membrane utilized analysis of membrane behavior during aspiration of membrane into glass micropipettes. Figure 6 illustrates the technique in one such micropipette experiment. Evans (25) has shown that the elastic tension tensor for a two-dimensional, hyperelastic material,

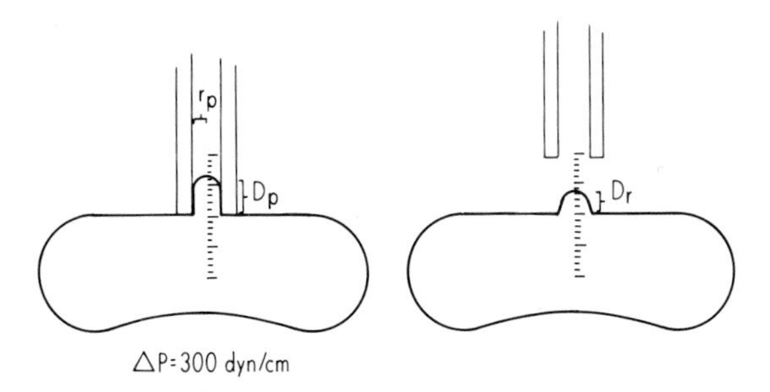

FIG. 6. Micropipette technique for controlled deformation of the erythrocyte membrane. *Left,* Negative pressure ΔP, in the small glass micropipette having radius r_p, causes aspiration of a portion of the membrane distance D_p into the micropipette without inducing tension throughout the entire membrane. A calibrated micrometer allows measurement of D_p and r_p. *Right,* When force is removed ($\Delta P \rightarrow 0$), the deformed region of membrane relaxes, or if plastic deformation has occurred, a residual deformation D_r may be observed. In a large micropipette, the pressure to cause the entire cell to enter the micropipette is a measure of total cellular deformability. Cellular deformability depends on membrane viscoelastic properties, and factors extrinsic to the membrane/cell area/volume relationship (shape) and the properties of the cell contents.

such as the erythrocyte membrane, which has large resistance to area change, is given by

$$T_{ij}^{e} = (\gamma + K\alpha)\delta_{ij} + \mu\epsilon_{ij}$$

where δ_{ij} is the identity matrix; i,j represents the in-plane membrane coordinates, γ is the interfacial free-energy density of hydrophobic interaction, K is the area compressibility modulus, α is the area strain ($\alpha = \lambda_1\lambda_2^{-1}$, where λ_1 and λ_2 are extension ratios = final dimension/initial dimension in the principal directions) and μ corresponds to the shear modulus.

As the area compressibility is extremely small, i.e., $K \geq \mu$, the material can be considered two-dimensionally incompressible and the first term may be replaced by isotropic tension:

$$T_{ij}^{e} = -P_{m}\delta_{ij} + \mu\epsilon_{ij}$$

where P_m is the membrane locally isotropic tension and represents energy storage when resisting dilation or compression of surface area, and ϵ_{ij} is the Lagrangian strain tensor.

In application of this concept to the analysis of uniaxial extensions produced by membrane aspiration into the micropipette (25,26) $\hat{\epsilon}_s$, the strain at the pipette tip for large deformations is given by

$$\hat{\epsilon} = \frac{\lambda_s^2 - 1}{2}$$

derived from the stress–strain law for the principal extensions

$$T_s = -P_m + \frac{\mu}{2}(\lambda^2 - 1)$$

where T is the stress resultant.

It can be shown (25,26) that the extension ratio λ at the pipette tip is a function of the distance D_p; the erythrocyte is aspirated into the micropipette; thus, D_p reflects the elastic constant of the membrane. D_p, shown in Fig. 7, may be related to pipette diameter r_p to permit comparison of data derived from pipettes of different sizes. Utilizing this analysis, the elastic constant μ has been found to be 7×10^{-3} dyn/cm (25,26).

η_e, the viscous material constant, is a constant which may relate viscous shear tension, $T_{ij}{}^v$ and the rate of deformation tensor, V_{ij}; $T_{ij}{}^v$ is the expression for viscous shear tension in the model for viscoelasticity where elastic tension and viscous shear tension are superposed (27):

$$T_{ij} = T_{ij}{}^e + T_{ij}{}^v$$

Evans and Hochmuth (28) have shown that a value for η_e may be derived from the relaxation of a membrane deformity upon removal of deforming force, as depicted in Fig. 6, right panel. The time constant T_{50}, time to relax to 50% of the initial deformation, $= 3\eta_e/\mu$. The T_{50} value for normal cells is $\sim$0.3 sec; thus $\eta_e \simeq 10^{-3}$ dyn-sec/cm.

In further application of the concept to the erythrocyte membrane undergoing extrusion of microtethers as result of fluid shear stress, Evans and Hochmuth have calculated η_p, membrane surface viscosity, from equations derived from a constitutive equation relating plastic behavior with yield tension and surface viscosity in tensor relationships between membrane tension and rate of deformation (28). $\eta_p \simeq 8 \times 10^{-3}$ dyn-sec/cm and T_0, yield tension in shear $= 2$ to 8×10^{-2} dyn/cm.

It should be noted that this surface viscosity value is 3 orders of magnitude greater than surface viscosities of lipid membrane components, indicating the dominant role of the proposed protein elastomer network as a determinant of the character of the two-dimensional membrane (29).

ELASTIC PROPERTIES OF PATHOLOGIC, AND MODIFIED NORMAL MEMBRANES

The force required to deform the membrane of lysolecithin-induced echinocytes (Fig. 7) in a 1.0 μm internal diameter glass micropipette, recorded in Table 1, is greater than for control membranes. Stomatocytes (Fig. 7) produced by chlorpromazine may require a slight, but not statistically significant, increment of pressure for this membrane extension. Homozygous S sickle cells at normal Po_2 were not different from normal if irreversibly sickled forms were excluded. The latter require significantly greater forces for deformation. Hereditary spherocytes are, in fact, stomatocytic in shape, as demonstrated by the scanning electron micrograph (Fig. 8); however, the

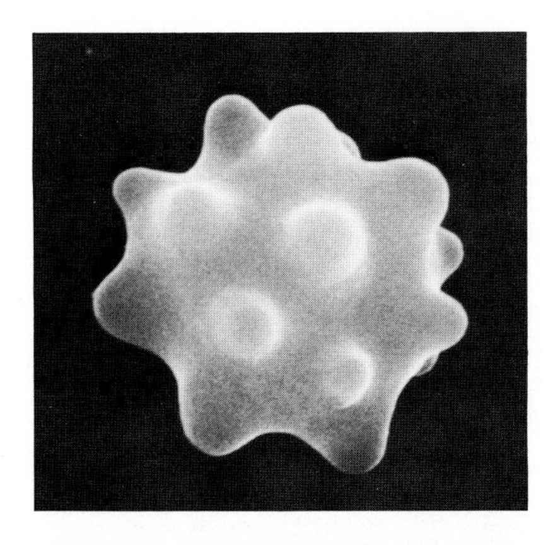
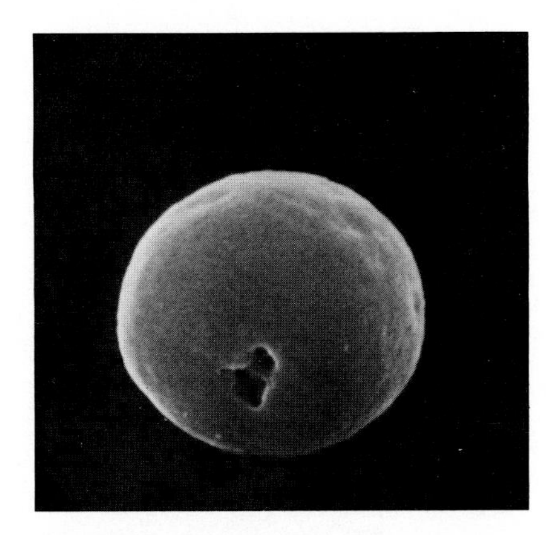

FIG. 7. Left: The echinocyte, produced by lysolecithin, has maintained its original volume and has regularly spaced spicules. **Right:** The spherostomatocyte, produced by chlorpromazine, has lost the typical cup shape of the stomatocyte; however, the location of the uniconcavity is still evident. (SEM.) The membrane deformability of echinocytes and stomatocytes is recorded in Table 1.

majority of such cells have normal membrane deformation characteristics, indicated by data for cells from five patients having hereditary spherocytosis. Occasional HS cells lysed when these deforming forces were applied, an observation not noted for normal erythrocytes. Cells from patients having thalassemia major required greater mean force for deformation, and in some poikilocytes (Fig. 9), variation of deformability over the surface was observed.

It should be noted that, despite normal membrane deformability, both HS cells and thalassemia cells may have significantly abnormal *cellular* deformability, i.e., the force required to cause such cells to enter and transit a 2.8–3.0-μm microcapillary is elevated (30).

When greater extensions, D_p/r_p, were related to stress, ΔP (Fig. 10), differences between normal erythrocytes and pathologic cells were apparent. Even heterozygous sickle cells (Hb = SA) appear different in this experi-

TABLE 1. *Pressure for deformation* ($D_p = r_p$) *of discocytes, echinocytes, stomatocytes, and pathologic cells*

	Δ Pressure (dyn/cm^2) mean $\pm$ SE
Control cell	140 ± 4
Echinocyte	292 ± 14
Stomatocyte	186 ± 8
SS cells (Po_2 = 100 mm Hg)	164 ± 26
Hereditary spherocytes	152 ± 6
Thalassemia major cells	204 ± 5

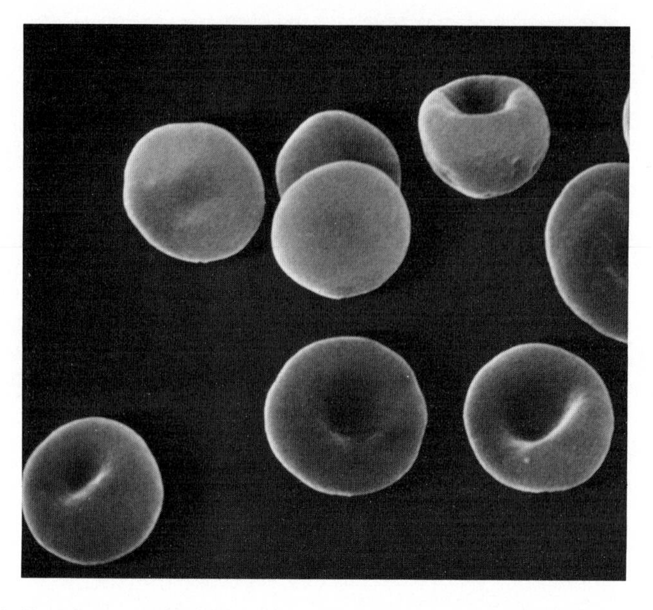

FIG. 8. Hereditary spherocytes. The dominant shape is stomatocytic. (SEM.)

ment, and thalassemia major erythrocytes were abnormal at higher extensions. Behavior of homozygous sickle cells clearly diverged from normal at greater extensions. In the case of senescent erythrocytes, more spherical in shape, a slight but not significant deviation from normal was found consistently at low deforming force (Fig. 11). Senescent cell membranes required less force for fragmentation than normal membranes.

Experimental modification of normal erythrocyte membrane lipid or protein components was undertaken to document the potential contribution of each membrane component to membrane deformability and total cellular deformability. Total cellular deformability was defined by the pressure re-

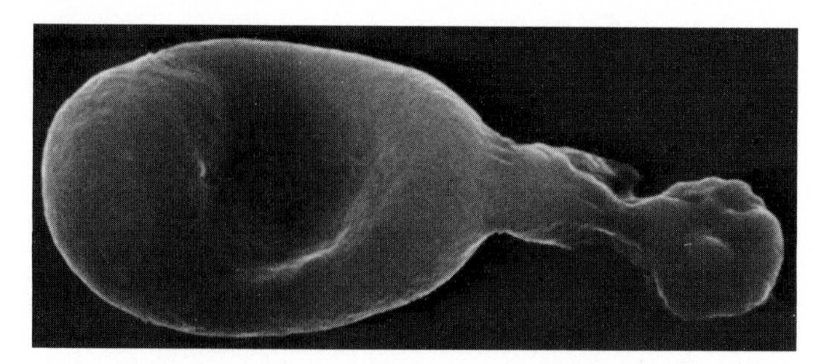

FIG. 9. Poikilocyte from thalassemic blood. The membrane projection on the right presumably contains relatively rigid Heinz bodies. (SEM.)

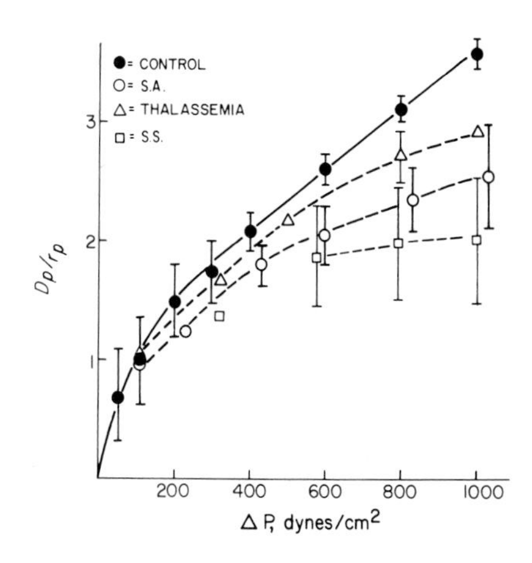

FIG. 10. Membrane extension, D_p/r_p, of normal, HbSS, HbSA, and thalassemic erythrocytes as a function of deforming pressure ΔP. D_p = length of membrane extension in the pipette, r_p = micropipette radius. Five samples of each cell type are recorded; $N = 40$ cells in each datum point.

quired to cause cells to traverse a 2.8-μm microcapillary or for filtration through a 2.8-μm aperture size low-pore-density polycarbonate filter under standard conditions. The results of such experiments are recorded in Table 2. Erythrocyte membrane lipids were altered by incubation of cells in media containing proportions of cholesterol to phospholipid, which resulted in increases of the membrane cholesterol or phospholipid. Some volume increase occurred in cholesterol-rich cells, but in these, as well as phospholipid-rich cells, membrane and total cellular deformability were normal when compared to control cells incubated without lipid addition.

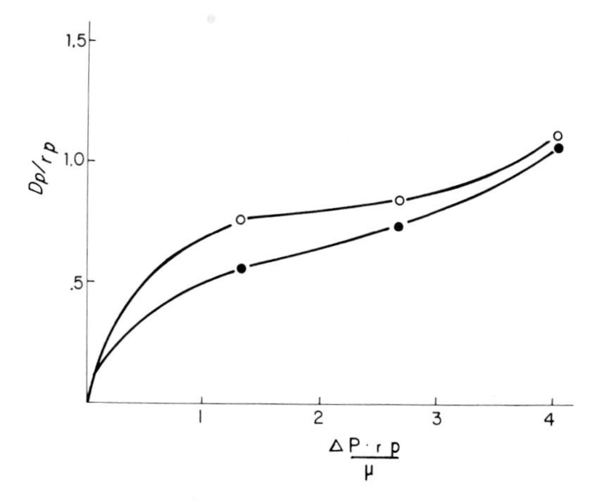

FIG. 11. An experiment in which membrane extension, D_p/r_p, for senescent erythrocytes (○) is compared to control cells from the same population. The senescent cells were obtained by a centrifugation technique (16). $(\Delta P \cdot rp)/\mu$ is dimensionless; ΔP is the deforming pressure, r_p is the micropipette radius, and μ is the elastic constant. The value for elastic constant $\simeq 7 \times 10^{-3}$ dyn/cm (27). Some difference in elasticity of senescent cells is suggested by this experiment. ($N = 20$ cells for each point.)

TABLE 2. *Effect of modifying lipid content and membrane proteins on normal erythrocyte deformability*

Test cell	Volume (μm^3)	Force for deformation (dyn/cm^2) Membrane ($D_p = r_p$)	Force for deformation (dyn/cm^2) Total cellular	Filterability (% filtered)
Control	85.5	120 ± 3	642 ± 28	95.7
20-hr incubated control	86.6	188 ± 16	980 ± 64	88.4
Cholesterol-rich	96.7	124 ± 4	734 ± 34	90.6
Phospholipid-rich	87.6	108 ± 6	592 ± 16	94.9
PCMB 0.5 m*M*	—	312 ± 18	7×10^4	—
NEM 0.1 m*M*	—	376 ± 25	1.47×10^5	—

Parachloromercuribenzoic acid (PCMB) and *N*-ethylmaleimide (NEM) causes obvious change in deformability, indicating the dominant role of membrane protein in the membrane material properties.

Further evidence to indicate the minimal contribution of the lipid component of the erythrocyte membrane to membrane elasticity is given in Fig. 12, where normal membrane extension is plotted against stress, and the failure point of a lipid bilayer is indicated. Clearly a lipid membrane fails when the deformation is small, indeed the calculated extension ratio at the micropipette tip is only 1.2, whereas the normal erythrocyte membrane tolerates extension ratios greater than 3. Figure 13 records deformation of cholesterol-rich cells (=lipid-altered erythrocytes) and erythrocytes from a patient having high serum cholesterol in cirrhosis. Normal deformability

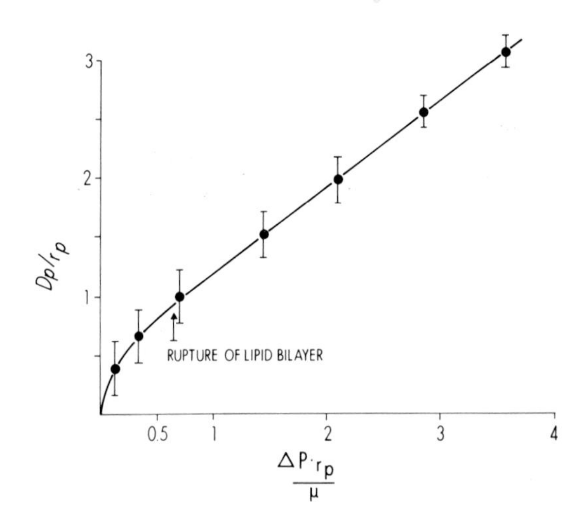

FIG. 12. Relation of membrane extension, D_p/r_p, to deforming force in normal erythrocyte membranes, illustrating the nearly linear relationship. *Arrow,* Point at which relatively inelastic lipid membrane bilayers disrupt at a calculated extension ratio $\lambda \simeq 1.2$.

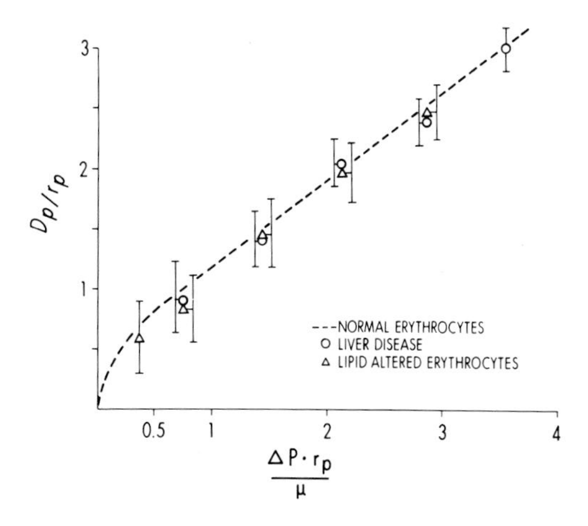

FIG. 13. Membrane extension of cholesterol-rich (lipid-altered), erythrocyte membranes, and membranes from a patient having hemolysis in liver disease compared to normal erythrocyte membranes. The value of the elastic constant μ was 7×10^{-3} dyn/cm.

is evident, indicating normal elastic properties, and thus normal rheologic characteristics. On the basis of their rheologic properties, a normal intravascular life span would be predicted, and the shortened life span associated with lipid abnormalities would appear to decrease by other mechanisms which compromise the membrane integrity.

OBSERVATIONS CONCERNING MEMBRANE VISCOSITY, PLASTIC DEFORMATION, AND TENSION AT HEMOLYSIS

When the time constants for relaxation of micropipette-induced deformation of oxygenated sickle cells and hereditary spherocytes were compared to normal cells, no significant differences were found in preliminary experiments. Up to twofold increases were observed in some poikilocytes from patients having thalassemia major, suggesting that η_e, the viscous material constant, may be significantly different from normal in these cells.

No direct values of η_p, the membrane surface viscosity in plastic flow, have been obtained. However, experiments concerning plastic deformation or permanent deformation as a function of the duration of applied force have produced evidence that normal erythrocyte membranes exhibit time-related change (27). This has been interpreted to indicate semisolid behavior determined by network topology, possibly represented by combinations of structural arrangements such as crosslinking, generation of new crosslinks with either time or crystallization, or both, at large extensions. The information in Table 3, expressed as residual extension ratios of membrane deformations produced by the micropipette method illustrated in Fig. 7,

TABLE 3. *Time-dependent residual deformation of normal, thalassemic, and sickle cell membranes*

Duration of stress	Cell type	λ, Extension ratio	
		With stress	Residual[a]
10 min	Normal	2.4	1.3
	Thalassemic	2.3	1.4
	Sickle	2.1	1.6
15 min	Normal	2.6	1.7
	Thalassemic	2.5	1.9
	Sickle	2.4	2.1

[a] Measured 2 min following removal of stress.

shows the increased tendency to permanent deformation in sickle cells (SS) and thalassemia major cells.

ACKNOWLEDGMENTS

This work was supported by US PHS Research Grants HL16421-04 and HL06241-14 and is based in part on work performed under contract with the USERDA at the University of Rochester and has been assigned Report No. UR-3490-685.

REFERENCES

1. Jolly, J. (1923): *Traité Technique d'Hematologie,* Vol. 1. Maloine, Paris, p. 64.
2. Teitel-Bernard, A. (1932): *Arch. Roum. Pathol. Exp. Microbiol.,* 5:989.
3. Ponder, E. (1948): *Hemolysis and Related Phenomena,* p. 10. Grune and Stratton, New York.
4. Weed, R. I., La Celle, P. L., and Merrill, E. W. (1969): *J. Clin. Invest.,* 48:795.
5. Bessis, M., and Lessin, L. S. (1970): *Blood,* 36:399.
6. Bessis, M., and Weed, R. I. (1973): *Adv. Biol. Med. Phys.,* 14:35.
7. Deuticke, B. (1968): *Biochim. Biophys. Acta,* 163:494.
8. Bessis, M., and Brecher, G. (1971): *Nouv. Rev. Fr. Hematol.,* 11:305.
9. Lichtman, M. A., and Gordesky, S. E. (1974): *Nouv. Rev. Fr. Hematol.,* 14:5.
10. Féo, C. (1972): *Nouv. Rev. Fr. Hematol.,* 12:757.
11. Hoffman, J. F. (1952): *Biol. Bull.,* 103:303.
12. Weed, R. I., and Chailley, B. (1972): *Nouv. Rev. Fr. Hematol.,* 12:775.
13. Nakao, M., Nakao, T., and Yamazoe, S. (1960): *Nature,* 187:945.
14. Hoffman, J. F. (1972): *Nouv. Rev. Fr. Hematol.,* 12:771.
15. Chailley, B., Weed, R. I., Leblond, P. F., and Maigné, J. (1973): *Nouv. Rev. Fr. Hematol.,* 13:71.
16. La Celle, P. L., Kirkpatrick, F. H., Udkow, M. P., and Arkin, B. (1972): *Nouv. Rev. Fr. Hematol.,* 12:789.
17. Avruch, J., and Fairbanks, G. (1974): *Biochemistry,* 13:5507.
18. La Celle, P. L., and Kirkpatrick, F. H. (1975): In: *Erythrocyte Structure and Function, Proceedings of the Third International Conference on Red Cell Metabolism and Function,* edited by G. Brewer, p. 535. Liss, New York.

19. Balzer, H., Makinose, M., and Hasselbach, W. (1968): *Arch. Pharmakol. Exp. Pathol.,* 260:444.
20. Schatzman, H. J. (1970): In: *Calcium and Cellular Functions,* edited by A. W. Cuthbert, p. 85. Macmillan, London.
21. Weed, R. I., and Bessis, M. (1973): *Blood,* 41:471.
22. Bull, B. S., and Brailsford, J. D. (1973): *Blood,* 41:833.
23. Skalak, R., Tozeren, A., Zarda, R. P., and Chien, S. (1973): *Biophys. J.,* 13:245.
24. Evans, E. A. (1973): *Biophys. J.,* 13:926.
25. Evans, E. A. (1973): *Biophys. J.,* 13:941.
26. Evans, E. A., and La Celle, P. L. (1975): *Blood,* 45:29.
27. Evans, E. A. (1975): In: *Comparative Physiology. 2. Structural Materials, Contractile Mechanisms and Energy Demands,* edited by L. Bolis, E. H. P. Maddrell, and Schmidt-Neilson. North-Holland, Amsterdam. (*in press*).
28. Evans, E. A., and Hochmuth, R. M. (1975): *Biophys. J.,* 16:1.
29. Evans, E. A., and Hochmuth, R. M. (1975): *Biophys. J.,* 16:13.
30. La Celle, P. L. (1975): *Blood Cells,* 1:269.

Membranes and Disease, edited by L. Bolis, J. F. Hoffman, and A. Leaf. Raven Press, New York, © 1976.

Energy-Dependent Endocytosis in White Erythrocyte Ghosts

Gordon A. Plishker, Lucy Vaughan, Harry W. Jarrett, Thomas Reid, John D. Roberts, and John T. Penniston

Kenan Laboratories of Chemistry and Dental Research Center, University of North Carolina, Chapel Hill, North Carolina 27514

The use of thoroughly washed, white erythrocyte ghosts for the study of energy-dependent endocytosis in red cell membranes has several advantages. These include the simplicity of the system, the accessibility of the inside of the membrane, the easier control of experimental parameters, and the ability to compare the white ghost system with the intact erythrocyte and the resealed erythrocyte. Therefore, the results reported in this chapter and elsewhere (1–5) may be viewed as complementary to the data obtained with whole cells and resealed ghosts (6–8).

The white erythrocyte ghost system is a highly purified erythrocyte membrane; not only has the huge amount of hemoglobin present in the erythrocyte been removed, but a very large number of soluble enzymes have also been removed. Despite this extreme purification, the white erythrocyte membrane shows shape changes that are very similar to those displayed by intact erythrocytes. The relative simplicity of the pure erythrocyte membrane system allows the researcher to focus on those aspects of the membrane structure which are directly responsible for the shape changes observed; complicating effects due to the presence of cytoplasmic enzymes are eliminated. The use of purified erythrocyte membranes should allow the separation of extraneous phenomena from those aspects of shape control that are intrinsic to the membrane itself.

The accessibility of the inside of the cell in these membranes, which have been made leaky by preparation in the presence of EDTA, greatly simplifies experimentation. Reagents added at the beginning of the experiment, or shortly after the experiment starts, can reach their site of action easily, regardless of whether this site is on the inside or the outside of the membrane. This is extremely advantageous in the performance of experiments, because reagents can simply be added and their effects observed without the intervention of long washing procedures. The accessibility of the inside also allows the direct observation of the enzymic activities responsible for endocytosis. For example, routine measurements of ATPase activity are much easier than in resealed ghosts.

The accessibility of the inside at the beginning of the experiment provides a greater degree of control of the parameters in the experiment. It is possible to vary the experimental parameters over a wider range than is possible in resealed cells; for example, media of very low osmotic pressure can be utilized, which would cause lysis in resealed cells. The internal concentration of Ca^{2+} or other ions can be varied at will simply by changing the external concentration.

One of the most important values of a purified ghost system is the new light that may be thrown upon the phenomena observed with it when they are compared with the phenomena observed in intact cells or resealed cells. As is discussed later in this chapter, differences in behavior may be due to the different conditions on the inside of the membrane which exist in leaky ghosts, or to differences in the properties of the membrane themselves. Such comparisons may very well allow identification of important controlling features in the behavior of the whole red cells.

The purpose of this chapter is twofold: because Ca^{2+} is reported to stimulate vacuolation in resealed erythrocytes, it is important to present carefully the data on the effect of Ca^{2+} on purified ghosts; extensive data on Ca^{2+} effects are presented here. The second aim is to discuss the shape changes observed upon incubation of purified ghosts; these are from the echinocytic form via the discocytic and stomatocytic to the spherendocytic form. We propose that the final form of the ghosts, which contains vacuoles (usually spherical) within the cell membrane be called a *spherendocyte,* from the Greek phrase, *αἱ σφαῖραι αἱ ἔνδον κύτεος*, connoting spheres within a cell. In this case, *sphere* refers to the vacuoles, *endo* to within, and *cyte* to the cell membrane. Consideration of the average curvature of the membrane allows the placing of these complex shape changes on a common scale and data pertaining to this are also presented in this chapter (*vide infra*).

METHODS

Ghosts were prepared as previously described (1), except that the data described in Figs. 4 and 5 were obtained on ghosts prepared by the method described by Penniston and Green (2). Human erythrocyte ghosts prepared from freshly drawn blood were utilized, except that ghosts from porcine erythrocytes were used to obtain the data reported in Fig. 5 and ghosts from rabbit erythrocytes to obtain the data reported in Fig. 4. Calcium determinations were carried out as previously described (1).

Endocytosis was normally carried out in a system which was 1 mg ghost protein/ml, 3 m*M* in $MgCl_2$, 3 m*M* in Tris-ATP, and 50 m*M* in TES–TEA, pH 7.4. Incubation was 30 min at 37°C. The data shown in Fig. 4 were obtained by incubation at 37° in a medium which was 100 m*M* in NaCl; 20 m*M* in KCl; 20 m*M* in TES, pH 7.1; 4 m*M* in $MgCl_2$ and 6 m*M* in ATP. The data shown in Fig. 5 were obtained by incubation at 37°C in a medium which was 160 m*M* in NaCl; 40 m*M* in KCl; 20 m*M* in TES, pH 7.3; 10 m*M* in $MgCl_2$; and 6 m*M* in ATP.

The amount of membrane taken inside was measured by estimating the hydrolysis of acetylthiocholine. This was done in the presence of Ellman's reagent. The reduced SH group released by the hydrolysis of acetylthiocholine was reoxidized by Ellman's reagent, releasing the colored 2-nitro-5-thiobenzoate, which was determined spectrophotometrically. The acetylthiocholine esterase is on the surface of the membrane, and is taken inside the vacuoles as endocytosis proceeds. Although the cells are leaky at the beginning of the experiment, the acetylthiocholine esterase, which is on the inside of the vacuole, is separated from the added substrates by *two* membranes, the inner of which may have become resealed in the course of the incubation (9), and thus is inaccessible to its substrates. Therefore, the acetylthiocholine esterase activity goes down as vacuolation occurs, and this has proved to be a reliable estimate of the amount of membrane taken in upon vacuolation. A further report on this method is in preparation.

Measurement of the Curvature of Ghost Membranes

In general, a curved surface in three dimensions has two radii of curvature. If one draws a line perpendicular to the surface at the point for which the curvature is being determined and then draws a plane through the surface and containing this line, one of the radii of curvature can be obtained from the curve describing the intersection between the plane and the surface. The second radius of curvature is obtained from the intersection between a second plane and the surface. This second plane is drawn perpendicular to the first plane, but also containing the line that was originally drawn perpendicular to the surface. In both cases the curve of intersection between the plane and the surface gives the radius of curvature, which is the radius of a circle that is tangent to the line at the point at which the perpendicular to the surface was drawn.

When a membrane is sectioned so that the plane of the section is perpendicular to the surface of the membrane then one of the radii of curvature of the membrane can be obtained by finding the circle that will be tangent to the membrane. If a small angle, θ, occurs between the perpendicular to the membrane and the plane of the section the observed radius of curvature is related to the actual radius of curvature by the equation $R' = R \cos \theta$. By a consideration of the parameters involved in the sections of red cell membranes, it is possible to show that only small errors are introduced by moderate values of θ. This can be done in the following way: The apparent thickness of the membrane and the attached spectrin when viewed in a section in which the membrane is perpendicular to the plane of the section is about 20 nm, whereas the average thickness of a section is 80 nm. When the angle θ is 30°, the apparent thickness of the membrane will be 3 times the thickness seen for $\theta = 0$, and this amount of tilting of the membrane relative to the plane of the section is easily detectable. Under this condition, cos $30° = 0.87$, and R' will be 87% of R; therefore, if we discard all sections of membranes that are more than 3 times the thickness of a well-defined membrane, the observed radii of curvature will vary from a correct reading to one that is understated by only 13%. Thus, the average observed radius of curvature will probably be about 5% too small when this method of omitting the most oblique sections is utilized.

The actual measurement of the radius of curvature was carried out utilizing a transparency containing a set of circles of appropriate sizes with calibrated circumferences. The circles were fitted to the curved region of the ghost membranes on the electron micrographs, and the length and radius of curvature of each region recorded. In most measurements, oblique sections of thickness more than 3 times the thickness of the thinnest membrane were omitted, but in some measurements both perpendicular and oblique sections were included.

Although the actual curvature of a three-dimensional surface is $(1/R_1 + 1/R_2)$, the measurement of a large number of radii of curvature in a population of ghosts sectioned at random angles assures that a representative set of curvatures has been obtained. Also, in most cases, the surface is sufficiently close to a sphere that one radius of curvature is not very different from the other.

RESULTS

Effect of Ca^{2+} on Endocytosis

As the ghosts utilized in these experiments were prepared in the presence of EDTA, the Ca content of the ghosts was quite low. The total Ca content of 15 different preparations of ghosts is summarized in Table 1. Since the concentration of ghosts in our assays was 1 mg protein/ml, the concentration of Ca^{2+} due to the ghosts was less than 1 μM. This low level of Ca^{2+} was due to the preparation of the ghosts in EDTA, and this method assured that the effects of added Ca^{2+} could be observed without the presence of appreciable amounts of Ca^{2+} in the membrane.

TABLE 1. *Ca content of ghosts*

	Protein (nmoles/mg)
Average of 15 preparations	0.41
Highest value	0.73
Lowest value	0.24
SD	±0.12
SEM	±0.03

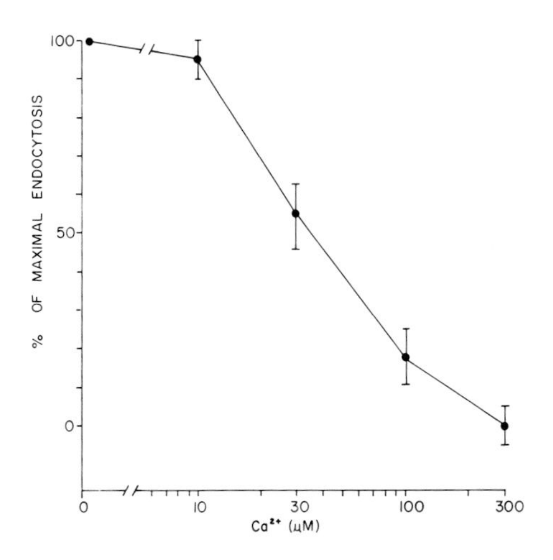

FIG 1. Effect of added Ca^{2+} on endocytosis. Values are the average of 15 separate experiments on 8 different ghost preparations. The error bars represent the standard error of the mean. Endocytosis was carried out in a system that was 1 mg ghost protein/ml, 3 m*M* in $MgCl_2$, 3 m*M* in Tris-ATP, and 50 m*M* in TES–TEA, pH 7.4. Incubation was 30 min at 37°C.

A large number of measurements of the dependence of endocytosis upon Ca^{2+} concentrations have been made utilizing ghosts with the low Ca content shown in Table 1. A summary of these measurements is shown in Fig. 1, which represents the average of 15 separate curves of endocytosis versus Ca^{2+} concentration, done utilizing 8 different preparations of ghosts. In some cases, the curves were done on the same preparation of ghosts at different times after the ghosts were prepared. There seemed to be no correlation between the age of the ghosts and the Ca inhibition curve. In only one of the 15 cases was any stimulation seen at low levels of Ca^{2+}, and in that one a 50% stimulation of endocytosis appeared at 10 μM Ca^{2+}. The circumstances surrounding that measurement suggested that this apparent stimulation was, in fact, caused by a depressed level of endocytosis at low Ca^{2+}, since the same preparation of ghosts showed more endocytosis when observed on the following day.

In order to control more carefully the actual level of free Ca^{2+} when no Ca^{2+} was added, an experiment was also done in which the total Ca present in the system was measured, and in which EGTA was present to obtain the point at zero Ca^{2+} concentration. The data obtained in this way are summarized in Fig. 2 and Table 2. In Fig. 2 the percent of the membrane area which was taken into vacuoles was estimated by the acetylthiocholine esterase assay and the Ca^{2+} concentrations shown correspond to the amount of Ca^{2+} added. The total Ca in the assay medium plus ghosts was measured by atomic absorption spectroscopy, the results of which are shown in Table 2. The 2.3 μM Ca present in the absence of added Ca^{2+} presumably originated

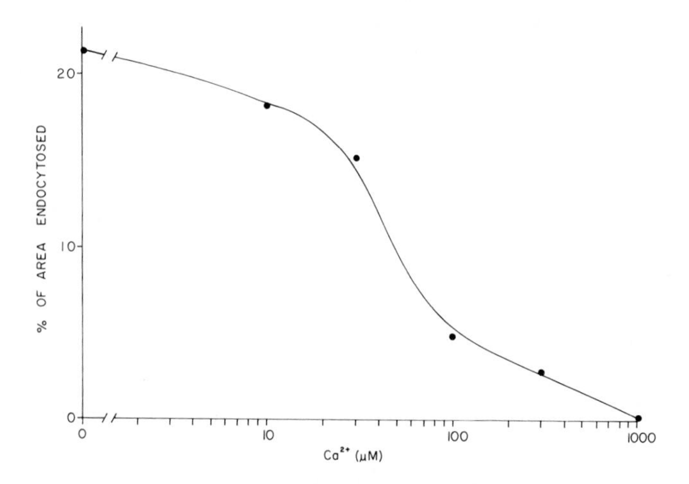

FIG. 2. Effect of added Ca^{2+} on endocytosis. Concentration of Ca^{2+} *added* is plotted versus endocytosis as measured by acetylthiocholinesterase disappearance. Conditions were the same as in Fig. 1. The samples to which no Ca^{2+} was added also contained 0.5 m*M* Mg-EGTA, to assure an extremely low free Ca^{2+} level.

from the ghosts, the ATP or the buffer. The presence of EGTA, with its 100-fold preference for Ca^{2+} over Mg^{2+}, acted to reduce the level of free Ca^{2+}. It would chelate a substantial portion of the total Ca^{2+}; hence, the level of Ca^{2+} that was available to these membranes was very low indeed.

The ionic content of the medium strongly influenced the rate and extent of endocytosis. Figure 3 shows the influence of Na^+ and K^+ on the extent of endocytosis. Under the conditions used, Na^+ showed a strong inhibition of endocytosis, whereas K^+ showed an equally strong stimulation. The effects of Na^+ and K^+ appear to be additive to those of Ca^{2+}, since they

TABLE 2. *Total Ca present at low amounts of added $CaCl_2$*

$CaCl_2$ added (μM)	Total Ca present (μM)
0	2.3 ± 0.0
10	12.9 ± 0.6
30	36.1 ± 0.6

These measurements were made on the same samples used for the endocytosis results shown in Fig. 1. Total Ca was measured by atomic absorption, and is the average of determinations on two samples. The error shown is the standard error of the mean.

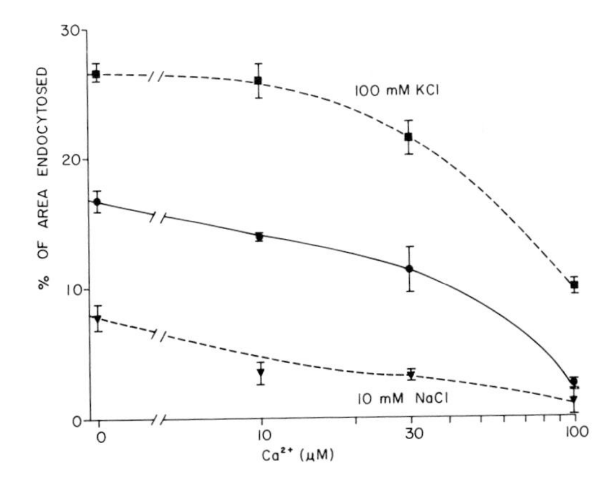

FIG. 3. Effect of K^+ and Na^+ on endocytosis. Error bars represent the extent of endocytosis, measured by acetylthiocholinesterase disappearance varied between experiments. In order to pool the results, the data were expressed relative to the maximal endocytosis in each experiment. Conditions were the same as in Fig. 1; ●, no further additions; ■, 100 m*M* in KCl; and ▲ 10 m*M* in NaCl.

remained effective in the presence of Ca^{2+}. These strong effects of the ions commonly present in erythrocytes were unexpected and their implications are discussed below.

Membrane Curvature Versus Time

When the shape of erythrocyte ghosts is plotted as a function of time, a complex pattern is seen, as depicted in Fig. 4. The echinocytic or crenated shapes rapidly disappear, being displaced mainly by disc shapes; the

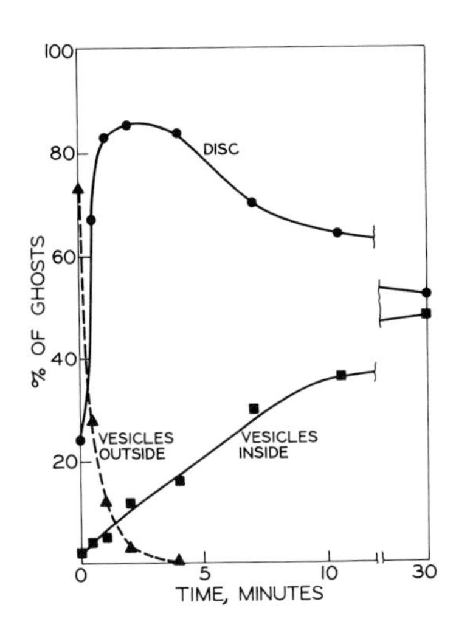

FIG. 4. Shapes of ghosts vs time of incubation. The reaction was started by addition of ghosts to a concentration of 1 mg protein/ml. Samples were fixed, stained, embedded, sectioned, and were examined by electron microscopy. The number of ghosts of each kind were counted on the electron micrographs.

spherendocytic forms then gradually appear. In some cases the appearance of the spherendocytes occurred after a lag of 5 or 10 min, although in the experiment shown here they appeared immediately. The analysis of this complex series of events may be simplified if the average curvature of the electron microscopic section of ghosts is used as a measurement of their shape. This was done for an experiment similar to that shown in Fig. 4 and the results are displayed in Fig. 5. The complex set of shape changes seen in the ghosts can be rationalized on the basis of a steady change in the curvature of the membrane. Curvature was defined as positive when the radius of curvature pointed toward the interior of the cell, and negative when the radius of curvature pointed toward the outside of the cell. The many

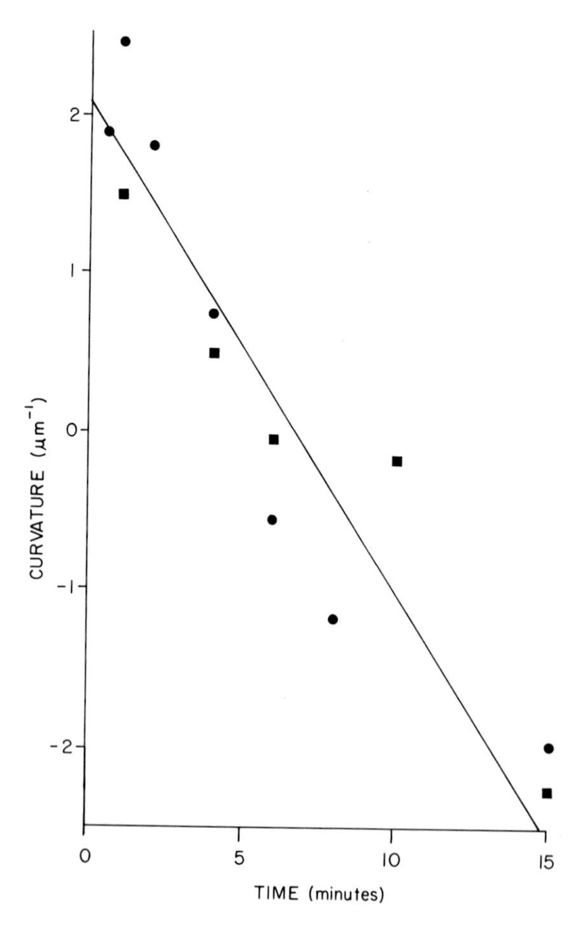

FIG. 5. Average curvature of ghosts vs time of incubation. Membrane curvature was obtained as described in the methods section of this paper. ■, Samples for which oblique sections were included in calculating curvature; ●, samples in which oblique sections were not used.

small spicules of the echinocytic shape gave it a strongly positive curvature, while the positive curvature of the rim of the discocyte was offset by the negative curvature of the concavities. In the spherendocyte, the radius of curvature was negative since the radius pointed toward the inside of the vacuole, which was the outside of the original red cell. This negative curvature was sufficient to overcome the positive curvature remaining in the envelope which bounded these vacuoles, with the results shown, a final negative curvature of greater than 2 μm^{-1}.

DISCUSSION

The data presented here show clearly that there was a negligible amount of Ca^{2+} in purified erythrocyte membranes, when care was taken to remove it. No level of Ca^{2+} in these membranes caused a stimulation of endocytosis, and higher levels of Ca^{2+} caused complete inhibition. Two possible explanations for this inhibition of endocytosis by Ca^{2+} may be proposed: Ca^{2+} may cause diversion of the metabolic energy produced by ATP hydrolysis, or it may interact with the membrane or vacuolating system in such a way as to make the membrane too rigid for vacuolation to occur.

The first suggestion implies that vacuolation and the Ca^{2+} transporting enzyme share a common source of metabolic energy. If this were so, the tapping of this source of energy by the Ca^{2+} transporting enzyme would divert the available chemical free energy and prevent it from being used for membrane movement. The situation would then be analogous to that which occurs when ion transport and oxidative phosphorylation compete for the central high-energy state in mitochondria. In that case also, the transporting of Ca^{2+} leads to prevention of the other energized functions of the mitochondria, particularly oxidative phosphorylation. Supporting this hypothesis is the fact that the inhibition of endocytosis occurs at the point at which Ca^{2+} transport begins to tap the energy supply significantly, as is indicated by the Ca^{2+} stimulation of ATP hydrolysis.

The presence of Ca^{2+}, and of other divalent cations, is known to decrease the fluidity of artificial bilayers (10). High levels of Ca^{2+} may similarly make the erythrocyte membrane more rigid, and thus make endocytosis difficult or impossible. Such a mechanical effect of Ca^{2+} would have no direct connection with its effect on the Ca^{2+} ATPase, except that it may be necessary to overload the Ca^{2+} transporting system in order to get sufficient Ca^{2+} to the appropriate membrane locations. Even though the membrane is sufficiently leaky that $Ca^{2+}{}_{out} \cong Ca^{2+}{}_{in}$, the Ca^{2+} concentration *at* the membrane may be kept low by an active pump, until the pump is overloaded by high levels of Ca^{2+}.

Having established the behavior of erythrocyte ghosts, it becomes important to explain the differences between the behavior of pure erythrocyte ghosts and of resealed ghosts (7). It is at this point that a careful experimental comparison would be valuable. Such a comparison has not yet been carried out, but the data shown in Fig. 3 suggest that the exact experimental conditions of incubation may strongly affect the details of the endocytosis observed. Thus, under our conditions, the exposure of the membrane to 100 m*M* KCl causes a very strong stimulation of endocytosis, while exposure of the membrane to a much lower concentration of NaCl causes an equally great inhibition. This suggests that the establishment of a high level of internal K^+, as well as a low level of Ca^{2+}, would be necessary for optimal endocytosis. The situation in resealed ghosts involves the redistribution of not only Ca^{2+} but also other substances which are moved across the membrane. The observed requirement for Ca^{2+} in resealed ghosts thus may be related to a complicated system of molecular redistributions.

Since resealed ghosts contain a number of compounds capable of chelating Ca^{2+} or Mg^{2+}, the effect of these chelators on endocytosis must also be taken into account. At a given total Ca concentration, the free Ca^{2+} would be much lower in resealed than in white ghosts. One of these chelators might inhibit endocytosis unless combined with Ca^{2+}, thereby explaining the requirement for Ca^{2+}. Clearly, further comparison of resealed ghosts, whole erythrocytes, and purified membranes will be necessary in order to obtain insight into the reasons for the behavior of this system under different circumstances.

The entire sequence of shapes seen during the vacuolation series can be explained simply as the equilibrium shapes necessary to obtain the lowest energy for a given relationship between the surface tension of the inner and outer halves of the membrane bilayer. Thus, if the inner half of the bilayer is shrunken relative to the outer half, the curvature of the membrane must become very positive. The red cell achieves this highly curved structure by the creation of many small protruding crenations of high positive curvature. If the inner half of the bilayer is expanded, whether by the operation of some expansile protein system or through a change in the surface properties of the phospholipids, the cell will gradually reduce its curvature by eliminating the crenation and become spherical. If the inner half of the bilayer expands still further, a shape of nearly zero curvature can be achieved; this is approximately true of the discocyte shape. If the expansion of the inner half of the bilayer continues, the cell is forced to create inward-directed vacuoles, whose curvature is strongly negative, until the spherendocytic shape is achieved. Thus, the series echinocyte–discocyte–stomatocyte–spherendocyte can be rationalized on the basis of the average curvature of the membrane. As predicted from this line of thought, a plot of the average curvature of the membrane versus time showed a smoothly varying curvature of the membrane as the curvature went from strongly positive at the beginning of the experiment to strongly negative when spherendocytes were created.

One can relate this to the fact that the energy is made available in a linear way, as the hydrolysis of ATP is linear with time. For a bilayer, the curvature of the bilayer is directly related to the energy input in the form of the difference between the surface tensions of the inside and outside of the bilayer (11). Thus, the curvature of the membrane may directly reflect the amount of energy put into the inside half of the bilayer in the form of an increase in its area, i.e., a decrease in surface tension. Clearly, a more sophisticated analysis than this will be necessary for accurate description of the system, because when a supporting network is present underneath the bilayer, the energy is no longer directly proportional to the curvature, but depends upon the exact shape of the vesicle. In any case, the smooth decrease in the average curvature with time indicates that curvature of the membrane is directly related to energy input as is suggested by this model.

We have attempted, in this chapter, to put into perspective the effects of Ca^{2+} on endocytosis in erythrocyte membranes and the relationship between energy input and erythrocyte membrane shape. Clearly, much more extensive comparisons between erythrocyte membranes and whole erythrocytes will be necessary before final conclusions can be drawn with regard to the effects of Ca^{2+} and the relationship of curvature to energy input. However, we believe it is clear that the purified erythrocyte membranes are uniquely useful for analyzing the effects of ions and the relationship of energy metabolism to shape, because of their purity, the easy access of the inside of the membrane, and the capability of a comparison between the simplified system and the more complicated whole erythrocyte system.

REFERENCES

1. Hayashi, H., Plishker, G. A., Vaughan, L., and Penniston, J. T. (1975): *Biochim. Biophys. Acta,* 382:218–229.
2. Penniston, J. T., and Green, D. E. (1968): *Arch. Biochem. Biophys.,* 128:339–350.
3. Penniston, J. T. (1972): *Arch. Biochem. Biophys.,* 153:410–412.
4. Hayashi, H., and Penniston, J. T. (1973): *Arch. Biochem. Biophys.,* 159:563–569.
5. Hayashi, H., and Penniston, J. T. (1974): *Biochem. Biophys. Res. Commun.,* 61:1–7.
6. Ben-Bassat, I., Bensch, K. G., and Schrier, S. L. (1972): *J. Clin. Invest.,* 51:1833–1844.
7. Schrier, S. L., Junga, I., and Seeger, M. (1973): *Proc. Soc. Exp. Biol. Med.,* 143:565–567.
8. Ginn, F. L., Hochstein, P., and Trump, B. F. (1969): *Science,* 164:843–845.
9. Kant, J. A., and Steck, T. L. (1972): *Nature* [*New Biol.*], 240:26–28.
10. Schnepel, G. H., Hegner, D., and Schummer, U. (1974): *Biochim. Biophys. Acta,* 367:67–74.
11. Evans, E. A. (1974): *Biophys. J.,* 14:923–931.

Membranes and Disease, edited by L. Bolis, J. F. Hoffman, and A. Leaf. Raven Press, New York, © 1976.

Endocytosis in Resealed Human Erythrocyte Ghosts: Abnormalities in Sickle Cell Anemia

Stanley L. Schrier and Klaus G. Bensch

Stanford University School of Medicine, Stanford, California 94305

Endocytosis has been shown to occur in resealed and unsealed erythrocyte ghosts (1,2). The process appears to involve an inward buckling of the membrane, then a narrowing of the neck of the invaginated pouch, and finally a membrane fusion which restores the continuity of the erythrocyte plasma membrane and results in the formation of a sealed endocytic vacuole (3). The substrate for ghost endocytosis appears to be Mg-ATP (2–4), and there is evidence associating ghost endocytosis with activation of the membrane-associated Ca, Mg-ATPase and its linked Ca efflux pump (3). The evidence is based on a similarity in substrate requirements of Ca, Mg-ATPase, Ca efflux, and ghost endocytosis, with all three phenomena showing requirements for Mg-ATP as substrate (3). There is also a biphasic response to Ca, such that increments of added Ca initially intensely stimulate and then profoundly inhibit the three phenomena (3,5) (Fig. 1). An inhibitor of Ca efflux and Ca, Mg-ATPase also inhibited ghost endocytosis (3).

In the studies to be described, the effect of a variety of membrane expanding agents (6) on ghost endocytosis was tested and the results related to the capacity of these agents to inhibit Ca, Mg-ATPase, and Ca efflux. The uniqueness of Mg-ATP as a substrate for endocytosis was retested (5). Ghost endocytosis was then measured in sickle disease, and when it was observed to be profoundly reduced, the capacity of sickle ghosts to extrude ^{45}Ca was measured (7).

METHODS

The methods for resealing human erythrocyte ghosts were previously published (3). The only change in these studies is that the ratio of washed red blood cells (RBC) to hemolysis solution was 1 : 10 rather than 1 : 20. The methods for measuring Ca, Mg-ATPase activity, ^{45}Ca efflux, and endocytosis were exactly as previously described (3).

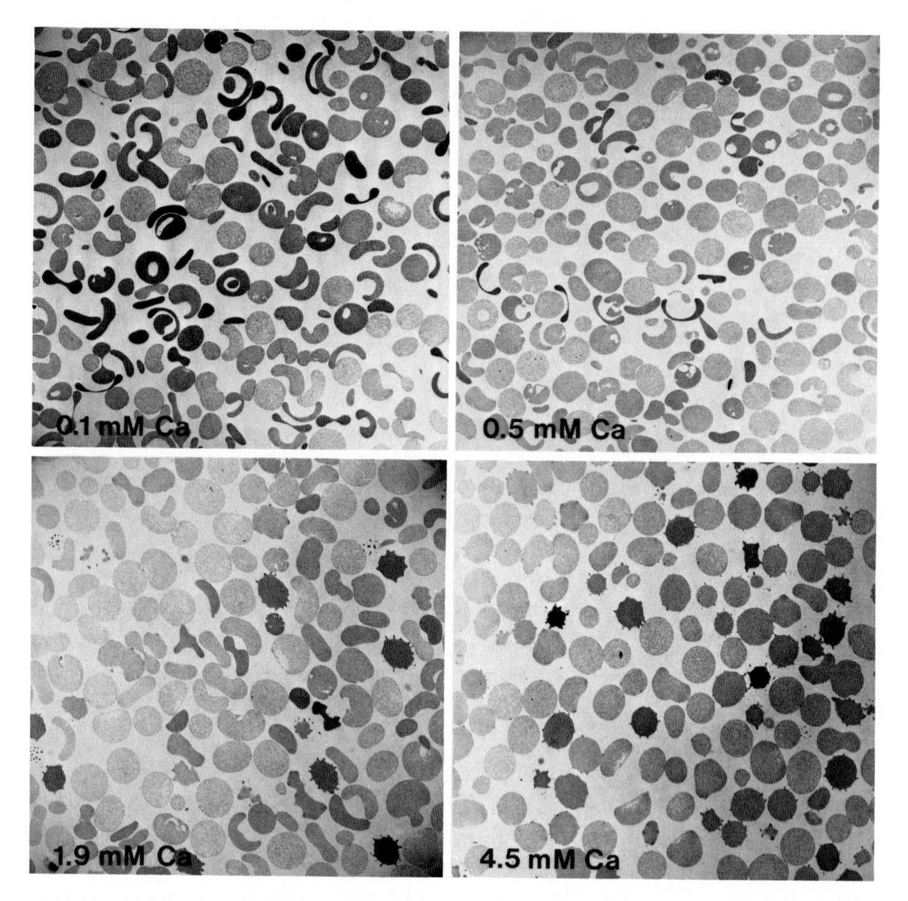

Fig. 1. Transmission electron microscopy of endocytosis in ghosts related to the actual calcium content of ghosts (nmoles Ca/10^{10} ghosts). The numbers superimposed on the different electron microphotographs indicate the calcium content of ghosts calculated from ^{45}Ca measurements performed on the ghosts isolated at the beginning of the incubation period.

RESULTS AND DISCUSSION

Effect of Other Membrane-Expanding Agents on Endocytosis in Resealed Human Erythrocyte Ghosts

The membrane-expanding agents primaquine, hydrocortisone, vinblastine, and chlorpromazine can induce endocytosis in intact human erythrocytes as distinguished from resealed ghosts (2). The effect of these agents on ghost endocytosis was tested in parallel with measurement of the effect of these agents on ^{45}Ca efflux and Ca, Mg-ATPase activity (Table 1). Endocytosis was measured morphologically and recorded as an arbitrary score relating

TABLE 1. *Effect of Drugs on resealed erythrocyte ghosts*

	^{45}Ca efflux	Ca, Mg-ATPase	Endocytosis index
Control	100[a]	100[a]	3.2
Chlorpromazine[b]			
0.225 m*M*	29	26	—
0.45 m*M*	20	16	0.15
Primaquine			
2.0 m*M*	95	84	4.3
Vinblastine			
0.5 m*M*	104	98	4.6
Hydrocortisone			
12.5 m*M*	102	82	5.6

[a] Control value normalized to 100.
[b] The drugs were added during the hemolysis procedure along with the ATP, Mg, and Ca.

numbers of endocytic vacuoles to numbers of ghosts sectioned in photographic enlargements of transmission electron-microscopic preparations. Only chlorpromazine inhibited ghost endocytosis and it was the only agent to inhibit Ca, Mg-ATPase and ^{45}Ca efflux with an apparent I_{50} of 0.15 m*M*.

These observations buttress the conclusions obtained with ruthenium red (3), in that an agent such as chlorpromazine, known to inhibit ^{45}Ca efflux and Ca, Mg-ATPase activity, also inhibits ghost endocytosis. Several points can be drawn from the observations: (i) The relationship between Ca, Mg-ATPase, ^{45}Ca efflux is emphasized (8). (ii) Inhibitors of Ca, Mg-ATPase and ^{45}Ca efflux block endocytosis. (iii) Caution must be used in analyzing lists of agents with a known capacity to produce membrane expansion and stomatocytosis in intact erythrocytosis (9). All of the agents tested in Table 1 produced endocytosis in intact erythrocytes and also produce stomatocytic changes. However, these drugs have other actions. Primaquine is a potent oxidant, and chlorpromazine, in the concentrations used, is a selective inhibitor of Ca, Mg-ATPase, ^{45}Ca efflux, and it alone inhibits endocytosis in ghosts.

Uniqueness of ATP as a Substrate for Endocytosis

It has previously been reported (4,5) that ATP is a uniquely required substrate for endocytosis. However, it had been reported that other nucleotide triphosphates could energize ^{45}Ca efflux (10). We therefore repeated these experiments (5), using nucleotide triphosphates other than ATP, to determine if they could energize Ca efflux with or without parallel induction of endocytosis. In these experiments, morphologic observation of ghost endocytosis was substituted for the radioisotopic method previously used. Table 2 indicates that ATP, CTP, ITP, and GTP energized ^{45}Ca efflux equally under the conditions used. Omission of added nucleotide resulted

TABLE 2. *Nucleotide substrate requirements for Ca efflux and endocytosis*

Ghosts resealed in the presence of 1 m*M* ^{45}Ca + 2.5 m*M* Mg Cl_2 +	^{45}Ca efflux (nmoles ^{45}Ca/min per 10^{10} ghosts)
2.5 m*M* ATP	84
2.5 m*M* CTP	74
2.5 m*M* ITP	82
2.5 m*M* GTP	64

Abbreviations: ATP, adenosine triphosphate; CTP, cytidine triphosphate; ITP, inosine triphosphate; GTP, guanosine triphosphate.

in a background ^{45}Ca efflux of 32 nmoles Ca/min per 10^{10} ghosts, presumably energized by residual ATP in the ghosts. Figure 2 shows that transmission electron microscopy revealed endocytosis only in ATP-containing ghosts.

Therefore, while Ca, Mg-ATPase activity and ^{45}Ca efflux are required for ghost endocytosis, ^{45}Ca efflux is not sufficient *per se* to induce endocytosis. ATP must also be present at some critical but undetermined level. Hence, ATP may have at least two roles in ghost endocytosis—one in providing substrate for Ca, Mg-ATPase and ^{45}Ca efflux, and the other in providing substrate for another aspect of the endocytosis phenomenon—perhaps involving a protein kinase.

Endocytosis in Sickle Disease

Because endocytosis in resealed ghosts involves a plastic property of the membrane, modulated by energy considerations expressed as the interaction of ATP with the membrane, we studied endocytosis in resealed sickle ghosts (Fig. 3). At all time intervals studied, endocytosis was decreased in SS ghosts and in the example shown in Fig. 3 at 15 min, SS ghosts produced one-eighth the normal number of endocytic vacuoles. A patient with autoimmune hemolytic anemia and comparable reticulocytosis had no comparable impairment of ghost endocytosis (Fig. 4).

Because it had been shown that SS erythrocytes contain more Ca than do normal erythrocytes and might have an impairment of Ca extrusion (11), we studied Ca efflux in SS resealed ghosts, looking for an explanation for the depression in endocytosis. Initial rates of ^{45}Ca efflux were normal in SS ghosts (Table 3), and in the reticulocyte matched controls, which included one patient with iron-deficient anemia, one patient responding to vitamin B_{12} therapy, one patient responding to folic acid therapy, one patient with autoimmune hemolytic anemia, and one patient with myeloproliferative disease and hemolysis. Ca, Mg-ATPase activity was normal in SS and provided a normal Ca/P ratio. However, when the ^{45}Ca extrusion from ghosts was followed over a prolonged period of time (Fig. 5), there was distinct retention of ^{45}Ca in resealed sickled ghosts in the order of 10 times the normal controls.

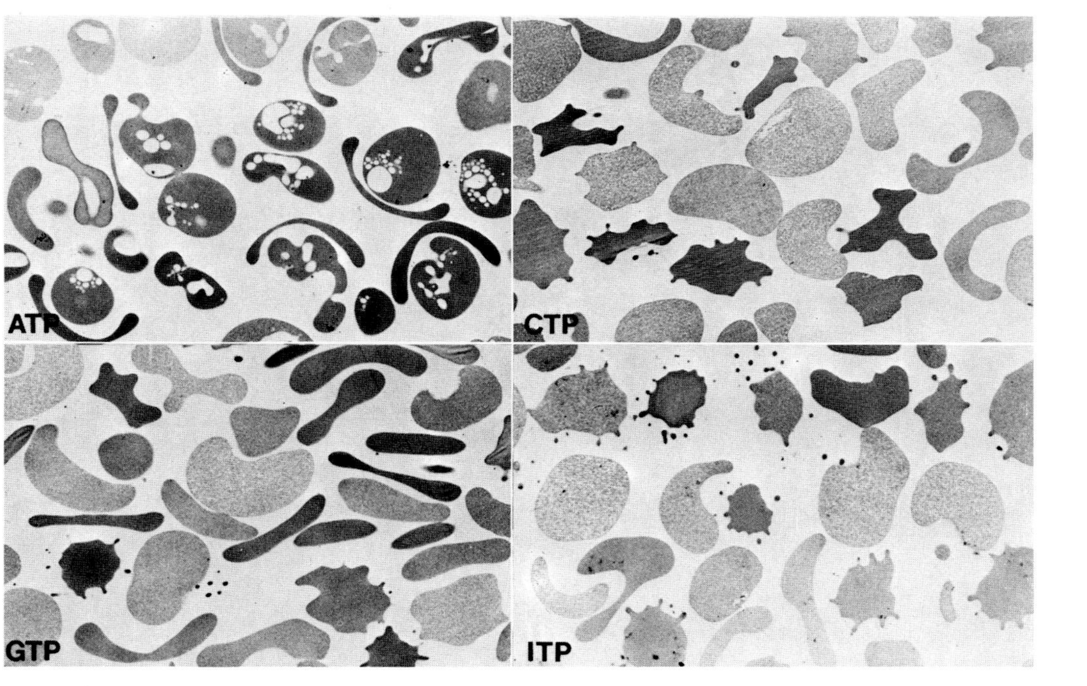

FIG. 2. Transmission electron microphotographs obtained from ghosts resealed in the presence of 1 m*M* calcium, 2.5 m*M* magnesium, and the following nucleotide triphosphates: ATP, CTP, GTP, and ITP. These samples were obtained after 15 min of incubation at 37°C.

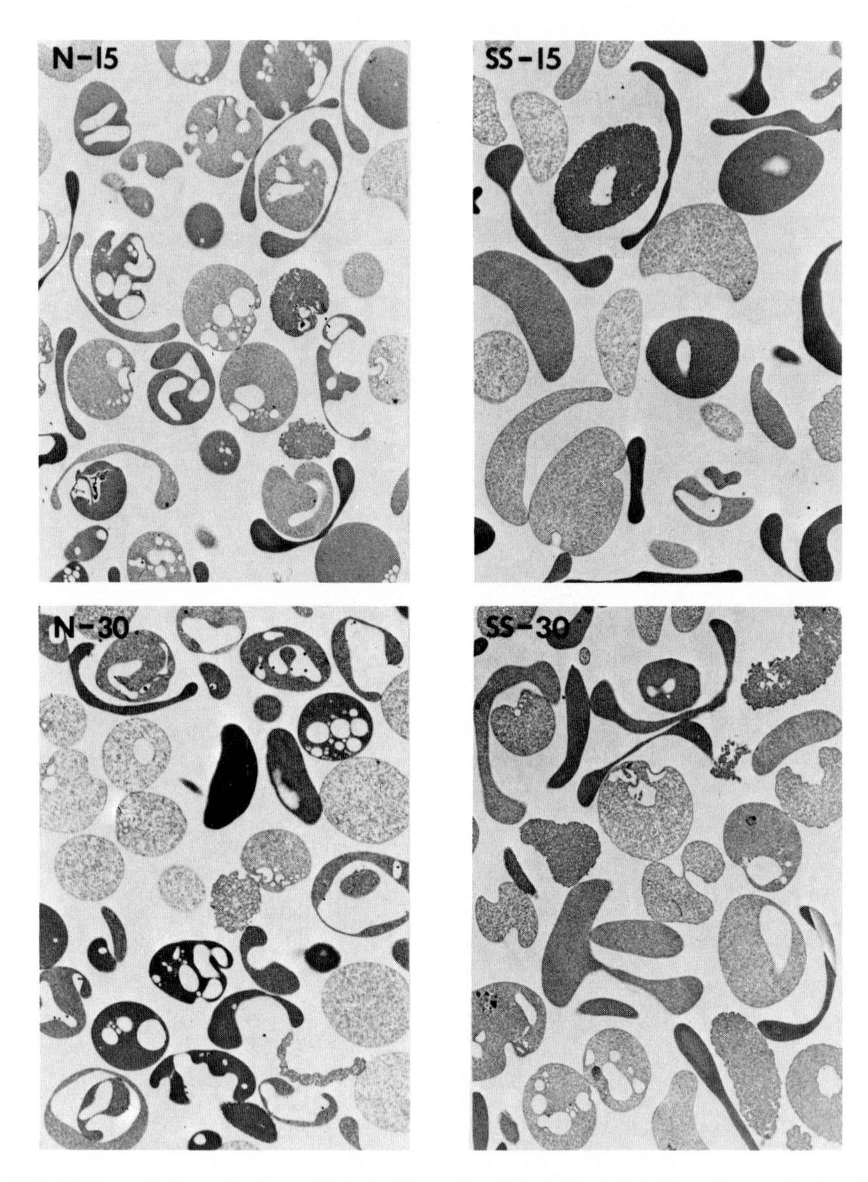

FIG. 3. Transmission electron microphotographs of normal and sickle cell anemia ghosts resealed in the presence of 1 m*M* calcium, 2.5 m*M* magnesium, and 2.5 m*M* ATP and then incubated. N-15 and SS-15 refer to normal ghosts and SS ghosts incubated for 15 min at 37°C. N-30 and SS-30 refer to normal and SS ghosts incubated at 37°C for 30 min.

When studied systematically, ghosts from reticulocyte-rich patients retained $1\frac{1}{2}$ times as much ^{45}Ca as the normal controls, but SS ghosts retained approximately 7 times as much Ca as the normal controls and 4 times as much ^{45}Ca as the reticulocyte-rich preparations (Table 4).

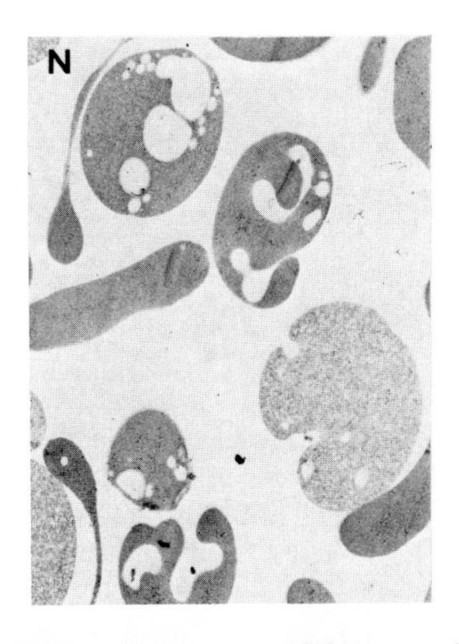

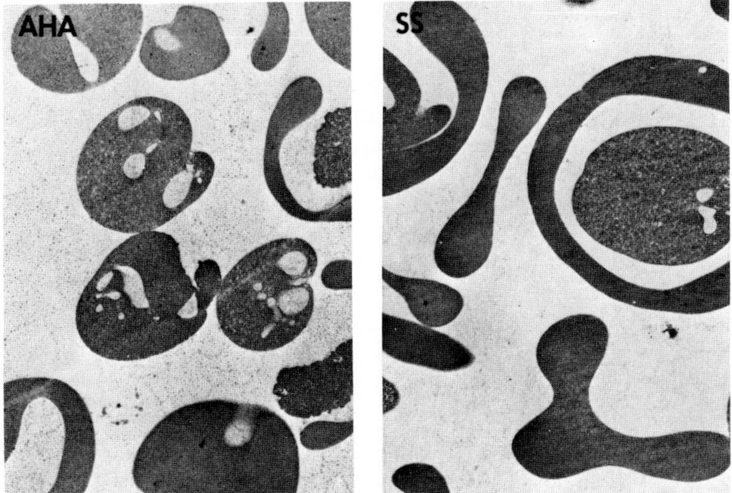

FIG. 4. Transmission electron microphotographs of normal ghosts (N) sickle cell anemia ghosts (SS) and ghosts obtained from a patient with autoimmune hemolytic anemia (AHA). All ghosts were resealed in the presence of 1 m*M* calcium, 2.5 m*M* magnesium, and 2.5 m*M* ATP, and incubated for 30 min at 37°C.

In order to determine whether there was sufficient ATP left at 15 and 30 min of incubation, $AT^{32}P$ was resealed into normal and SS ghosts and the Ca, Mg-ATPase activity and residual $AT^{32}P$ at designated time intervals was measured (Table 5). It can be seen that Ca, Mg-ATPase activity decreased after 15 min when most of ^{45}Ca had been extruded, confirming our prior observations (3) that Ca addition results in enhanced Ca, Mg-ATPase

TABLE 3. *Ca efflux from resealed ghosts*

Subjects tested	Mean ± SEM	Ca/P ratio ± SEM[a]
Normal controls	35 ± 2.6	0.92 ± 0.12
Reticulocyte-rich patients	36.9 ± 2.67	—
AS patient	47	—
SS patients	34 ± 3.3	1.1 ± 0.16

Values expressed as nanomoles ^{45}Ca per minute per 10^{10} ghosts.
[a] Relates moles of calcium extruded to moles of ATP hydrolyzed to ADP and P_i.

and Ca depletion results in reduced Ca, Mg-ATPase. There is abundant $AT^{32}P$ left in SS ghosts at 15 and 30 min for energization of ^{45}Ca efflux.

Therefore, given the fact that Ca, Mg-ATPase is normal in SS ghosts, that initial rates of ^{45}Ca efflux and Ca/P ratios are normal in SS, and that there is adequate ATP available in SS ghosts, the residual ^{45}Ca in SS ghosts must be at sites not accessible to an otherwise normally functioning Ca, Mg-ATPase and Ca pump. We propose that this additional Ca is in sites which

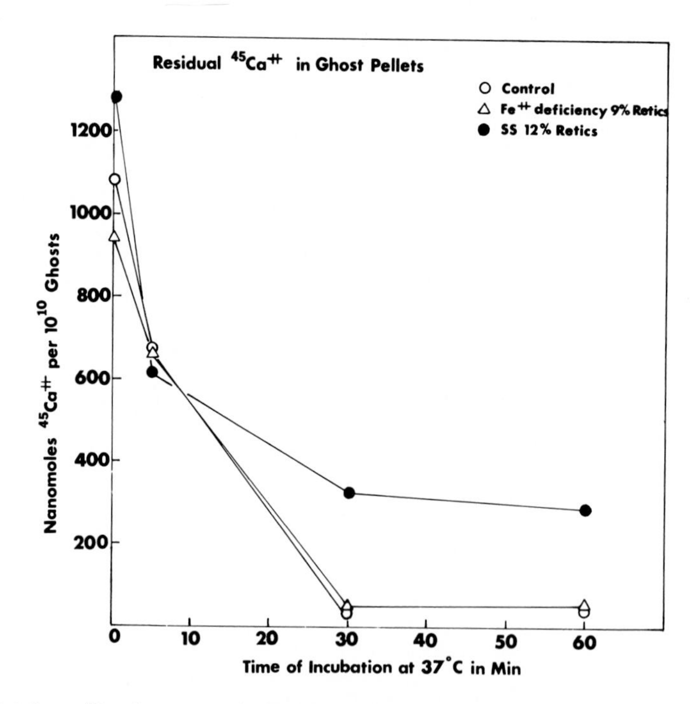

FIG. 5. Calcium efflux from resealed human erythrocyte ghosts. The efflux is recorded as the residual ^{45}Ca remaining in the ghost pellet at varying periods of incubation at 37°C (abscissa). The ordinate indicates the nanomoles ^{45}Ca per 10^{10} ghosts. The normal control ghosts are indicated by the open circles, the reticulocyte-rich control is indicated by the open triangle, and the patient with sickle cell disease is indicated by the closed circle.

TABLE 4. *Residual Ca in ghosts*

	After 30 minutes[a]		After 60 min[a]	
Subjects tested	Mean ± SEM	Range	Mean ± SEM	Range
Normals (10)	31 ± 4	16–40	27 ± 3	12–39
Reticulocyte-rich patients (5)	49 ± 7	35–67	44 ± 2[b]	42–48
AS patient	16		18	
AC patient	20		15	
SS patients (6)	150 ± 32	58–330	+155 ± 39[c]	78–296

Values expressed as nanomoles ^{45}Ca per 10^{10} ghosts.
[a] Incubation time at 37°C.
[b] Different from control $p < 0.02$.
[c] Different from control $p < 0.01$.

TABLE 5. *Calcium, ATP content, and Ca,Mg-ATPase of resealed ghosts*

Content	0	5 min	15 min	30 min[a]
Ca				
Control	739	284	20	18
SS	890	510	127	114
ATP				
Control	2,048	1,935	1,613	1,516
SS	1,862	1,615	1,446	1,354
Ca,Mg-ATPase activity expressed as Δ ^{32}Pi generated				
Control	–	339	581	613
SS	–	354	600	708

All values expressed as nanomoles per 10^{10} ghosts.
[a] Incubation time at 37°C.

adversely modulate plastic properties of the sickle ghost, rendering it rigid and incapable of endocytosis. How sickle hemoglobin conditions the cytoplasmic surface of the membrane so that it binds additional Ca in a site not accessible to the Ca pump is not known. Elsewhere in this volume, Dr. J. Palek has presented data with intact sickle erythrocytes showing that they bind excessive amounts of Ca. Our data using resealed sickle ghosts are in complete agreement with those of Palek. However, Palek found no differences between SS and resealed reticulocyte-rich controls. Our data here are in conflict and require resolution.

SUMMARY

Endocytosis in resealed ghosts requires Ca, Mg-ATPase activity and Ca efflux. Agents that interfere with Ca, Mg-ATPase also block endocytosis.

ATP must be present within the resealed ghost. SS ghosts bind 7 times as much Ca as controls and this Ca is not accessible to a normally functioning Ca, Mg-ATPase and Ca pump. This Ca may produce a membrane rigidity which blocks endocytosis. The role of Ca in the irreversible sickled cell *in vivo* remains to be determined.

ACKNOWLEDGMENT

This work was supported by a grant from the National Institutes of Health, AM-13682.

REFERENCES

1. Penniston, J. T., and Green, D. E. (1968): *Arch. Biochem. Biophys.,* 138:339–2315.
2. Ben-Bassat, I., Bensch, K. G., and Schrier, S. L. (1972): *J. Clin. Invest.,* 51:1833–1844.
3. Schrier, S. L., Bensch, K. G., Johnson, M., and Junga, I. (1975): *J. Clin. Invest.,* 56:8–22.
4. Hayasai, H., Plishker, G. A., Vaughn, L., and Penniston, J. T. (1975): *Biochem. Biophys. Acta,* 382:218–229.
5. Schrier, S. L., Junga, I., and Seeger, M. (1973): *Proc. Soc. Exp. Biol. Med.,* 143:565–567.
6. Seeman, P. (1972): *Pharmacol. Rev.,* 24: 583–655.
7. Schrier, S. L., Bensch, K., Junga, I., Johnson, M. (1975): *Clin. Res.,* 23:406a.
8. Schatzmann, H. J. (1970): In: *Calcium and Cellular Function,* edited by A. W. Cuthbert, pp. 85–95. Macmillan, London.
9. Deuticke, B. (1968): *Biochem. Biophys. Acta,* 163:494.
10. Lee, K. S., and Shin, B. C. (1969): *J. Gen Physiol.,* 54:713–729.
11. Palek, J. (1974): In: *Proceedings of the First National Symposium of Sickle Cell Disease,* edited by J. I. Hercules, A. N. Shechter, W. A. Eaton, and R. E. Jackson, pp. 219–221. DHEW Publication #NIH 75-723. Bethesda, Maryland.

Membranes and Disease, edited by L. Bolis, J. F. Hoffman, and A. Leaf. Raven Press, New York, © 1976.

Transmembrane Movements and Distribution of Calcium in Normal and Hemoglobin S Erythrocytes

J. Palek, A. Church, and G. Fairbanks

Hematology Research Laboratory, Division of Hematology, The Saint Vincent Hospital and University of Massachusetts Medical School, Worcester, Massachusetts 01610, and The Worcester Foundation for Experimental Biology, Shrewsbury, Massachusetts

It is well recognized that calcium is excluded from normal erythrocytes through the action of an active transport system (1–3) and that the accumulation of this cation in cells undergoing ATP depletion is associated with dramatic changes in red cell membrane deformability, permeability, and shape (4–13). On the other hand, there is only limited information on the mechanism of these alterations (4,14–16) and on movements of calcium and its role in altered physical properties of abnormal erythrocytes (16–18). The subject of this chapter is (i) to review our data on the movements and distribution of radioactive calcium in fresh and ATP-depleted red cells and the relationship of calcium accumulation to the binding of the complex of fibrillar proteins, termed "spectrin," to the inner surface of the membrane, and (ii) to demonstrate alterations in the membrane calcium permeability of red cells containing hemoglobin S, and to document that these alterations are related to the sickling phenomenon.

THE EXCHANGE OF RADIOCALCIUM BETWEEN PLASMA AND FRESH RED CELLS

Figure 1 depicts relationships between the entry of radiocalcium and the net calcium content during 12 hr incubation of fresh human erythrocytes. The open circles depict the amount of radiocalcium entering cells expressed in micromoles per liter of cells, calculated from the specific activity of ^{45}Ca in the supernatant plasma. Since the net red cell calcium content, as measured by atomic absorption spectrophotometry (AAS, solid line) remained constant (16 ± 5 μmoles/liter), the rise in radiocalcium represented an exchange of radiocalcium with an exchangeable red cell calcium pool. This pool, when estimated after steady state has been achieved after 4 hr of incubation, was 3.4 ± 1.2 μmoles/liter cells, that is, approximately 20% of total red cell calcium. It is unlikely that the initial uptake represented binding of ^{45}Ca to the external cell surface as the radiocalcium

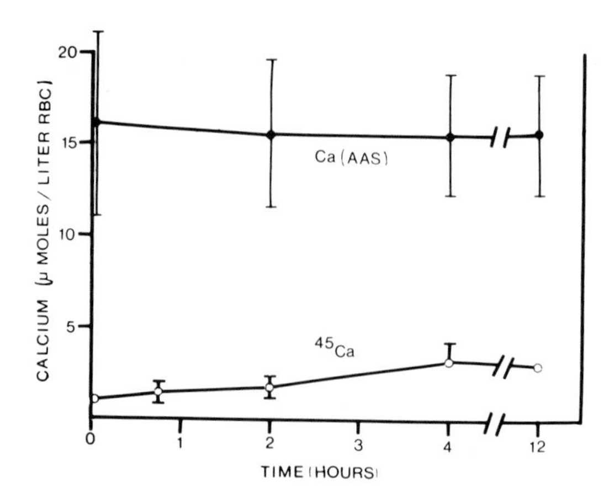

FIG. 1. Relationship between red cell calcium content and the uptake of ^{45}Ca during 12 hr incubation at 37°C. Twenty percent red cell suspension in plasma was equilibrated with 5% CO_2 in air and incubated with ^{45}Ca (specific activity mCi/mg of calcium). At times indicated, cells were washed 3 times in isotonic saline solution, ^{45}Ca determined by liquid scintillation counting (55), and the calcium content determined by atomic absorption spectrophotometry (19).

was not removed by washing cells in isotonic saline solution containing 5 m*M* EGTA.

In order to determine if the exchangeable red cell calcium was present principally in the cytosol or in the membrane, red cells previously incubated for 4 hr with ^{45}Ca were hemolyzed in 5 m*M* Tris HCl buffer (pH 6.9), unwashed ghosts and supernatants were separated by centrifugation and their radioactivity determined (Table 1). The radioactivity per unit volume of unwashed ghosts was only slightly higher than that of the supernatant and 79.2% of the total exchangeable red cell calcium was present in red cell

TABLE 1. *Distribution of calcium in human red cells*

	Ghosts	Supernatant
^{45}Ca		
dpm/100 μl	633	489
% dpm recovered	20.8	79.2
Total Ca (AAS)		
μmoles/liter	16.8	<1.0
% recovered	86 ± 8%	–

20% red cell suspension in plasma incubated for 4 hr (37°C) with ^{45}Ca. Red cells were hemolyzed in 5 m*M* Tris HCl buffer (pH 6.9), ghosts and supernatants were separated at 40,000 × *g* for 30 min in a Sorvall RC-5 centrifuge and their net calcium content and ^{45}Ca determined as described in Fig. 1.

cytosol. In contrast, most of the total red cell calcium as measured by atomic absorption spectrophotometry (AAS) was firmly associated with the membrane as previously shown (5,19,20), and 96 ± 12% was recovered in red cells washed in isotonic saline solution containing 5 m*M* EDTA. These data suggest that there are at least two compartments of calcium in *fresh, normal* erythrocytes: a "nonexchangeable" calcium which can be recovered in red cell membranes, representing approximately 80% of total cellular calcium and an exchangeable calcium, present in the cytosol (3.4 ± 1.2 μmoles/liter cells), presumably in the form of Ca–ATP complex (21).

THE LOCATION OF CALCIUM IN ATP-DEPLETED RED CELLS

Figure 2 shows the relationship between the net calcium content and radiocalcium (in μmoles/liter cells, as calculated from the specific activity in the supernatant plasma) in cells which have depleted their ATP content. In this case, the radiocalcium entering cells approximated a net gain in cellular calcium. We have studied the distribution of this accumulated radiocalcium between unwashed ghosts and supernatants made from ATP-depleted red cells by hypotonic hemolysis (Table 2). The radiocalcium per unit volume of ghosts and ghost-free hemolysates approached equilibrium, suggesting no preferential partitioning of radiocalcium to the membranes of ATP-depleted cells. When such ghosts were subsequently washed without a complexing agent (Fig. 3), they retained 3.8 ± 1.2 μmoles of residual

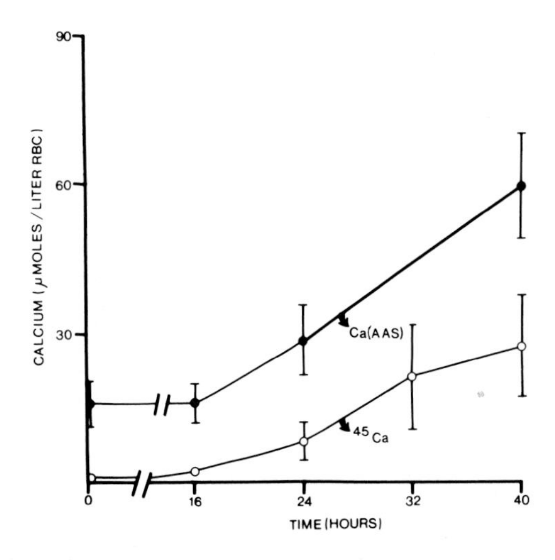

FIG. 2. Relationship between changes in net calcium content and the uptake of radiocalcium during incubation of ATP-depleted red cells at 37°C. ATP at 16 hr <0.18 μmoles/ml RBC. Forty percent red cell suspension in plasma was incubated with ^{45}Ca. At times indicated, cells were washed and their ^{45}Ca and net calcium content determined as described in Fig. 1.

TABLE 2. *Distribution of ^{45}Ca in ATP-depleted red cells*

	Ghosts	Supernatant
dpm/100 μl	1369	1136
% dpm recovered	19.4	80.6

40% red cell suspension in plasma was incubated for 40 hr (37°C) with ^{45}Ca. Red cells were hemolyzed and ghosts and supernatants separated as described in Table 1.

radiocalcium per liter ghosts which was not removed from ghosts by additional washing. This represented 6.8% of the radioactivity of the initial washed red cells. In contrast, >99.9% of the initial red cell radioactivity was removed when 5 m*M* EGTA was added to the washing solution. It can be seen that ghosts washed without a complexing agent exhibited a marked shape distortion and increase in density, whereas ghosts from which most of calcium was removed by EGTA were smooth and flat, similar to

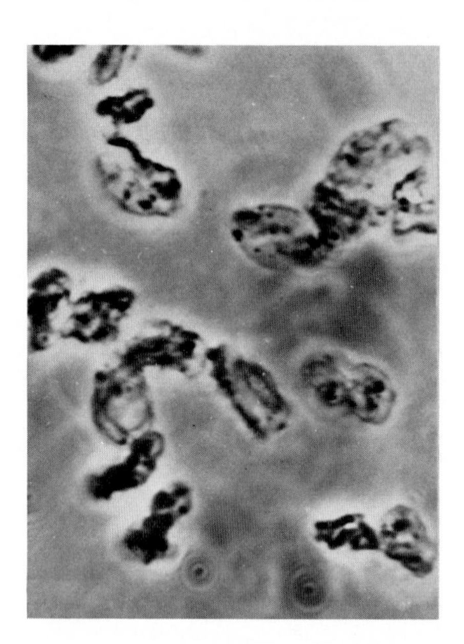

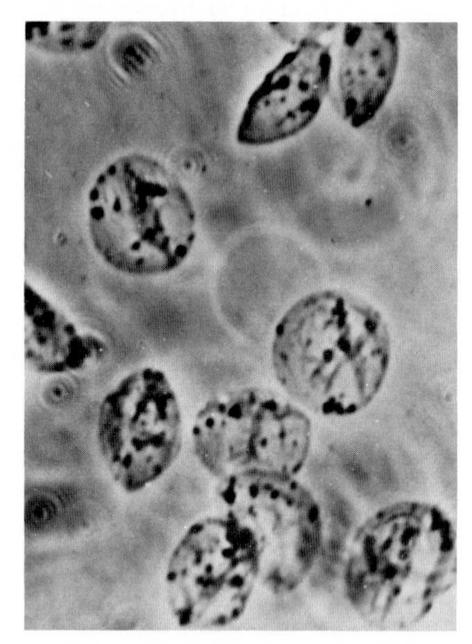

FIG. 3. Effect of EGTA on the shape and calcium content of ghosts from ATP-depleted erythrocytes. A 40% red cell suspension in plasma was incubated at 37°C for 44 hr in the presence of ^{45}Ca. Red cells were homolyzed in 30 vol of 5 m*M* Tris HCl buffer (pH 6.9) with or without EGTA (5 m*M*) and washed 3 times in the same buffer. **Left:** Tris (5 m*M*, pH 6.9); Ca (3.8 ± 1.2 μmoles/liter). **Right:** Tris + EGTA (2 m*M*), Ca (0.2 ± 0.8 μmoles/liter). Subsequently, ghosts were separated by high-speed centrifugation (40,000 × *g*, 60 min at 0°C on Sorval RC-5 centrifuge). The radiocalcium was determined by liquid scintillation counting (55) and ghosts examined in a phase-contrast microscope after fixation with osmium tetroxide as described (15).

ghosts of fresh erythrocytes. This suggested that a binding of calcium in the amounts of approximately 3×10^{-6} moles/liter ghosts to one of the membrane components was associated with these alterations of ghost shape. It was shown previously that similar calcium concentrations were required for echinocytic transformations and decrease in volume of ghosts from fresh red cells, and that this effect was inhibited by a SH-blocking agent (14,15, 20).

THE ROLE OF CALCIUM IN THE SPECTRIN–MEMBRANE INTERACTION

Ghosts from red cells depleted in ATP have an increased content of nonhemoglobin protein (4,22,23). We asked, first, if the increase in nonhemoglobin protein is due to increased retention of spectrin, the high molecular weight, fibrillar protein attached to the inner surface of the membrane (24–27,32); and, second, whether the calcium that accumulates in ATP-depleted cells is bound to this protein.

The first question is addressed in Fig. 4, which shows sodium dodecyl sulfate (SDS) polyacrylamide gel electrophorograms of red cell membrane proteins. These gels were scanned with a densitometer to determine the integrated intensity of Coomassie blue stain in bands I, II, and III. The

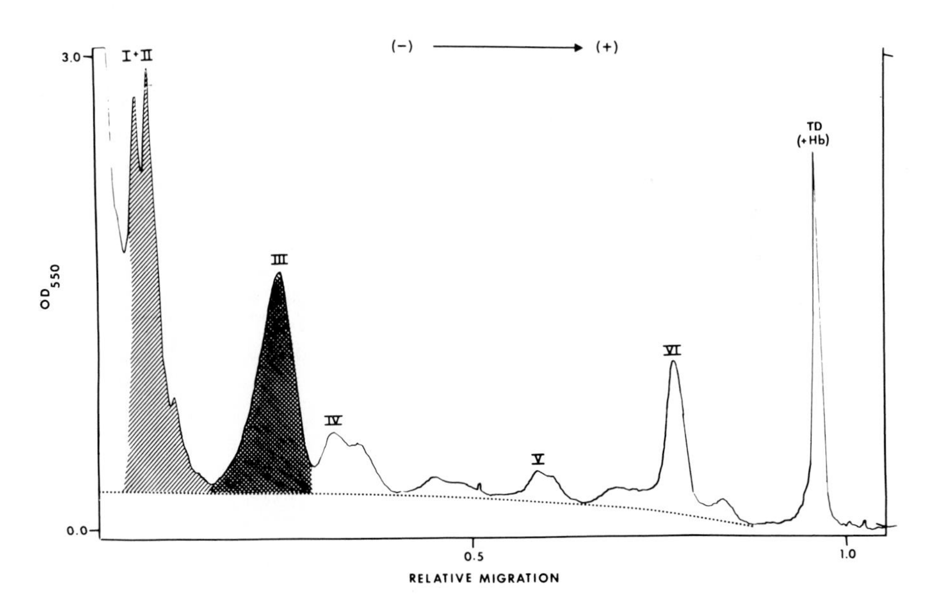

FIG. 4. A densitometric scan of a gel stained for protein with Coomassie blue after electrophoresis of ~2–3 μg of ghost protein, according to ref. 29. The areas under the spectrin peaks I and II, and under the peak of the tightly bound membrane protein III, were determined.

content of spectrin was then estimated from the ratio of the spectrin bands, I and II, to the major tightly bound membrane protein, III (Fig. 5). It can be seen that the ratio of spectrin to band III is significantly higher in ghosts of ATP-depleted red cells than in ghosts from fresh cells hemolyzed under the same conditions. This suggests that ATP depletion of the intact cells results in increased spectrin retention by the ghost membranes. Similar results were recently reported by Lux and John (28).

In order to investigate the locus of calcium binding in membranes of ATP-depleted cells, we have carried out the selective elution of spectrin (plus component V) by dilution into warm 1 m*M* Tris HCl (pH 8), as previously described (30). The supernatant and the pellet containing membrane vesicles were separated by centrifugation at 40,000 × *g* for 60 min (0°). Both fractions were then assayed for radiocalcium and examined on SDS polyacrylamide gels (Fig. 6). We observed that most of the spectrin and component V were extracted at this very low ionic strength both in the presence and absence of 1 m*M* EGTA. However, when the elution was performed in the absence of the calcium chelator, 90–95% of the radiocalcium remained associated with the membrane. (It was interesting to note that this radioactivity migrated in SDS gels in a diffuse band with an apparent average molecular weight of about 50,000.) When, on the other hand, spectrin elution and vesiculation was carried out in the presence of EGTA, more than 95% of the radioactivity was recovered in the supernatant. (In this case, the counts migrated faster than the tracking dye, presumably as an EGTA–calcium complex.) These data suggested that most of the calcium that accumulates in ATP-depleted erythrocytes is not associated with the elutable fraction, either the high molecular weight spectrin components or the 45,000-dalton actin-like polypeptide V.

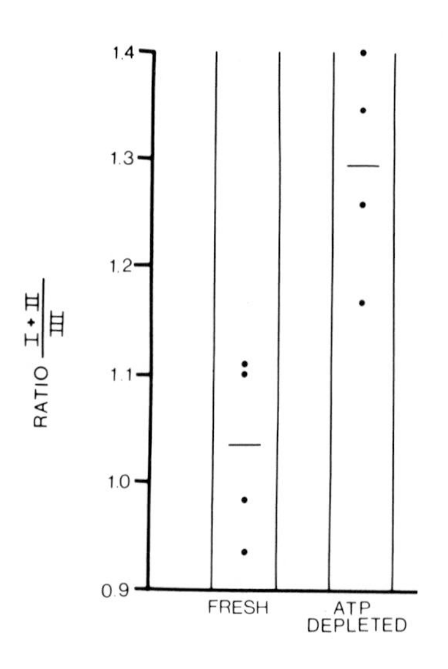

FIG. 5. Retention of spectrin by ghosts from erythrocytes depleted in ATP by incubation of heparinized blood at 37°C for 48 hr without glucose. Washed red cells were hemolyzed in 30 vol of 10 m*M* Tris HCl (pH 8), washed twice in the same buffer, red cell membrane proteins were separated electrophoretically, and stained gels were scanned on a densitometer (Fig. 4). The integrated staining intensities in bands I, II, and III were measured and the ratios of bands I + II to band III were calculated.

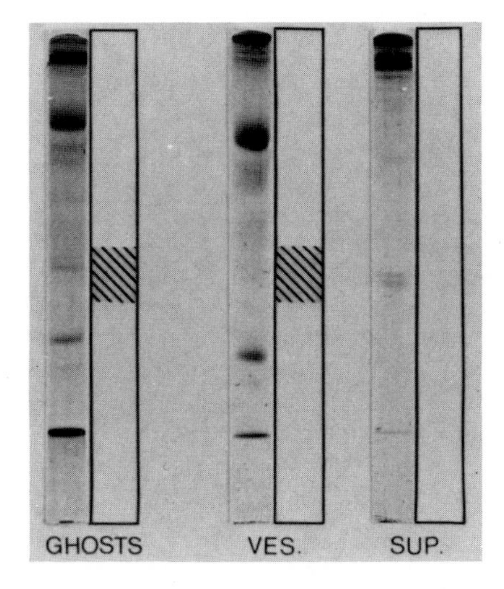

FIG. 6. The extraction of spectrin and ^{45}Ca from ghosts of ATP-depleted red cells by m*M* Tris HCl (pH 8), according to ref. 30. The supernatant and the pellet containing membrane vesicles were separated by centrifugation at 40,000 × *g* for 60 min (0°C), and both fractions were examined on SDS-polyacrylamide gels. For each fraction, the pattern of Coomassie blue staining is shown at the *left,* and the position of ^{45}Ca in autoradiographs of unstained gels is diagrammed at the *right.*

This conclusion was strongly supported by the complementary experiment, in which we extracted ghosts with the nonionic detergent, Triton X-100. This procedure selectively extracts the principal integral protein, III, together with the sialoglycoprotein and other major components, including VI (31). A meshwork of spectrin, component V, and part of the membrane lipid is recovered as a gelatinous precipitate. Extraction under these conditions solubilized 81% of the radiocalcium, leaving less than 20% associated with the spectrin–component V complex. Again, the results indicate that the calcium that accumulates in ATP-depleted cell membranes is not bound primarily to the elutable actomyosin-like proteins of the membrane inner surface, but to other, yet-to-be-identified, components in the insoluble matrix of integral membrane proteins and lipids. (We emphasize that although the migration of radiocalcium in SDS gels is attractive and intriguing, it probably represents an artifact rather than association with a calcium-binding protein, since similar diffuse bands are seen in electrophorograms of radiocalcium in the absence of added membrane proteins.)

Our findings in this line of investigation suggest that the effect of calcium on red cell rigidity and discocyte–echinocyte transformation is not mediated by direct binding of the cation to spectrin or polypeptide V. This is also supported by recent observations (33) that a spectrin-containing aggregate, produced by incorporation of Ca (>1 m*M*) into ghosts, did not bind any ^{45}Ca. Although it is known that about 80% of membrane-bound calcium is associated with membrane protein (34,35), the mode of this association and the calcium binding proteins in red cell membranes remain to be elucidated. At this stage, we can only speculate as to what the mode of association may be. It is clear, however, that the binding site for the extra bound calcium in ATP-depleted cells need not be a major membrane component, because the binding of only about 3×10^{-19} calcium ions per ghost is associated with the induction of echinocytic shape. Calcium at this locus might (i) reduce the flexibility of the membrane without participation of spectrin, (ii) establish additional binding sites for the fibrillar protein, or (iii) alter the conformation of the polymer indirectly by modulating the activity of a regulatory enzyme, such as a phosphoprotein phosphatase.

NET CHANGES IN CALCIUM CONTENT OF HEMOGLOBIN S RED CELLS

Our interest in hemoglobin S red cells was stimulated by earlier studies of Tosteson (36–38), who observed that the sickling phenomenon is associated with increased passive movements of sodium and potassium across the sickle cell membranes. We have investigated if similar increases in membrane permeability do occur in the case of calcium.

Table 3 shows the net calcium content as determined by AAS. It can be seen that while the AS heterozygotes had a normal calcium content, the patients with sickle cell anemia (HbSS) exhibited a statistically significant elevation of cellular calcium which was considerably higher than in normal red cells and in patients with other hemolytic anemias, as was also recently observed by others (16–18). Furthermore, the rise in cellular calcium could not be accounted for by increase in reticulocytes, as documented by absence of correlation between the reticulocyte count and red cell calcium content (not shown). However, the calcium content of red cells, separated on the basis of their density (44), was significantly higher in the bottom, ISC-rich fraction of packed HbSS red cells (64 ± 12 μmoles/liter; 64 ± 8% ISC, mean ± SD of 6 determinations) than in the top fraction rich in reticulocytes, but low in ISC (42 ± 8 μmoles/liter, ISC < 2%, $p < 0.01$).

TABLE 3. *Red cell calcium content (mean ± SD) of normal individuals, heterozygotes for sickle cell hemoglobin (HbAS), homozygotes for hemoglobin S (HbSS), and hemolytic anemias*

Red cells	Calcium (μmoles/liter)	*p*
Control	19 ± 6	
HbAS	17 ± 4	>0.5[a]
HbSS	50 ± 21	<0.01[a] <0.02[b]
Other hemolytic anemias	25 ± 18	

Calcium determined by AAS (19).

[a] Statistical difference from control red blood cells.

[b] Statistical difference from red cells of other hemolytic anemias.

TABLE 4. *Changes in red cell calcium content during incubation of 20% red cell suspensions under oxygenated (95% O_2 + 5% CO_2) and deoxygenated (95% N_2 + 5% CO_2) conditions*

		Calcium (μmoles/liter red cells)			
	Fresh	Oxygenated[a]	p	Deoxygenated[a]	p
Controls	18 ± 5	20 ± 6		18 ± 6	
HbAS	16 ± 4	17 ± 4	>0.05	16 ± 6	>0.05
—[b]		16 ± 5	>0.05	17 ± 6	>0.05
HbSS	46 ± 18	49 ± 20	>0.05	102 ± 36	<0.01
Other hemolytic anemias	26 ± 12	28 ± 12	0.05	22 ± 36	>0.05

Mean ± SD of 6 determinations. Value p is a statistical difference between oxygenated and deoxygenated sample.

[a] Incubated for 4 hr at 37°C.

[b] HbAS red cells were incubated under 20% O_2 + 80% CO_2 or 20% N_2 + 80% CO_2.

In order to evaluate the cause of the elevated calcium content of fresh HbSS erythrocytes, we have measured net changes in cellular calcium during incubation of cells under oxy- or deoxygenated conditions. Table 4 summarizes the net changes in cellular calcium content during incubation at 37°C of 20% red cell suspensions under oxygenated (95 O_2 + 5% CO_2) or deoxygenated (95% N_2 + 5% CO_2) conditions. HbSS red cells incubated under deoxygenated conditions exhibited a statistically significant net gain in cellular calcium. On the other hand, there was no increase in calcium content of these erythrocytes when incubated under oxygen. Likewise, there was no net increase in cellular calcium in a control group of patients with other hemolytic anemias, or in hemoglobin AS individuals when examined both under oxygenated or deoxygenated conditions. Becase sickling *in vivo* appears to exhibit similar relationships to oxygen saturation as that *in vitro* (39), it can be proposed that a gain in calcium similar to that *in vitro* occurs *in vivo* in HbSS cells during repeated episodes of sickling, thereby accounting for the elevated calcium content of fresh HbSS erythrocytes.

^{45}CA INFLUX INTO HEMOGLOBIN S ERYTHROCYTES

In order to evaluate the mechanism of calcium accumulation during incubation of deoxygenated HbSS erythrocytes, we have studied the unidirectional calcium movements using ^{45}Ca. Figure 7 shows influx of ^{45}Ca into red cells from supernatant plasma as represented by changes in the ratio of red cell disintegrations per minute to plasma disintegrations per minute (RBC ^{45}Ca/plasma ^{45}Ca) during 6 hr of incubation at 37°C. In contrast to normal erythrocytes, HbSS cells which have transformed in >90%

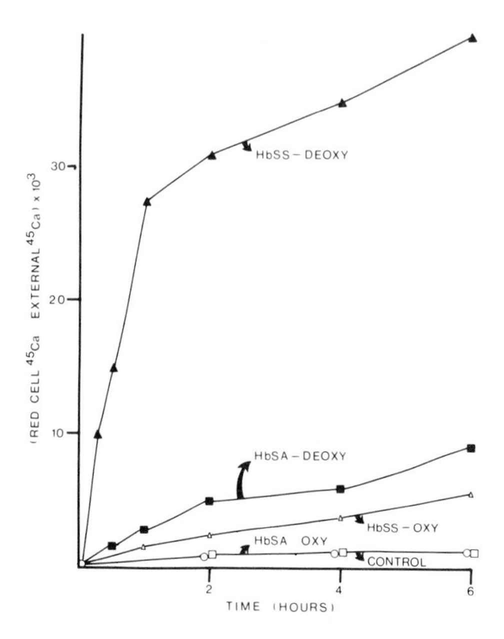

FIG. 7. Influx of radiocalcium into red cells of controls, HbSS, and HbAS individuals. Control or HbSS red cell suspensions containing traces of ^{45}Ca (specific activity 0.03 mCi/mg plasma Ca) were incubated at 37°C under 95% N_2 + 5% CO_2 (▲), or 95% O_2 + 5% CO_2 (○, △). HbAS red cell suspensions were incubated under 80% CO_2 + 20% N_2 (■), or 80% CO_2 + 20% O_2 (□). The ratio of red cell ^{45}Ca/plasma ^{45}Ca was determined at times indicated. Controls (○) are mean of 15 observations. HbSS patients (△) are means of 12 determinations in 8 HbSS individuals. HbAS individuals (□) are means of 4 determinations. (From J. Palek: Changes in calcium content and transmembrane calcium movements of hemoglobin S red cells *in vitro. Submitted for publication.*)

into sickled shape during incubation under deoxygenated conditions exhibited a marked increase in radioactivity indicative of a marked influx of calcium into these cells. On the other hand, considerably lower amounts of radiocalcium entered these cells when they were incubated under oxygenated conditions. Patients with other hemolytic anemias containing HbA as their major hemoglobin were in a similar low range as the oxygenated HbSS cells both when incubated under oxy- or deoxygenated conditions (not shown).

The increase in ratio of RBC ^{45}Ca/plasma ^{45}Ca during incubation of deoxygenated HbSS blood could be divided into two kinetic components (Fig. 7). The first component was represented by a rapid initial uptake of ^{45}Ca by the cells during the first hour of incubation, indicating influx of calcium into these cells. At the end of this rapid kinetic component the amount of calcium entering cells as calculated from the specific activity of ^{45}Ca in the supernatant plasma was 64 $\pm$ 8 μmoles/liter cells. The net red cell calcium content determined by AAS was 62 $\pm$ 4 μmoles/liter cells at the onset of incubation and 71 ± 6 μmoles/liter cells at the end of the first rapid component of ^{45}Ca uptake (1 hr of incubation). The corresponding specific activities of ^{45}Ca in red cells at 1 hr of incubation was 0.028 mCi/mg Ca, while the specific activity of plasma ^{45}Ca was 0.032 mCi/mg Ca (means of 4 determinations). Therefore, approximately 80% of cellular calcium in fresh HbSS erythrocytes exchanged with ^{45}Ca, and the elevation of cellular calcium content in these cells resulted principally from the increase in the exchangeable fraction of cellular calcium. After approximately 1 hr, the net gain of calcium and a further increment of ^{45}Ca paralleled each other, suggesting that the second kinetic component of increase in red cell ^{45}Ca represented a net gain of calcium, which occurred after an isotopic equilibrium had been achieved between external radiocalcium and the exchangeable internal calcium. Nearly identical kinetics of radiocalcium entry into red cells were observed when ^{45}Ca was added to hemoglobin S cells 2 hr after the onset of deoxygenation when 90% or more of red cells have transformed into a sickled shape.

In contrast to patients with sickle cell anemia, red cells from heterozygotes for hemoglobin S (HbAS), when incubated under deoxygenated conditions, exhibited no appreciable uptake of ^{45}Ca. However, under such conditions, only 45% or less of cells transformed into sickled shape (not shown). When the percentage of sickling was increased to 86 ± 6% by deoxygenating cells with 20% N_2 and 80% CO_2 to decrease the pH, there was again an increase in RBC ^{45}Ca/plasma ^{45}Ca ratio (Fig. 7) which, however, was of considerably lower magnitude than in cells from HbSS homozygotes, shown on Fig. 7. (The incubation of HbSS red cells under the same conditions, i.e., under 20% N_2 and 80% CO_2 resulted in only slightly higher ^{45}Ca uptake than in HbSS cells incubated under 95% N_2 + 5% CO_2, not exceeding 18% of the ^{45}Ca uptake under the latter conditions.) The increased radiocalcium uptake was absent when HbAS red cell suspensions were incubated at the same pH under oxygenated conditions (20% O_2 + 80% CO_2). Thus, the markedly lower radiocalcium influx into HbAS cells than into HbSS cells suggests that the ^{45}Ca influx is related to the percentage of hemoglobin S in the cells.

The conclusion that the influx of calcium into deoxygenated HbSS red cells undergoing sickling resulted from the membrane alterations associated with sickling is further supported by an experiment, in which we have exposed HbSS red cells to carbon monoxide and subsequently incubated under 95% N_2 + 5% CO_2. The formation of carboxyhemoglobin effectively prevented sickling as well as influx of radiocalcium (not shown).

LOCALIZATION OF CALCIUM IN HEMOGLOBIN S ERYTHROCYTES

In order to study the localization of radiocalcium in HbSS red cells which have transformed into the sickled shape, we have compared the distribution of radiocalcium between HbSS ghosts and the supernatants after hypotonic

TABLE 5. *Distribution of ^{45}Ca in HbSS erythrocytes*

	^{45}Ca (dpm × 10^3/ml)	Hemoglobin (g/dl)
Supernatant	4.9	4.8
Ghosts	19.6	5.9
$\frac{\text{Ghosts}}{\text{Supernatant}} \times 100$	400 ± 58	123 ± 12

HbSS red cells suspensions were incubated for 4 hr at 37°C under 95% N_2 + 5% CO_2, and subsequently chilled to 0°C and oxygenated (2 min). Washed HbSS red cells were hemolyzed and ghosts and ghost-free supernatants were separated as described in Table 1 and the distribution of ^{45}Ca was compared to that of hemoglobin. (Mean ± SD of 4 experiments).

hemolysis of HbSS erythrocytes that were previously deoxygenated in the presence of radiocalcium (Table 5). It can be seen that in HbSS ghosts, there was approximately 4 times more radioactivity per unit volume than in the supernatant hemolysates, suggesting preferential accumulation of radiocalcium in HbSS membranes. In contrast, the concentration of hemoglobin in ghosts and supernatant hemolysates was close to an equilibrium.

EXTRUSION OF CALCIUM FROM HbSS RED CELLS

Since human erythrocytes contain a highly active ATP-dependent calcium pump extruding large amounts of calcium from cells into the external medium against a high concentration difference (1–3), we have investigated if the net gain in calcium during deoxygenation of HbSS erythrocytes was not in part due to a concomitant decrease in calcium extrusion from the cells. Figure 8 demonstrates the extrusion of calcium from intact hemoglobin S red cells under oxygenated conditions. These cells have accumulated calcium during previous deoxygenation for 4 hr at 37°C under 95% N_2 + 5% CO_2. They were subsequently rapidly cooled to 0°C, oxygenated, cells separated and resuspended in a plasma containing no ^{45}Ca. We have observed a rapid extrusion of the radioactive tracer from cells associated with corresponding increase in radioactivity of the supernatant medium. As a control to this experiment we have used normal red cells into which calcium has been introduced by a brief mixing of cells with a hypotonic medium containing 2 m*M* $CaCl_2$ with traces of ^{45}Ca. After this treatment, reconstituted normal red cells had retained over 70% of initial levels of intracellular magnesium and ATP. (Since the amount of calcium introduced into cells depended on the osmolarity and calcium concentration in the hypotonic medium, we were able to introduce into normal cells similar amounts of calcium as those found in HbSS cells exposed to previous deoxygenation.)

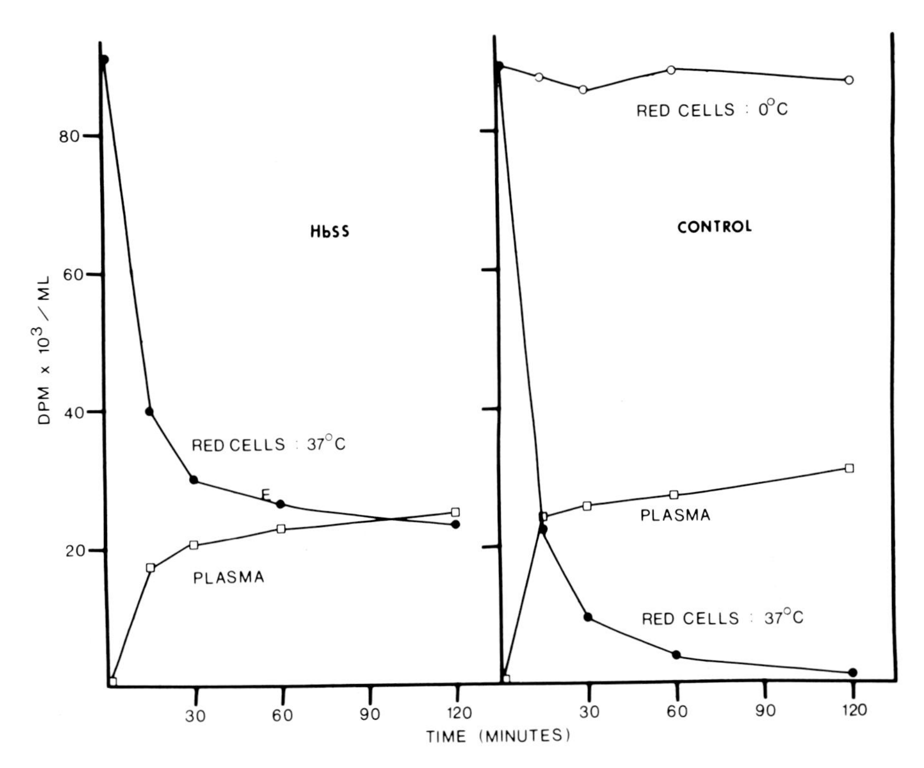

FIG. 8. Extrusion of radiocalcium from HbSS and normal erythrocytes. *Left,* HbSS red cell suspensions were incubated for 4 hr at 37°C under 95% N_2 + 5% CO_2 with traces of ^{45}Ca. The suspensions were then cooled to 0°C, oxygenated (2 min), red cells separated and resuspended in plasma containing no radiocalcium to the hematocrit of 20 ± 1%, and incubated at 37°C. Red cell and plasma ^{45}Ca (dpm × 10^3/ml) are plotted against the time of incubation. Data are means of 4 experiments. The mean red cell and plasma calcium content (by AAS) at the onset of incubation was 106 ± 22 μmoles/liter red cells and 2.2 ± 0.2 mmoles/liter plasma. *Right,* Normal red cells were loaded with calcium by suspending them at 0°C in hypotonic NaCl buffered with glycylglycine (10 m*M,* pH 7.4), containing $CaCl_2$ (2 m*M*) with traces of ^{45}Ca. Red cells were separated from ghosts, restored to isotonicity and resuspended in plasma without ^{45}Ca. Means of 5 experiments in which the initial cell calcium content (by AAS) (92 ± 18 μmoles/liter red cells) and ^{45}Ca were in a similar range as in the deoxygenated HbSS erythrocytes.

There was a rapid extrusion of the radioactive tracer from these cells which was not seen if these cells were either incubated at 0°C (Fig. 8) or depleted in ATP prior to the loading phase in the hypotonic medium (not shown). However, after 2 hr of incubation, HbSS red cells extruded significantly less ^{45}Ca (74 ± 6% the initial radioactivity) than normal erythrocytes (95% ± 7, $p < 0.02$).

We have also investigated the extrusion of calcium from reconstituted ghosts (Fig. 9) enriched in calcium, magnesium, and ATP (2 m*M*) during hemolysis. There were no differences in calcium extrusion from hemoglobin S red cell ghosts and ghosts from red cells of other hemolytic anemias

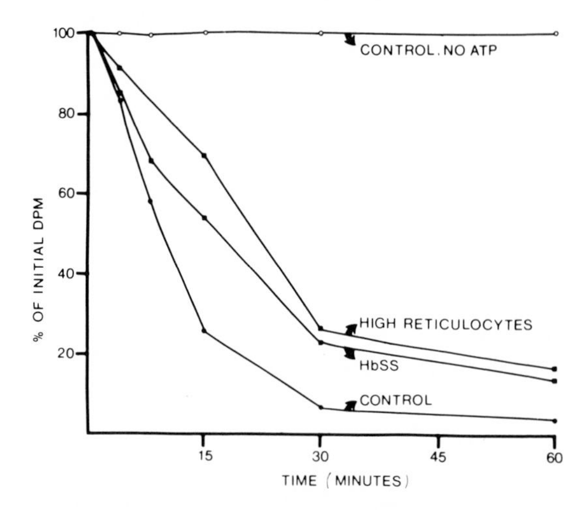

FIG. 9. Extrusion of calcium from reconstituted ghosts enriched in Ca^{2+}, Mg^{2+}, and ATP. Control and high reticulocyte (12–18%) red cells (from 1 patient with paroxysmal nocturnal hemoglobinuria and 2 patients with immunohemolytic anemia) were hemolyzed at 0°C in 5 m*M* Tris HCl buffer, pH 6.9, containing 2 m*M* $CaCl_2$ (with traces of ^{45}Ca) $MgCl_2$ and Na salt of ATP. The hemolysates were restored to isotonicity by 1.25 *M* KCl, 0.25 *M* NaCl solution, ghosts separated at 40,000 × *g* for 30 min at 0°C, resuspended in prewarmed (37°C) isotonic NaCl containing 2 m*M* $CaCl_2$, 5 m*M* KCl, 10 m*M* glycylglycine buffer (pH 7.4), and incubated at 37°C. At times indicated, ghosts were separated at 40,000 × *g* for 30 min at 0°C, and their radioactivity counted by liquid scintillation counting. Means of 3 experiments. The samples labeled "control, no ATP" were kept at 0°C throughout the incubation.

with similar increases in reticulocytes, although more rapid extrusion was observed in ghosts from normal red cells. However, considerable decreases in calcium extrusion may occur during deoxygenation of intact red cells when the increased binding of ATP to hemoglobin (40,41) may decrease the concentration of free red cell ATP, which may become a rate-limiting factor in calcium extrusion.

Therefore, it can be concluded that the accumulation of calcium in HbSS erythrocytes results from the influx of this cation into HbSS red cells which have transformed into sickled shape causing the opening of pathways allowing a free movement of calcium from plasma into cells according to the existing concentration difference.

THE ROLE OF CALCIUM IN THE DECREASED DEFORMABILITY OF HbSS ERYTHROCYTES AND FORMATION OF IRREVERSIBLY SICKLED CELLS (ISC)

Since red cell deformability depends on the ratio of calcium to ATP in the cells (4–6,15,21,41), it can be predicted that deoxygenated HbSS erythrocytes incubated without added glucose will exhibit increases in deformability earlier than the corresponding oxygenated HbSS red cells. This prediction is supported by measurements of filterability of fresh HbSS cells incubated under deoxygenated conditions with and without EDTA (5 m*M*). Results are shown in Fig. 10. At times indicated, cells were rapidly cooled to 0°C, reoxygenated for 2 min, and the flow rate of 0.2% red cell suspensions through 3-μm Nuclepore filters was determined. HbSS red cells previously incubated under deoxygenated conditions differed from HbSS cells incubated under oxygen in a longer filtration time, which was partly improved when deoxygenated HbSS red cells were incubated in the presence of EGTA (5 m*M*). This suggested that calcium, which accumulated in HbSS cells undergoing sickling, contributes to their altered deformability. Similar conclusions were previously made from viscosity measurements of incubated HbSS erythrocytes (55).

We have also investigated if the accumulation of calcium in HbSS erythrocytes contributes to the formation of irreversibly sickled cells (ISC). These cells constitute 4–44% of red cells in the oxygenated blood of patients with sickle cell anemia (42–44). They were shown to be formed *in*

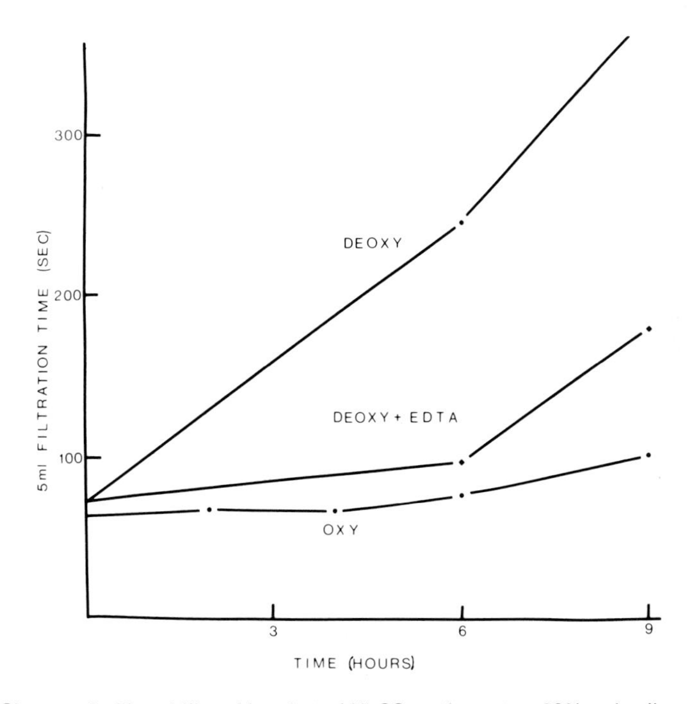

FIG. 10. Changes in filterability of incubated HbSS erythrocytes. 20% red cell suspensions in plasma were incubated at 37°C under 95% N_2 + 5% CO_2 or 95% O_2 + 5% CO_2 with or without EDTA (5 m*M*). At times indicated, samples were taken, rapidly cooled to 0°C, oxygenated (2 min), and diluted in isotonic saline solution containing 10 m*M* glycylglycine (pH 7.4) to make 0.2% red cell suspension. Ten ml of the suspension was put into Amicon cell containing a 3-μm Nuclepore filter and the time of flow of 5 ml of the suspension under a positive pressure of 10 cm H_2O was recorded. Representative data on 1 of 5 experiments.

vitro during prolonged incubation under deoxygenated conditions (45) and their irreversibly sickled shape was recovered in ghosts, suggesting a permanent membrane deformation (46). ISCs are rapidly removed from the circulation with a half-life of 2 days as compared with that of 5 days (^{51}Cr) of whole HbSS blood (42). ISCs also have lower mean corpuscular volume, higher density, decreased surface area-to-volume ratio, reduced cell potassium and water due to increased efflux of potassium from cells, elevated internal viscosity, reduced membrane flexibility, and they contain membrane-bound microaggregates consisting of hemoglobin (42,44,45,47–51).

Since ISCs were shown by others (16) and us to have elevated calcium content and their formation was shown by some (52), but not by others, to be stimulated by ATP depletion (51,53), we have studied the possibility that these cells result from calcium accumulation during sickling. Results shown in Table 6 indicate that the rate of formation of ISC was not influenced when ATP was maintained during incubation by adding adenine (0.54 m*M*) and inosine (12.7 m*M*), and when EGTA (5 m*M*) has been added to diminish calcium influx into cells. However, considerably more ISCs exhibited spiculated surface when incubated without the above additives, and the differentiation of these cells from unsickled echinocytes presented considerable difficulties (Fig. 11). Thus, although the cellular rigidity of HbSS blood depends on red cell calcium accumulation, ISC formation is not influenced by the accumulation of calcium. The elevated calcium content of ISC (16) appears to be a consequence, rather than cause, of membrane alterations leading to irreversible sickling. Conclusive evidence of the independence of ISC formation from calcium has been recently

TABLE 6. *Independence of ISC formation on calcium and ATP*

ISC (%)	Incub.: 1 hr		Incub.: 24 hr	
	Control	Ad–In–EGTA	Control	Ad–In–EGTA
Smooth				
I	11	11	6	22
I + II	21	21	9	43
Spiculated				
I	—	—	31	12
I + II	—	—	57	25
Total				
I	11	11	37	34
I + II	21	21	66	68

1% red cell suspension in plasma incubated at 37°C without or with adenine (Ad) (0.54 m*M*), inosine (In) (12.7 m*M*), and EGTA (5 m*M*). Cells were oxygenated for 5 min, fixed with glutaraldehyde and examined in phase-contrast microscope. ISCs were differentiated on the basis of their surface characteristics (smooth, spiculated), and dimensions (type I: elongated cells with length equal or greater than doubled width; type II: other permanently sickled form, as shown in Fig. 11). Means of 3 experiments.

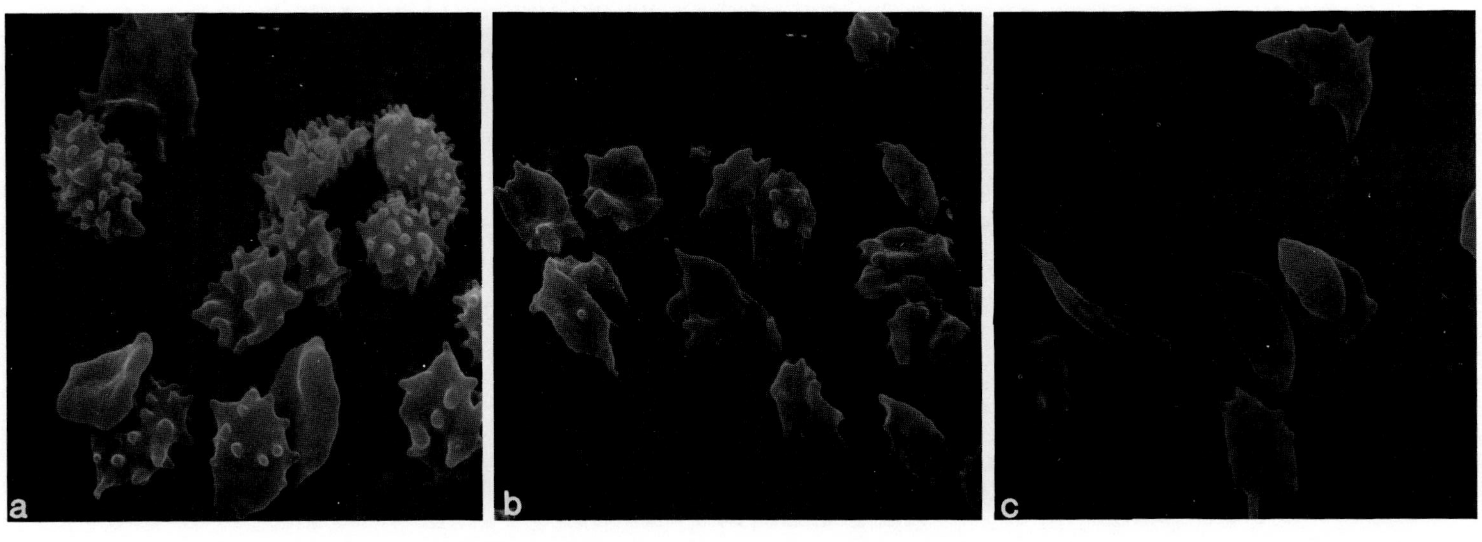

FIG. 11. Scanning electron micrographs of oxygenated HbSS red cells, which were previously incubated under deoxygenated conditions (95% N_2 + 5% CO_2) without or with adenine, inosine, and EGTA. Detailed conditions are in Table 6. After reoxygenation for 5 min at room temperature, cells were fixed according to ref. 56. **a:** 10 hr of incubation, no additives. **b:** 10 hr of incubation, with adenine, inosine, and EGTA. **c:** 22 hr of incubation with adenine, inosine, and EGTA.

provided by studies on normal reconstituted red cell ghosts into which hemoglobin S has been introduced under hypotonic conditions (54), and which gave rise to irreversibly sickled forms even in the absence of calcium and in the presence of EGTA.

CONCLUSIONS

1. In normal red cells, 20% (3.4 ± 1.2 μmoles/liter) of red cell calcium exchanged with ^{45}Ca of the supernatant plasma after 4 hr at 37°C; 79% of the exchangeable red cell calcium was recovered in red cell cytosol.
2. Of the calcium which accumulated in ATP-depleted red cells, 80% was recovered in red cell cytosol. However, ghosts from these cells washed in the absence of EGTA exhibited echinocytic configuration and retained 3.8 ± 1.2 μmoles/liter of residual radiocalcium. ^{45}Ca removal by EGTA (2 m*M*) was associated with restoration of discoid shape.
3. When spectrin was extracted from ghosts of ATP-depleted red cells by 1 m*M* Tris HCl (pH 8, 37°C), 90–95% of the ghost radiocalcium was recovered in the residual vesicles and only 5–10% was associated with spectrin and the actin-like polypeptide (band V), suggesting that the role of calcium in cellular rigidity is not mediated through direct interaction of calcium with these proteins.
4. Red cells of patients with sickle cell anemia (HbSS) exhibited elevated calcium content, and their transformation into sickled shape was associated with net gain in calcium and a marked influx of radiocalcium from plasma into cells.
5. Calcium that accumulated in sickled HbSS cells was preferentially located in red cell membranes.
6. Extrusion of calcium from HbSS ghosts enriched in calcium, magnesium, and ATP did not differ from that of ghosts of reticulocyte-rich red cells of patients with other hemolytic anemias.
7. Oxygenated HbSS red cells which accumulated calcium during previous deoxygenation for 6–9 hr exhibited lower filterability than corresponding cells incubated under oxygenated conditions or deoxygenated conditions in the presence of a complexing agent.
8. The percentage of cells which transformed into irreversibly sickled shape (ISC) was not influenced by the addition of adenine, inosine, and EGTA. However, considerably more ISCs exhibited spiculated surface when incubated without the above additives.
9. Calcium which accumulates in hemoglobin S red cells contributes to decreases in cellular deformability, but does not increase number of cells which transform into irreversibly sickled cell shape.

ACKNOWLEDGMENTS

This study was supported by NIH grants 7-R01-HL17955-01, 5-R01-AM15914-02, 5-R01-AM16914-02, and by Boston Sickle Cell Center. We want to thank Ms. Donna Ozog, Mr. Peter O'Connell, Mr. Martin Thomae, and Mr. George Njoku for skillful technical assistance, Mr. Francis Kiernan for preparation of scanning electron micrographs, and Mrs. Betty Stevenson for typing and organizing the manuscript.

REFERENCES

1. Schatzman, H., and Vincenzi, F. (1969): *J. Physiol. (Lond.)*, 201:369.
2. Lee, K., and Shin, B. (1969): *J. Gen. Physiol.*, 54:713.
3. Olson E., and Cazort, R. (1969): *J. Gen. Physiol.*, 53:311.
4. Weed R., LaCelle, P., and Merrill, E. (1969): *J. Clin. Invest.*, 48:795.
5. Lichtman, M., and Weed, R. *Nouv. Rev. Fr. Hematol.*, 12:799.
6. Palek, J., Curby, W., and Lionetti, F. (1972): *Blood*, 40:261.
7. Nakao, M., Nakao, T., and Yamozoe, S. (1961): *Nature*, 187:945.
8. Gardos, G. (1958): *Biochim. Biophys. Acta*, 30:653.
9. Gardos, G. (1959): *Acta. Physiol. Acad. Sci. Hung.*, 15:121.
10. Lepke, S., and Passow, H. (1960): *Pfluegers Arch.*, 271:473.
11. Lepke, S., and Passow, H. (1967): *J. Physiol. (Lond.)*, 191:39.
12. Romero, P., and Whittam, R. (1971): *J. Physiol. (Lond.)*, 214:481.
13. Lew, V. (1971): *Biochim. Biophys. Acta*, 233:827.
14. Palek, J., Curby, W., and Lionetti, F. (1971): *Am. J. Physiol.*, 220:1028.
15. Palek, J., Stewart, G., and Lionetti, F. (1974): *Blood*, 44:583.
16. Eaton, J., Skelton, T., Swofford, H., Kolpin, C., and Jacob, H. (1973): *Nature*, 246:105.
17. Palek, J. (1973): *Blood*, 42:988.
18. Wiley, J., and Shaller, C. (1973): *Proc. 16th Ann. Mtg. Am. Soc. Hematol., Chicago, 1973.*
19. Harrison, D., and Long, G. (1968): *J. Physiol. (Lond.)*, 199:367.
20. Palek, J., Curby, W., and Lionetti, F. (1971): *Am. J. Physiol.*, 220:19.
21. LaCelle, P., Kirkpatrick, F., Udkow, M., and Arkin, B. (1973): In: *Red Cell Shape: Physiology, Pathology, Ultrastructure*, edited by M. Bessis, R. Weed, and P. Leblond, p. 67. Springer Verlag, New York, Heidelberg, Berlin.
22. Sears, D. (1973): *J. Lab. Clin. Med.*, 82:719.
23. Langley, G., and Axell, M. (1968): *Br. J. Haematol.*, 14:593.
24. Marchesi, S., Steers, E., Marchesi, V., and Tillack, T. (1969): *Biochemistry*, 9:50.
25. Marchesi, V., and Steers, E., Jr. (1969): *Science*, 159:203.
26. Steck, T. (1974): *J. Cell Biol.*, 62:1.
27. Juliano, R. (1973): *Biochim. Biophys. Acta*, 300:341.
28. Lux, S., and John, K. (1971): *Proc. 17th Ann. Mtg. Am. Soc. Hematol., Atlanta, Georgia, 1974*, p. 49.
29. Laemmli, V. (1970): *Nature*, 227:680.
30. Avruch, J., and Fairbanks, G. (1974): *Biochemistry*, 13:5507.
31. Yu, J., and Fischman, D. A. (1973): *J. Supramol. Struct.*, 1:233.
32. Nicolson, G., Marchesi, V., and Singer, S. (1971): *J. Cell Biol.*, 51:265.
33. Carraway, K., Triplett, R., and Anderson, D. (1975): *Biochim. Biophys. Acta*, 379:571.
34. Forstner, J., and Manery, J. F. (1971): *Biochem. J.*, 124:563.
35. Duffy, M., and Schwarz, V. (1973): *Biochim. Biophys. Acta*, 330:294.
36. Tosteson, D., Shea, E., and Carling, R. (1952): *J. Clin. Invest.*, 31:406.
37. Tosteson, D., Carlsen, E., and Dunham, E. (1955): *J. Gen. Physiol.*, 39:31.
38. Tosteson, D. (1955): *J. Gen. Physiol.*, 39:55.
39. Serjeant, G., Petch, M., and Serjeant, B. (1973): *J. Lab. Clin. Med.*, 81:850.
40. Bunn, R., Ransil, B., and Chao, A. (1971): *J. Biol. Chem.*, 246:5273.
41. Wong, M. C., and Seeman, P. (1971): *Biochim. Biophys. Acta*, 241:473.
42. Bertles, J. F., and Milner, P. F. (1968): *J. Clin. Invest.*, 47:1731.
43. Serjeant, G. R., Serjeant, B. E., and Milner, P. F. (1969): *Br. J. Haematol.*, 17:527.
44. Chein, S., Usami, S., and Bertles, J. F. (1970): *J. Clin. Invest.*, 49:623.
45. Shen, S. C., Fleming, E. M., and Castle, W. B. (1949): *Blood*, 4:498.
46. Jensen, W. A., Bromberg, P. A., and Barefield, K. (1969): *Clin. Res.*, 17:464.
47. Dobler, J., and Bertles, J. F. (1968): *J. Exp. Med.*, 127:711.

48. Jensen, W. N. (1969): *Am. J. Med. Sci.,* 257:355.
49. Murayama, M. (1966): *Science,* 153:145.
50. Glader, B., and Nathan, D. (1974): In: *Proceedings of 1st National Symposium on Sickle Cell Disease,* edited by J. I. Hercules, A. N. Schecter, W. A. Eaton, and R. E. Jackson. Washington, 1974.
51. Lessin, L., and Wallas, C. (1973): *Blood,* 42:978.
52. Jensen, M., Shohet, A., and Nathan, D. (1973): *Blood,* 42:835.
53. Palek, J. (1973): *Clin. Res.,* 21:562.
54. Clark, M. R., and Shohet, S. B. (1975): *Clin. Res.,* 23:402.
55. Palek, J. (1974): In: *Proceedings of the 1st National Symposium on Sickle Cell Disease, Washington, 1974,* edited by J. I. Hercules, A. N. Schecter, W. A. Eaton, and R. E. Jackson.
56. Bessis, M., and Weed, R. I. (1972): *Proc. 5th SEM Symp., Chicago, 1972,* p. 289.

Membranes and Disease, edited by L. Bolis, J. F. Hoffman, and A. Leaf. Raven Press, New York, © 1976.

Mechanisms of Red Cell Membrane Lipid Renewal

Stephen B. Shohet

Cancer Research Institute, School of Medicine, University of California, San Francisco

In this chapter, I will, first, briefly outline some of the known mechanisms for the renewal and maintenance of red cell lipids, and, then, try to illustrate them by means of two conditions where a disturbance of those mechanisms is associated with functional abnormalities of the red cell membrane.

Figure 1 depicts the approximate lipid composition of the red cell membrane. The bulk of the lipids consist of phospholipids, with a smaller proportion of cholesterol and glycerides—the so-called neutral lipids. The majority of this discussion will be confined to phospholipids, although the importance of cholesterol composition in the red cell membrane should not be minimized. Of the phospholipids, phosphatidylcholine is the metabolically most active species in the red cell. Phosphatidylethanolamine, which is present in nearly equal amounts, is considerably less active, and sphingomyelin has almost no detectable turnover in the mature cell.

Figure 2 shows one of the primary reactions involved in the maintenance of the phospholipids of the mature erythrocyte. On the left side, lysophosphatidylcholine is shown. This molecule, which is normally present in red cells in only very small amounts, is primarily bound to plasma albumin. However, this plasma lysolecithin undergoes a rapid passive exchange with the red cell membrane. The role of lysolecithin should be emphasized in that it is a strongly detergent molecule, which can induce membrane phase-state changes in model systems. In small concentrations, it causes hemolysis and, as will be discussed, its accumulation in red cells can produce marked changes in cell shape and membrane stability. On the right side of Fig. 2, phosphatidylcholine is shown. This compound, which is a major constituent of the red cell membrane, differs from the former only by the addition of one fatty acid. It is much more lipophilic and, although found in both the plasma and the red cell membrane, tends to exchange at a much slower rate. Within the membrane itself, an enzyme system dependent on ATP and magnesium and coenzyme A can convert lysophosphatidylcholine to phosphatidylcholine. This so-called "acylase" reaction is extremely important in the maintenance of red cell phospholipids, because when lysophosphatidylcholine is converted to phosphatidylcholine within the membrane, it is, in effect, trapped there owing to the differences in exchange rates. Since the red cell can get both its fatty acid and its lysolecithin

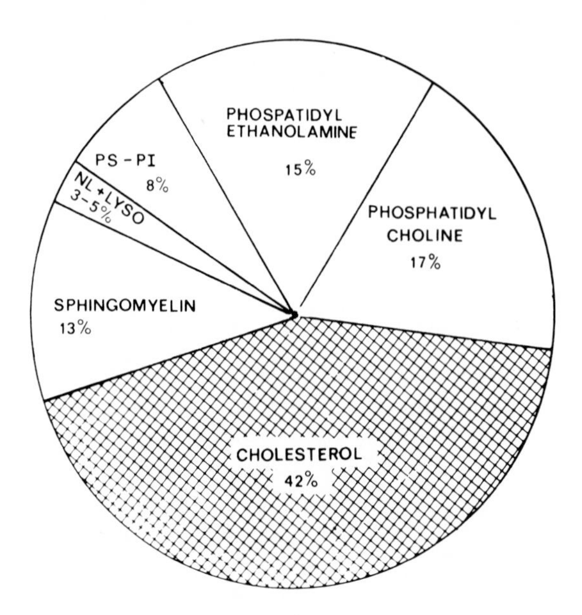

FIG. 1. Red cell lipid composition. Total erythrocyte lipids, approximate molar percentage. See ref. 1 for further details.

at no metabolic cost from the plasma, and since it only has to supply the enzyme and small amounts of ATP to entrap these comparatively complex molecules, this is a very economical procedure for the cell.

Figure 3 shows a more general diagram which emphasizes this ATP-mediated active reaction for the renewal of membrane phosphatides, and which also depicts some passive exchange pathways for the renewal of phosphatides and cholesterol in the red cell. If we confine ourselves initially to fatty acid pathways, which are indicated by the horizontal arrows, plasma free fatty acid bound to albumin is in rapid equilibrium with a superficial membrane pool of free fatty acid (3,4). This "F_1 pool" is defined as being

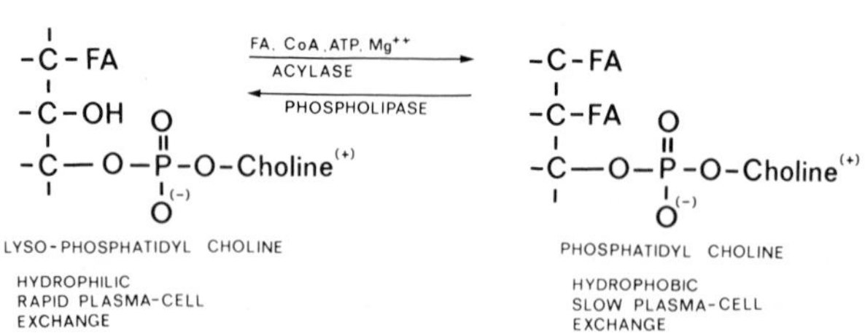

-C-FA
-C-OH
-C—O—P(=O)(O$^{(-)}$)-O-Choline$^{(+)}$

FA, CoA, ATP, Mg^{++}
ACYLASE →
← PHOSPHOLIPASE

-C-FA
-C-FA
-C—O—P(=O)(O$^{(-)}$)-O-Choline$^{(+)}$

LYSO-PHOSPHATIDYL CHOLINE
HYDROPHILIC
RAPID PLASMA-CELL EXCHANGE

PHOSPHATIDYL CHOLINE
HYDROPHOBIC
SLOW PLASMA-CELL EXCHANGE

FIG. 2. The acylation reaction fundamental to red cell membrane lipid renewal. A lysophosphatide is esterified with a free fatty acid to produce a complete phosphatide which has profoundly different physical properties and which tends to remain within the membrane. See refs. 5–7 and 9. (Reproduced with permission from the *New England Journal of Medicine,* 286:557–583, 1972).

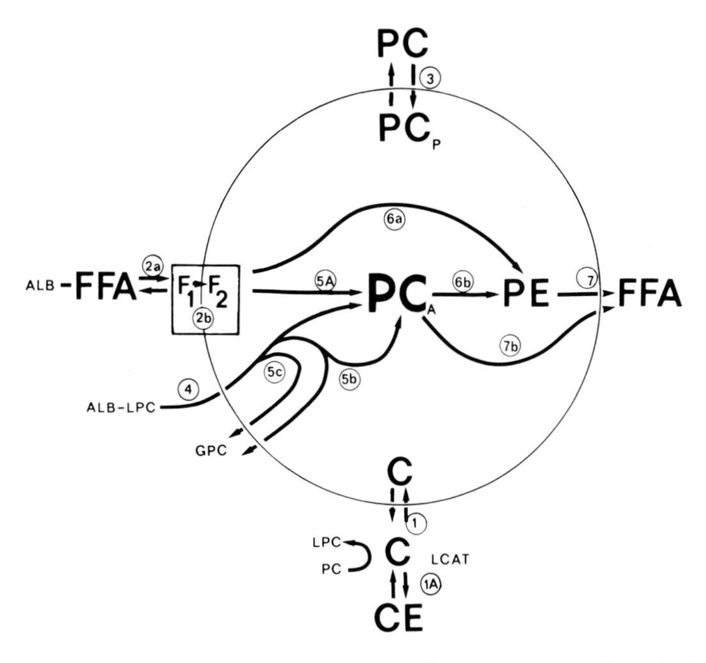

FIG. 3. Schema of the major exchange and metabolism pathways for lipids in the mature erythrocytes. Alb-FFA, albumin-bound free fatty acid; F_1, surface pool of freely exchangeable free fatty acid; F_2, "deeper" pool of free fatty acid used as a source of acyl groups for phosphatides within the membrane; PC_A, phosphatidylcholine actively synthesized within the membrane; PC_p, phosphatidylcholine passively acquired by exchange with the plasma by the membrane; PE, phosphatidylethanolamine; Alb-LPC, albumin-bound lysophosphatidylcholine which, together with Alb-FFA, serves as the precursor for PC_A; GPC, glycerolphosphorylcholine; C, cholesterol; CE, cholesterol ester; LCAT, lecithin cholesterol acyltransferase. See refs. 2, 8, 9, and 11. (Reproduced with permission from the *New England Journal of Medicine,* 286:638–644, 1972.)

superficial in that it is easily extracted from the cell by defatted albumin. Fatty acid delivered into this first pool can then be transposed into another pool of fatty acid, the "F_2 pool," which is much less extractable with defatted albumin. This second process requires metabolic energy. Fatty acid in the second pool, the F_2 pool, can then undergo the reaction described in Fig. 2; lysolecithin from the plasma can be acylated by this fatty acid in the presence of the appropriate enzyme, magnesium and ATP, to form phosphatidylcholine. This resultant phosphatidylcholine is labeled with a subscript, A, to indicate that it comes to the red cell actively. It represents the major elements of active phospholipid turnover in the mature erythrocyte. As shown by the upper curved arrow, F_2 fatty acid can also be incorporated into phosphatidylethanolamine in an analogous fashion to phosphatidylcholine, but endogenous lysophosphatidylethanolamine serves as substrate in this case. Both phosphatidylcholine and phosphatidylethanolamine can release their fatty acids as free fatty acids back to the plasma to account for the catabolic side of this metabolic pathway. In addition, a special pathway is available for the transfer of phosphatidylcholine fatty acid to phosphatidylethanolamine, and this pathway accounts for approximately one-third of the turnover of phosphatidylcholine fatty acid (10). Finally, as shown in the lower left portion of the figure, a series of phospholipase and transmutase reactions exist for the removal of excess lysophosphatidylcholine.

In addition to these active fatty acid incorporation reactions, the red cell also undergoes two important passive transfer reactions for the maintenance of its lipid constituents. At the top of Fig. 3 is shown the passive exchange of preformed, intact, plasma phosphatidylcholine and membrane phosphatidylcholine, which is distinct from, and does not appreciably mix with, the pool of phosphatidylcholine obtained by the active incorporation pathway. Likewise, at the bottom of the figure, the passive exchange of plasma cholesterol with membrane cholesterol is shown. This exchange involves only free cholesterol, not esterified cholesterol. Accordingly, the activity of lecithin cholesterol acyltransferase (LCAT), an enzyme which converts cholesterol esters to free cholesterol and which thereby regulates the level of plasma free cholesterol, is important in the regulation of red cell membrane cholesterol.

Abnormalities in any of these lipid renewal reactions can be associated with abnormalities in the red cell membrane. Dr. Murphy, Dr. Cooper, and other workers have shown that abnormalities in the plasma free cholesterol (and perhaps in plasma bile acids) can passively influence the levels of cholesterol in the membrane. This, in turn, results in abnormalities in osmotic fragility, and, rarely, in the flagrant hemolysis of so-called "spur cell" anemia. Two other conditions are instructive illustrations of abnormalities in active lipid metabolic pathways associated with membrane malfunction and instability.

The rare condition of high phosphatidylcholine hemolytic anemia (HPCHA) is associated with a metabolic block in the catabolic pathway for actively incorporated phosphatidylcholine. As shown in Table 1, these cells have elevated levels of phosphatidylcholine and decreased levels of phosphatidylethanolamine in their membranes.

Figure 4 demonstrates some of the abnormal fatty acid metabolic findings in these cells. Here it can be seen that, in comparison to the normals, the gross incorporation of fatty acid into phosphatidylcholine in these cells was markedly increased. In contrast, incorporation into phosphatidylethanolamine was decreased.

However, as shown in Fig. 5, when the acylation capacity of these cells was compared with normal, age-matched controls, no significant differences were seen for fatty acid incorporation capacity into either lysophosphatidylcholine or lysophosphatidylethanolamine.

This apparent paradox was explained by the data shown in Fig. 6, which show that the fractional turnover of actively acquired phosphatidylcholine

TABLE 1. *Red cell lipids in high phosphatidylcholine hemolytic anemia (HPCHA)*

	Erythrocyte lipids (γP/cc, ± SD)		
	Patients		Controls
Total PL γP/cc	164 ± 11		156 ± 9
PC γP/cc	60 ± 4	$P < 0.001$	45 ± 5
PE γP/cc	35 ± 3	$P < 0.01$	43 ± 3

Analyses are from 5 affected patients and 6 reticulocyte-matched controls. PL, phospholipid; PC, phosphatidylcholine; PE, phosphatidylethanolamine.

Reproduced with permission from *Blood,* 38:445–456, 1971.

was markedly reduced in these patients, whereas that in phosphatidylethanolamine was increased. In parallel experiments done with HPCHA cells passively labeled with phosphatidylcholine, no abnormalities in fractional turnover could be detected.

These observations were consistent with a block in the active transfer of phosphatidylcholine fatty acid to phosphatidylethanolamine and, as shown in Table 2, this was indeed so. When cells prelabeled with ^{14}C-linoleic fatty acid were washed and reincubated in serum for several hours, there was very little transfer of radioactivity from phosphatidylcholine to phosphatidylethanolamine in patient cells in comparison to normal control cells.

Figure 7, which is similar to Fig. 3, summarizes the fatty acid metabolic abnormality in these cells. Fatty acid enters the cell normally through the

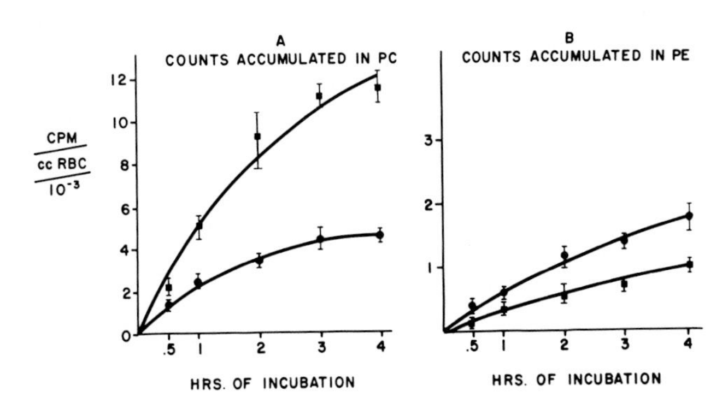

FIG. 4. Incorporation of ^{14}C-linoleic acid into phosphatidylcholine and phosphatidylethanolamine in red cells of HPCHA patients (squares) and reticulocyte-matched controls (circles). (Reproduced with permission from *Blood,* 38:445–456, 1971.)

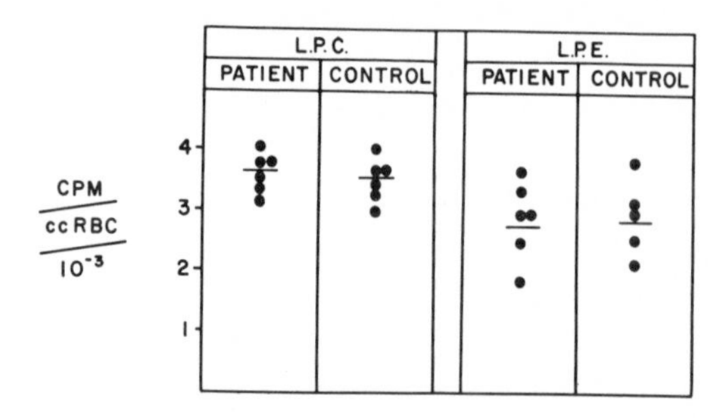

FIG. 5. Acylation capacity for lysophosphatidylcholine (LPC) and lysophosphatidylethanolamine (LPE) substrates in cells from patients with HPCA and reticulocyte-matched controls. Acylase capacity was determined on erythrocyte stroma in a system with excess ^{14}C-linoleic acid, ATP, coenzyme A, and the appropriate lyso compound. (Reproduced with permission from *Blood,* 38:445–456, 1971.)

surface pools, and then is transferred normally to the actively acquired pool of phosphatidylcholine. Likewise, some normal incorporation of fatty acid directly in phosphatidylethanolamine, as indicated by the dotted arrow, occurs. However, transfer of phosphatidylcholine fatty acid to phosphatidylethanolamine is blocked, and phosphatidylcholine levels rise and phosphatidylethanolamine levels fall. Passively acquired phosphatidylcholine, as shown by the arrows at the bottom of the figure, is unaffected. The net effect of this distortion in phospholipid metabolism is an abnormality in phospholipid composition. This, in turn, may well be related to an abnormality in membrane function, which is shown in Fig. 8.

As can be seen in this cation flux study (Fig. 8), HPCHA cells show

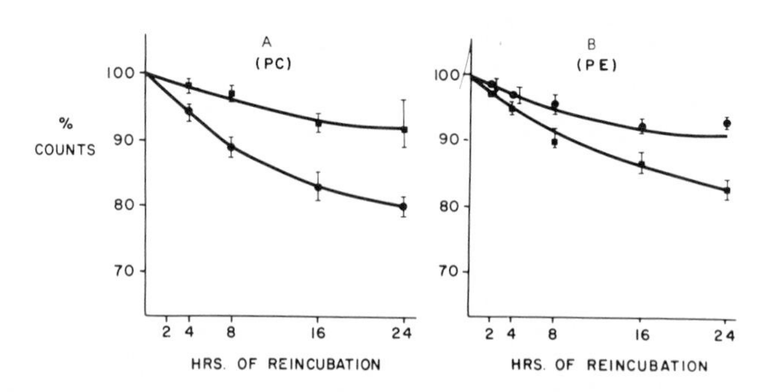

FIG. 6. Percent fractional turnover of actively labeled phosphatidylcholine fatty acid, and phosphatidylethanolamine fatty acid. **A:** Turnover of counts in PC. **B:** Turnover of counts in PE. Cells were preincubated for 4 hr with ^{14}C-linoleic acid and glucose to actively label the phosphatides. They were then washed free of media and surface fatty acid with saline and defatted albumin, respectively. They were then reincubated in fresh serum and aliquots removed at various times. The counts present zero time in each fraction were arbitrarily set at 100%. The squares represent the patients, the circles, the reticulocyte-matched controls. (Reproduced with permission from *Blood,* 38:445–456, 1971.)

TABLE 2. *Transfer of PC-FA to PE-FA in briefly prelabeled cells*

Reincubation time (hr)	Cumulative rise in PE counts as % of initial PC counts	
	Controls	Patients
4	5	1
8	12	1
16	16	2
24	8	4

Cells were prelabeled with high-specific active fatty acid for ½ hr and washed with saline and defatted albumin to remove media and surface fatty acid, respectively. They were then reincubated in fresh serum and sequential aliquots were removed for extraction and analysis. The cumulative increases in PE counts were calculated as percent of original PC counts and represent the net effect of continued incorporation of PC counts into PE minus the release of PE counts into the serum. (See Table 1 for abbreviations.)

Reproduced with permission from *Blood,* 38:445–456, 1971.

abnormalities in membrane permeability, whereby both sodium and potassium permeability is increased. At rest, the cation pump is nearly able to compensate for the increased leak of these ions. However, when the cation pump is inhibited by ouabain, marked abnormalities in cation flux occur with an abnormal efflux of cell potassium and an abnormal influx of medium

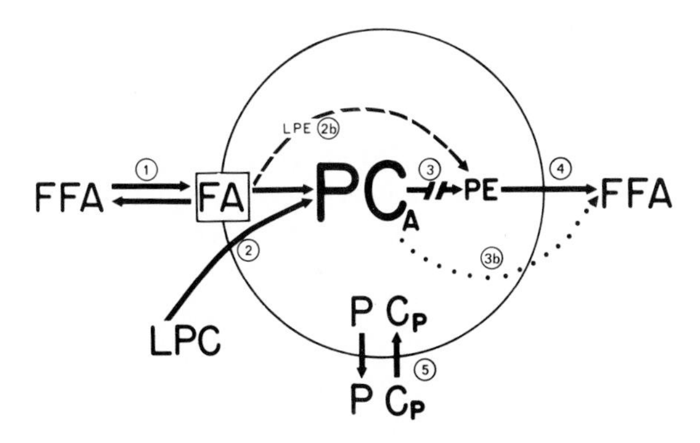

FIG. 7. Proposed phosphatide–fatty acid renewal pathway in HPCHA erythrocytes. FFA, free fatty acid; FA, surface fatty acid pools; LPC, lysophosphatidylcholine; LPE, lysophosphatidylethanolamine; PC_A, phosphatidylcholine accumulated through the active acylation pathway; PE, phosphatidylethanolamine; PC_p, phosphatidylcholine accumulated through the passive exchange pathway. There is an apparent block in the transfer of PC_A fatty acid to PE with a resultant reduction in PE levels and an increase in PC levels within the cell membrane. (Reproduced with permission from *Blood,* 38:445–456, 1971.)

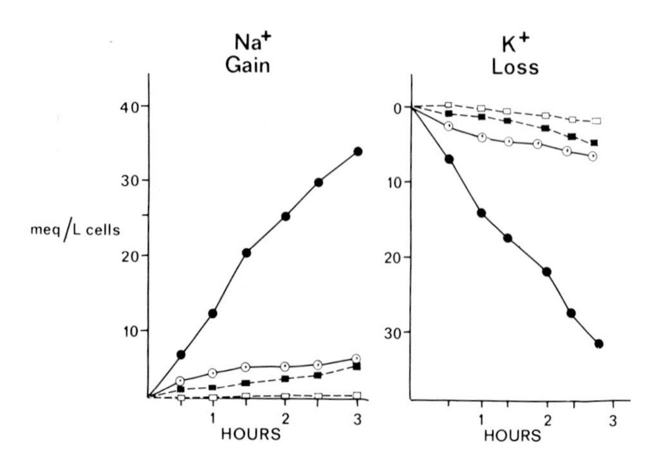

FIG. 8. Changes in sodium and potassium concentrations HPCHA cells during incubation at 37°C. Circles indicate patient cells, squares indicate reticulocyte-matched control cells. Open symbols indicate incubation in glucose alone, closed symbols indicate incubation with glucose and 10^{-4} *M* ouabain. (Reproduced with permission from *Blood,* 42:1–7, 1973.)

sodium. The clinically significant hemolytic anemia of these patients may be related to these changes.

In addition to this example of highly specific defect in a late transfer reaction of fatty acid metabolism, which appears to induce a membrane permeability defect, red cell membrane abnormalities may be induced by nonspecific early inhibition of fatty acid incorporation into red cell phosphatides. Figures 9–15 show an experiment in which such inhibition was induced by the reduction of red cell ATP stores by glucose starvation *in vitro.*

In the experiment depicted in Fig. 9, acylation activity, as measured by the incorporation of radioactive palmitic acid into phosphatidylcholine, was followed in cells which were incubated in fresh serum at low hematocrit without glucose. ATP levels began to fall at approximately 8 hr, and were below 25% of starting levels at approximately 16 hr of incubation. In contrast, acylation activity did not begin to fall until these low levels were reached. This is consistent with the fact that acylation reaction only uses a very small amount of total cell ATP, and also with the fact that the K_m for this reaction is quite low. However, once acylation activity did begin to fall, lysolecithin levels within the cells began to rise.

Figure 10 shows scanning electron micrographs taken with the help of Dr. James Haley which depict what happens to the erythrocytes under these conditions. In Fig. 10A, virtually normal cells with membrane lysophosphatidylcholine concentrations of about 0.1 m*M* are shown at the beginning of the incubation. In Fig. 10B, early echinocytic changes at 17 hr

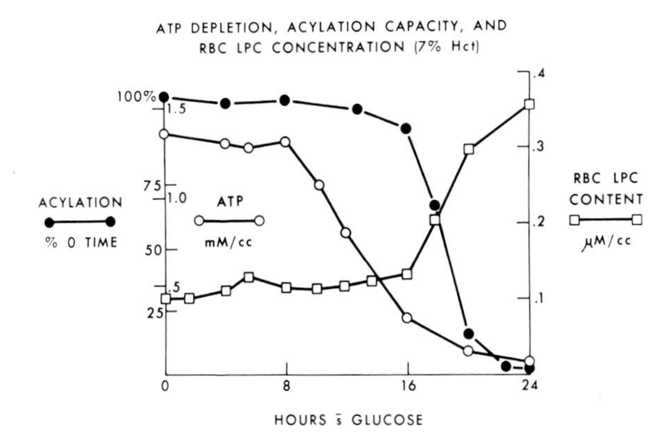

Fig. 9. Metabolic and lipid parameters of red blood cells (RBC) incubated in autologous serum for prolonged periods at 7% hematocrit without added glucose. Closed circles indicate acylation activity as percent of zero time; open circles, ATP concentration in millimoles per cubic centimeter packed cells. Open squares indicate lysophosphatidylcholine (LPC) concentration in micromoles per cubic centimeter packed cells. (Reproduced with permission from *Nouvelle Revue Francaise d'Hematologie,* 12: 761–770, 1972.)

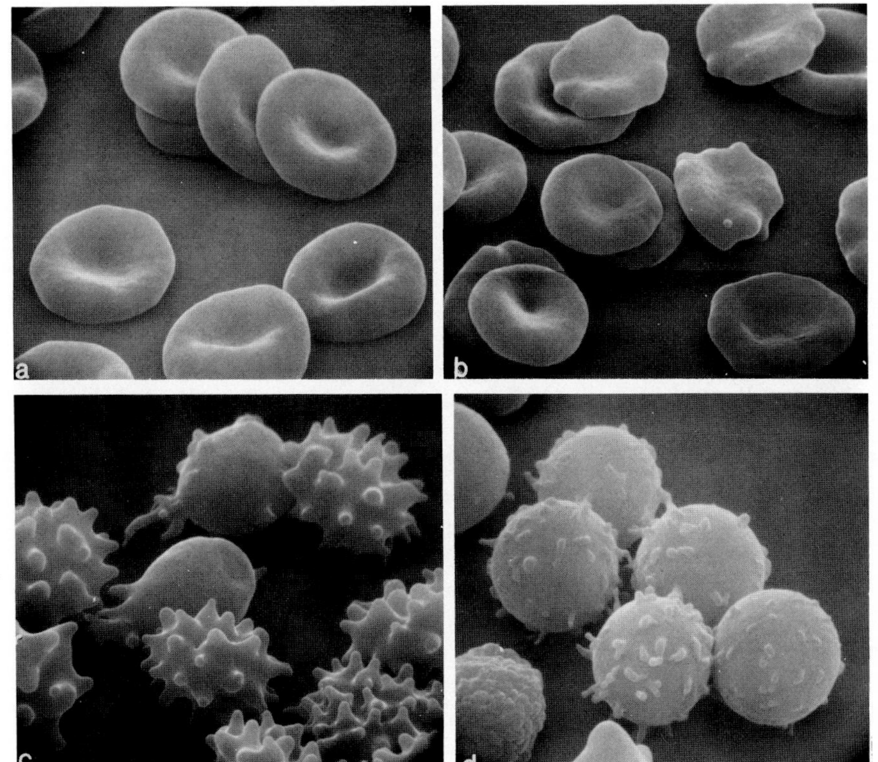

FIG. 10. Scanning electron micrographs of erythrocytes subjected to increasing concentrations of membrane lysophosphatidylcholine. **a:** Membrane LPC 0.12 $\mu M/cm^3$ cells. **b:** Membrane LPC 0.15 $\mu M/cm^3$ cells. **c:** Membrane LPC 0.30 $\mu M/cm^3$ cells. **d:** Membrane LPC 0.50 $\mu M/cm^3$ cells. The bulk of the changes seen in **A–C** could be reversed by washing the cells with defatted albumin. The changes seen in the red cells in **D** are irreversible, and represent, in part, membrane loss due to microvesiculation induced by the high LPC concentrations. (Reproduced with permission from *Nouvelle Revue Francaise d'Hematologie,* 12:761–770, 1972.)

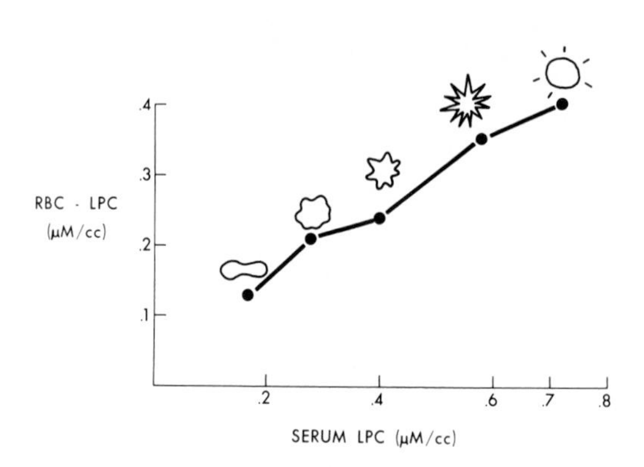

FIG. 11. Summary of morphologic changes associated with increased red cell lysophosphatidylcholine induced by the addition of exogenous lysophosphatidylcholine (4°C) to suspending serum.

of incubation with ~0.2 m*M* lysolecithin in the red cell membrane are shown. In Fig. 10C, marked echinocytic changes at 20 hr with ~0.3 m*M* lysolecithin in the red cell membrane are shown. In Fig. 10D, irreversible, spheroechinocytic changes at 26 hr and ~0.5 m*M* red cell lysolecithin are shown. This stage may, in part, represent membrane loss due to microvesiculation induced by the high lysolecithin concentration.

Although it is recognized that many other things may be happening to the red cells besides the accumulation of lysolecithin during this period, control experiments done with glucose-depleted serum do not show these morphologic changes, and all other cell lipids, including cholesterol, were identical at the various times of incubation in the depleted and the control cells. Total cell cation contents as well were not sufficiently disturbed to explain these changes on the basis of pump failure. Moreover, as shown in Fig. 11, analogous morphologic changes have been created in fresh cells with normal ATP levels by simply adding sufficient lysolecithin to the incubating serum to produce levels, in the fresh cell membrane, which are similar to those which were previously induced by glucose depletion.

These glucose-deprived lysolecithin-enriched cells are not simply peculiar looking. Figure 12 shows an assay system designed to study the stability of these membranes. Red cells are doubly labeled with chromium-51 in their hemoglobin moiety and with carbon-14 in their membrane phospholipids. They are then layered upon a monolayer of either human splenic or circulating macrophages from a blood type-compatible donor and allowed to incubate at 37°C for 3 hr without agitation. The red cells are then gently washed off the monolayer by dipping the plates through saline solutions, and the transfer of membrane lipids to the adherent monolayer is measured by radioactive counting. Corrections for erythrophagocytosis, rather than

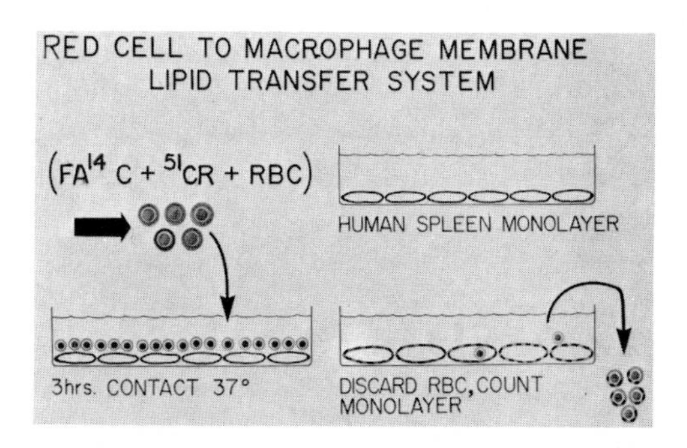

FIG. 12. Schema for technique to measure instability of erythrocyte membrane lipids *in vitro.* Red cells are doubly labeled with ^{14}C-fatty acid by active acylation in their membrane phosphatides and by ^{51}Cr in their hemoglobin. Monolayers of human macrophages are prepared and the doubly labeled red cells are incubated in contact with the monolayer for various periods of time. Nonadherent red cells are then gently removed by dipping the small petri dishes through wash solutions and the residual radioactivity in the monolayer is counted for radioactivity. ^{51}Cr radioactivity is taken as a measure of adherent or phagocytized whole cells or hemoglobin-containing red cell fragments, while ^{14}C radioactivity in excess of ^{51}Cr radioactivity is taken as a measure of red cell membrane lipid transfer to the monolayer.

membrane transfer, are made by examining the chromium vs the carbon counts.

Figure 13 shows the results of such an experiment conducted with fresh, doubly labeled, human red cells. Initially, there was a very rapid early transfer of a small amount of membrane radioactivity, which may reflect the equilibration of unesterified free fatty acids between the erythrocytes and the monolayer. Thereafter, there was a continuous, gradual transfer of

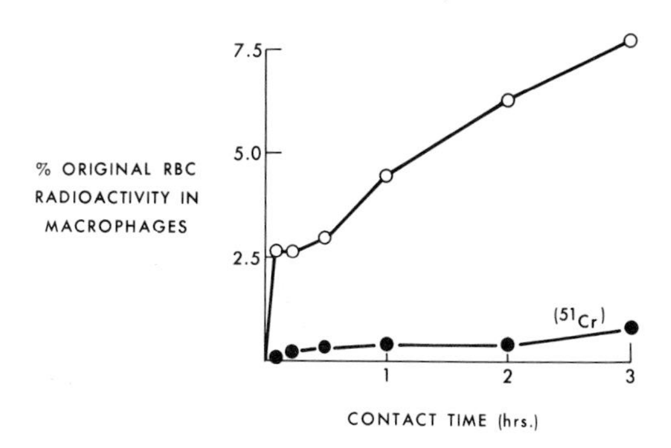

FIG. 13. Transfer of membrane label (^{14}C) to splenic macrophage monolayers (fresh cells). Open circles indicate membrane lipid (^{14}C) label; closed circles, hemoglobin (^{51}Cr) label. After an initial rapid equilibration, approximately 5% of original red cell radioactivity is transferred from fresh red cells to the macrophage monolayer. (Reproduced with permission from *Nouvelle Revue Francaise d'Hematologie,* 12:761–770, 1972.)

the membrane constituents to the macrophage layer. The latter transfer may represent an analog of normal human splenic function.

In Fig. 14, the same experiment is shown with echinocytic, metabolically depleted red cells which are bearing an excess burden of lysolecithin. Here, the membrane transfer is greatly accelerated and approximately 15% of the total lipid label is transferred within 3 hr. Similar results with lysolecithin-treated fresh cells or with ATP-reconstituted, adenosine-treated, aged cells suggest that this phenomenon is a function of the increased lysolecithin content per se, and not simply a reflection of the general depletion in ATP.

In Fig. 15, the percentage of original membrane phosphatide label transferred to the monolayer at 3 hr is plotted as a function of the lysolecithin content of red cell membranes whether the lysolecithin was increased by ATP depletion in aged cells or by simply adding lysolecithin to fresh cells. Again, the parallel effects of increasing membrane lysolecithin independent of the method used, or the metabolic state of the cells, strongly suggests

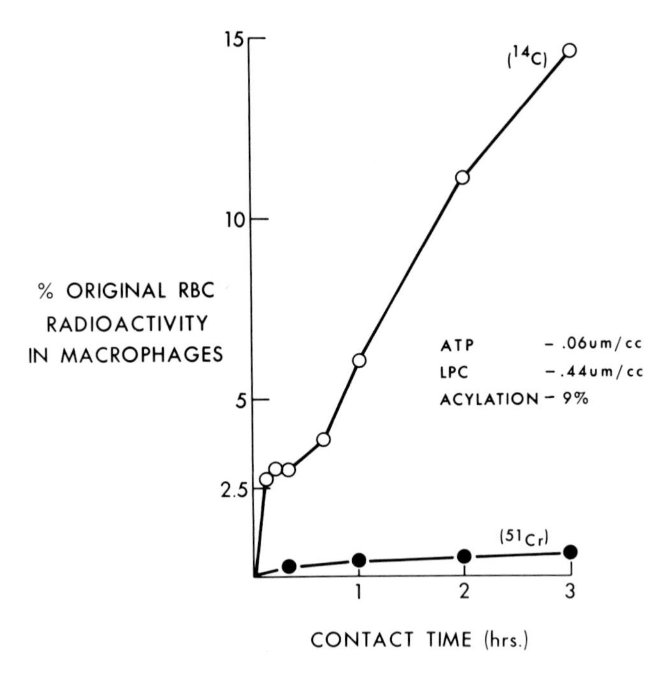

FIG. 14. Transfer of membrane label (^{14}C) to splenic macrophage cells (aged cells). The experimental design and the symbols are the same as in Fig. 13. In this case, however, red cells had been preincubated without glucose. ATP levels and acylation activity were depressed and lysophosphatidylcholine content was increased in the membranes. (See Fig. 9 at 24 hr.) After a similar initial, rapid equilibration, approximately 12% of membrane lipid label is transferred to the macrophage monolayer over a 3-hr period. (Reproduced with permission from *Nouvelle Revue Francaise d'Hematologie,* 12:761–770, 1972.)

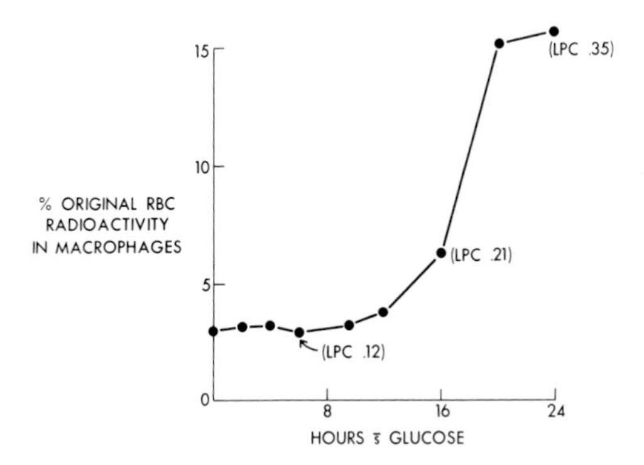

FIG. 15. Extent of membrane transfer to splenic macrophages (3 hr contract) as a function of time of preincubation without glucose and membrane lysolecithin concentration. Note that no significant change is noted until red cells have been incubated without glucose for at least 12 hr.

that the lysolecithin concentration per se is responsible for the observed membrane instability.

Since lysolecithin is known to induce phase-state changes in simple lipid bilayer systems, and since lysolecithin appears to be able to induce phase-state changes leading to cell fusion in more complex mammalian cell systems, it seems likely that lysolecithin here may be producing a phase-state change in the membrane, which renders the membrane unstable, and easily transferred to the monolayer.

The red cell membrane is a complex mixture of lipids and proteins. Though we are only beginning to analyze the proteins, at least the composition of the lipids is well characterized. Moreover, several of the metabolic pathways responsible for the maintenance and renewal of these red cell lipids have been described. Disturbances in some of these pathways produce abnormalities of the membrane which are associated with hemolysis and membrane instability. Perhaps, when we gain some understanding of the additional role of proteins and lipid–protein interactions in regulating these pathways, we will develop further understanding for the mechanism of these abnormalities of membrane function.

SELECTED REFERENCES

1. Ways, P., and Hanahan, D. J. (1964): Characterization and quantification of red cell lipids in normal man. *J. Lipid Res.,* 5:318.
2. Reed, C. F. (1968): Phospholipid exchange between plasma and erythrocytes in man and the dog. *J. Clin. Invest.,* 47:749.
3. Switzer, S., Eder, H. A. (1965): Transport of lysolecithin by albumin in human and rat plasma. *J. Lipid Res.,* 6:506.
4. Goodman, D. S. (1958): The interaction of human erythrocytes with sodium palmitate. *J. Clin. Invest.* 37:1729.

5. Oliveira, M. M., and Vaughan, M. (1964): Incorporation of fatty acids into phospholipids of erythrocyte membranes. *J. Lipid Res.,* 5:156.
6. Mulder, E., Van Deenen, L. L. M. (1965): Metabolism of red-cell lipids. I. Incorporation *in vitro* of fatty acids into phospholipids from mature erythrocytes. *Biochim. Biophys. Acta,* 106:106.
7. Mulder, E., Van Den Uerg, J. W. O., Van Deenen, L. L. M. (1965): Metabolism of red-cell lipids. II. Conversions of lysophosphoglycerides. *Biochim. Biophys. Acta,* 106:118.
8. Shohet, S. B., Nathan, D. G., et al. (1968): Stages in the incorporation of fatty acids into red blood cells. *J. Clin. Invest.,* 47:1096.
9. Shohet, S. B. (1970): Release of phospholipid fatty acid from human erythrocytes. *J. Clin. Invest.,* 49:1668.
10. Shohet, S. B. (1971): The apparent transfer of fatty acid from phosphatidylcholine to phosphatidylethanolamine in human erythrocytes. *J. Lipid Res.,* 12:139.
11. Brecher, G., and Bessis, M. (1972): Present status of spiculed red cells and their relationship to the discocyte-echinocyte transformation: A critical review. *Blood,* 40:333.

Membranes and Disease, edited by L. Bolis, J. F. Hoffman, and A. Leaf. Raven Press, New York, © 1976.

Impaired Reassemblance of Red Blood Cell Membrane Components in Hereditary Spherocytosis

Rupert Engelhardt

Medizinische Universitätsklinik, 78 Freiburg, West Germany

It is generally accepted that the shortened life span of the red blood cell (RBC) in hereditary spherocytosis (HS) is caused by its reduced deformability (1), reduced surface area:volume ratio, and increased permeability to sodium ions (2). However, the exact nature of the primary biochemical defect in this hemolytic disorder remains unknown: The total lipid content of the HS cell membrane is only slightly decreased, and the relative proportions of cholesterol and the various phospholipids are normal (3). On the other hand, the results of Jacob (4) favored the assumption of some alterations of the protein moiety of the membrane.

The experiments reported here are based on the assumption that the normal arrangement of the major membrane components, controlling its structure and permeability, is determined in part by electrostatic forces (5). Therefore, alterations of the charge of the membrane components possibly could cause alterations of the structure and function of the membrane. Partially solubilized RBC membrane proteins did not show striking differences in normal controls or other hemolytic disorders in electrophoresis (6). When membrane proteins were totally solubilized by sodium dodecyl sulfate, charge-dependent differences could not be detected. Therefore, we tried to find another approach to quantify the charge of the membrane proteins. Based on results of Rottem (7), the components of the disintegrated RBC membrane can be induced to reassemble at the boundary of a two-phase system. The reassemblance depends on the ionic strength as well as on the charge of the components involved.

METHOD AND RESULTS

White ghosts were disintegrated by deionization on a mixed bed ion exchange resin (Serva MB-1), and then dialyzed for 18 hr against a 1 m*M* Tris buffer, pH 7.4 at 4°C. Modifying the membrane protein extraction method given by Maddy (8) and Rega (9), aliquots of the dialyzed material were given in a series of plastic tubes: 1 part cold *n*-butanol was added to 2 parts sample, and then shaken vigorously for 20 sec. After centrifugation, the upper butanol phase contained 100% of the cholesterol and 80% of

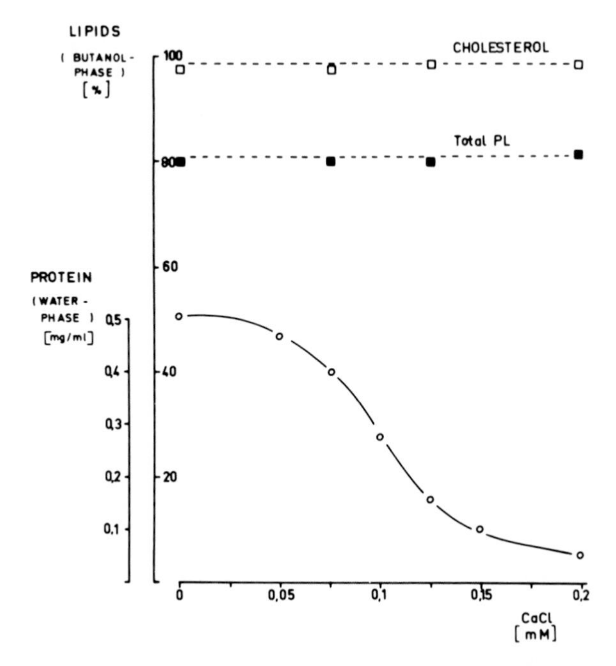

FIG. 1. The protein content of the water phase (open circles) decreases when the salt concentration is raised. Simultaneously the thickness of the membranous film at the boundary increases. 80% of total phospholipids (filled squares) and 100% of cholesterol (open squares) were found in the butanol phase. Only 5% of the phospholipids were recovered in the water phase and in the pellet, and the rest is supposed to take part in the recombination reaction.

the phospholipids, and the lower water phase, ~80% (77.26; 2σ 7.28; $n =$ 10) of the ghost protein (see Fig. 1). Decreasing amounts of protein were found in the water phase when the ionic strength was raised by adding mono- or divalent cations, with a dose–response curve sigmoidal in shape. The increasing moiety of the unsolubilized protein forms a tiny membranous film at the butanol–water interphase. Simultaneously, at the bottom of the tube a small pellet develops, containing up to 15% of the total protein moiety. The protein content of the butanol phase is about 5%, and does not depend on the salt concentration.

When the protein concentration was increased (as shown in Fig. 2), more salt was necessary to get, e.g., 50% of the protein out of the water phase into the interfacial "membrane," indicating a direct relationship between the amount of cations and the amount of membrane proteins involved. When the effect of monovalent cations, such as potassium or sodium, was studied, the dose–response curve was sigmoidal in shape as well (Fig. 2). However, approximately 10-fold salt concentration was necessary to induce a comparable effect. When studying the pH dependence (Fig. 3),

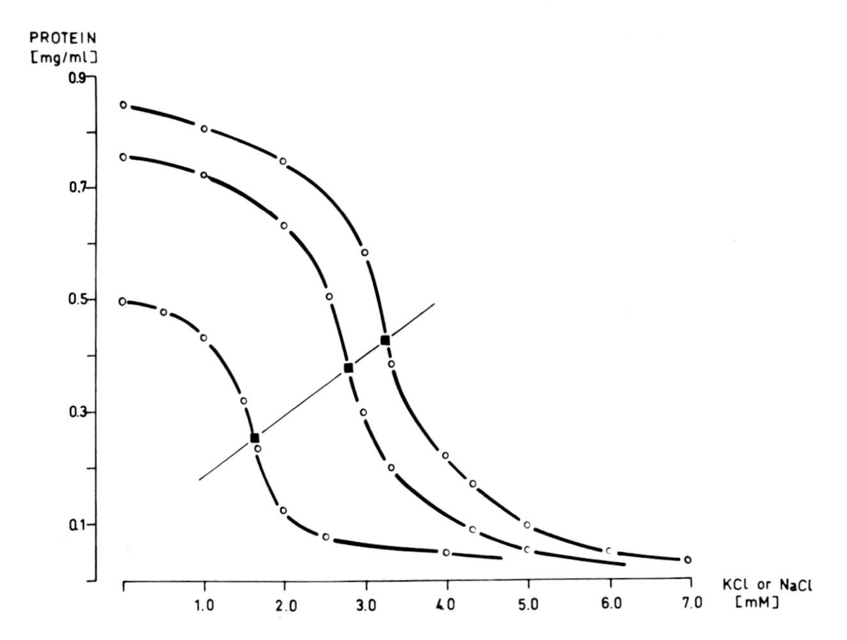

FIG. 2. The protein concentration, representing the part of the total protein not yet consumed for the recombination, decreases when the concentration of monovalent cations is raised. The dose–response curve is sigmoidal in shape as in the use of divalent cations. The 50% points (closed squares) of several curves, starting at different protein concentrations, can be connected by a straight line.

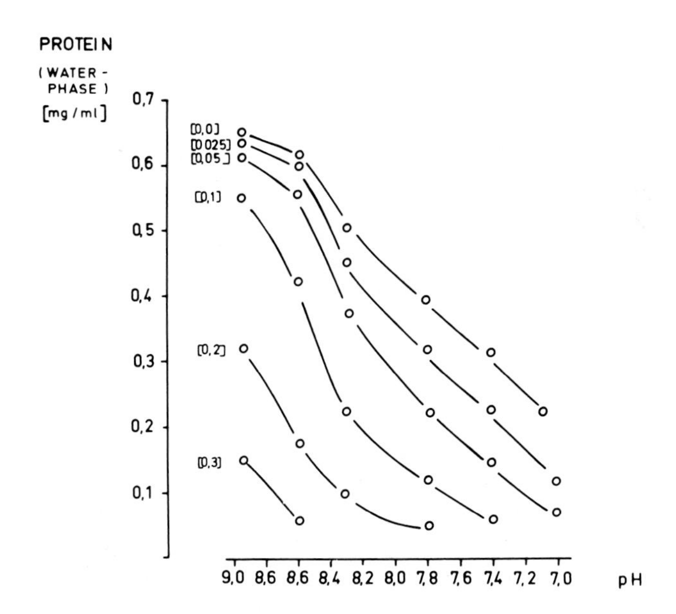

FIG. 3. pH dependence of protein consumption for the formation of the recombinant. At any calcium concentration the lowering of the pH value induces a loss of protein out of the water phase into the membranous film at the boundary of the two-phase system.

quite similar kinetics of protein consumption was found: At any calcium concentration the lowering of the pH value induces a loss of protein out of the water phase into the membranous film at the boundary. This result supports the charge dependence of the reaction based on the ampholyte character of the proteins.

In collaboration with W. Kreutz, Department of Biophysics, University of Freiburg, we studied the recombinant by electronmicroscopy. Figure 4 presents an electron micrograph of a sample we obtained upon addition of calcium. The micrograph shows filaments rather uniform in diameter. In some areas they are arranged in layers. The fibers themselves look like flat bands, which can be seen at those points where they are twisted. The formation of these filaments was then studied in HS.

Figure 5 gives the results of a typical experiment: Plotted against a logarithmic scale, the dose–response curve of a HS proposita is placed against a normal control of a comparable amount of reticulocytes. As confirmed in four other HS patients, the slope of the HS curve is flattened markedly, indicating that more cations are necessary to get the same amount of proteins into the interfacial film as in the control.

We did not find any shifting of this curve when we studied patients with

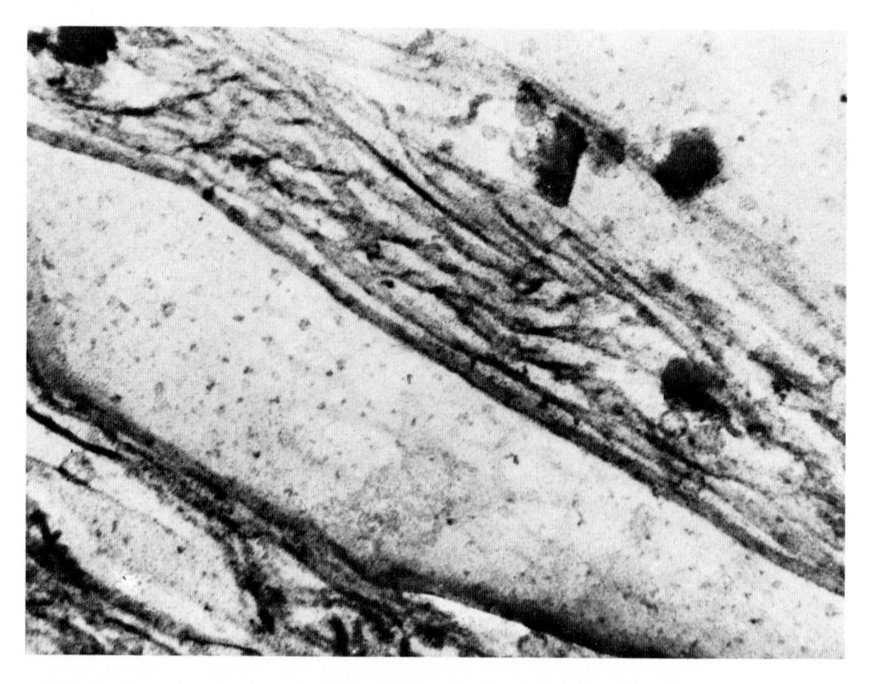

FIG. 4. Electronmicrograph of a recombinant in the presence of calcium chloride. Preparation by sectioning technique; ×30,000; fixation by osmium tetroxide; embedded in Araldite; double staining with uranylacetate and lead citrate.

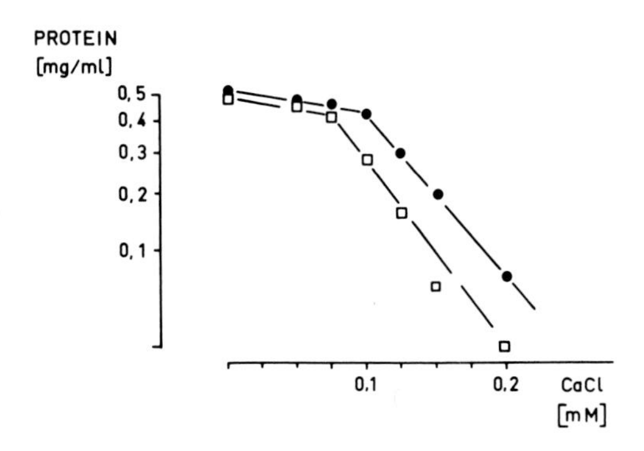

FIG. 5. The dose–response curve of a HS proposita (closed circles) is typically flattened when compared with a normal control (open squares).

other hemolytic disorders, e.g., hereditary elliptocytosis, hemolysis in Gaucher's disease, severe hemolysis in a man with erythroleukemia, in a case with Zieve syndrome, and one case of hemolytic anemia of unknown etiology.

DISCUSSION

The initial formation of complexes between the components of the disintegrated RBC membrane depends on electrostatic attraction. At the neutral or alkaline pH range, no interaction is possible since all the solubilized constituents possess negative charge. The cations added are necessary to reduce the electrostatic repulsion. Moreover, they induce, perhaps, the proper conformation of the proteins to allow the lipid–protein and protein–protein interaction.

Based on these principles, alterations of the dose–response-curve given above may indicate alterations of the charge of the components involved; more precisely, the alterations we found in HS would consist in an excess of negative charges of their membrane proteins compared with those of the normal controls.

ACKNOWLEDGMENT

This work was supported by Deutsche Forschungsgemeinschaft Grant AR 59/3.

REFERENCES

1. LaCelle, P. L. (1970): *Semin. Hematol.,* 7:355.
2. Bertles, J. F. (1957): *J. Clin. Invest.,* 36:816.
3. De Gier, J., van Deenen, L. L. M., Verloop, M. C., and van Castel, C. (1969): *Br. J. Haematol.,* 10:246.

4. Jacob, H. S., Ruby, A., Overland, E. S., and Mazia, D. (1971): *J. Clin. Invest.,* 50:1800.
5. Coleman, R. (1973): *Biochim. Biophys. Acta,* 300:1.
6. Gomperts, E. D., Metz, J., and Zail, S. S. (1972): *Br. J. Haematol.,* 23:363; (1973): 25:421.
7. Rottem, S., Stein, O., and Razin, S. (1968): *Arch. Biochem. Biophys.,* 125:46.
8. Maddy, A. H. (1964): *Biochim. Biophys. Acta,* 88:448.
9. Rega, A. F., Weed, R. I., Reed, C. F., Berg, G. G., and Rothstein, A (1967): *Biochim. Biophys. Acta,* 147:297.

Membranes and Disease, edited by L. Bolis, J. F. Hoffman, and A. Leaf. Raven Press, New York, © 1976.

The Molecular Basis of Membrane Elasticity and Strength

Martin Blank

Department of Physiology, Columbia University College of Physicians and Surgeons, New York, New York 10032

A volume devoted to understanding disease processes in terms of fundamental membrane properties should consider the divergent ideas that are frequently included in our current conceptions of the natural membrane. The many paradigms of the natural membrane are not always compatible, and in this chapter we shall examine the relation between the biochemical and the rheological views. Specifically, we shall consider if the properties of the known components of the membrane can account for its responses to mechanical stresses. These two paradigms of the membrane are compatible if we modify our ideas about the distribution of protein in a membrane. In so doing, we must also necessarily consider the effect of the protein distribution on ion transport processes. The interrelationships between the various properties pose great difficulties for those who are interested in developing ways of characterizing disease processes.

MEMBRANE PARADIGMS

Natural membranes differ considerably from one cell to another, and even in a single cell there are differences between the outer or plasma membranes and those of the organelles. Furthermore, the regular turnover of membrane components, the cumulative changes with cell age, and the reported diurnal variation of the activities of some membrane enzymes (1), all add to the multiplicity of structures covered by the single term. Yet we speak of a generalized natural membrane that is about 100 Å thick, because we believe that there is a basic structure (2) or unit membrane that is present in all cases.

Even though we acknowledge the above restrictions regarding our concept of the membrane, we still have different pictures in mind when we use the term. Table 1 indicates the wide diversity of the points of view of the various experimental approaches to the study of membrane properties. The different ways of looking at membranes are useful when they contribute to our understanding of membrane properties by providing complementary types of information. However, there are difficulties if the

TABLE 1. *Paradigms of the natural membrane*

Approach	Membrane composition	Arrangement of components	Function
Biochemistry	Polar lipids and polymers (e.g., proteins)	Chemically asymmetric bilayer	Matrix for enzymic reactions
Morphology	Components that react with heavy metals and oil soluble reagents	Asymmetric, ultrathin layer split by freeze fracture	Boundary of cell or organelle
Physiology (transport)	Semipermeable array with special molecules (e.g., pump enzyme, carrier)	Permeability barrier with selective structures (e.g., pore)	Separation and concentration of cell constitutents
Physiology (electrophysiology)	Resistance, capaci- tance, and ion channels with gates	Equivalent circuit of electrical elements	Electrical properties (e.g., excitation)
Electrochemistry	Dielectric material, ions, and dipoles	Phase boundaries, oriented dipoles, and asymmetric charge distribution	Charge transport and variations of membrane potential
Rheology	Elastic, viscous, and plastic (yield) elements	Equivalent circuit of rheological elements	Strain and flow under mechanical stresses; yield strength

models do not overlap when they refer to the same system. All membrane paradigms must eventually coincide, e.g., the membrane components of the rheologist must be the same as those described by the biochemist.

The attempt to account for a physical property of the natural membrane in terms of its molecular components and their arrangement is in line with a general goal of Biology today. (Many biological scientists go so far as to say that all one really needs to know is the structure of a system and an understanding of the function will follow.) In this chapter we show that many aspects of membrane rheology can be explained in terms of the known properties of the protein constituents when present in the form of monolayers, and that in this case the functions may be based directly on the structure.

Much appears to be known about the composition, structure, and physical properties of the erythrocyte membrane, and we shall use this system as the prototype of a natural membrane. Most investigators agree that the membrane consists of an asymmetric lipid bilayer (choline phospholipids and glycolipids on the external face and amino phospholipids on the cytoplasmic face), and an asymmetric distribution of membrane protein, with the major portion on the cytoplasmic side and some going through the bilayer (3). But how much protein is actually there? Investigators agree on the amount, approximately 10^{-12} g/cell, but not on how much membrane area this would cover. If one were to repeat the Gorter and Grendel experiment with these proteins, one would expect this amount to cover the membrane area about 7 times, based on the properties of protein monolayers (4). This would constitute a layer that is at least as thick as the bilayer, and it may even be thicker if the protein is not close packed, but in a looser linked chain-mail arrangement.

The quantitative aspects of the biochemical paradigm of the natural membrane lead us to the conclusion that there must be much more protein present in the structure than is usually depicted. This amount of protein must influence the properties of membranes, and in the next section we shall see that this layer could account for the observed rheological properties.

MEMBRANE RHEOLOGY

The ability of a membrane (especially of the erythrocyte) to remain stable when responding to mechanical stresses should be based upon its composition and structure. Recent experiments on the properties of protein monolayers suggest that the membrane proteins can account for these properties. The conformations of protein molecules in monolayers (5–7) are probably quite similar to the conformations of the various interfaces in membrane structures, so the physical properties should be related.

Protein molecules generally interact on a surface, and form a two-dimensional network when the film is at a relatively high surface concentration (8–10). As would be expected, the interactions between molecules in protein films are affected by factors such as age, degree of shear in the monolayer, subphase composition, and ionic strength. But it is the strong binding of multivalent cations to proteins that causes the major change in the rheological properties of protein films. In particular, there are large changes in the yield stress (11,12), a measure that is based on the pressure difference needed to start flow through a two-dimensional channel, and that does not involve a flow of subphase. The variation of the yield stress due to cations in the aqueous subphase, shown in Table 2, bears a strong resemblance to the variation of erythrocyte rigidity with the same ions. The surface yield stress of the protein monolayer can be converted to a comparable three-dimensional value by assuming a film thickness of 10^{-7} cm. The comparable bulk value, 10^{7}–10^{8} dyn/cm^2 shows that the strength of the network is similar to that of the metals aluminum, magnesium, and zinc, which have yield points about 10^{9} dyn/cm^2, and lead, which is about 10^{8} dyn/cm^2. The looser two-dimensional structures that form in protein films have considerably greater strength than expected.

The measured elasticities of some protein monolayers (13) are about 10 dyn/cm. When these values are converted to three-dimensional values, they are of the same magnitudes as determined for erythrocytes. The monolayer values are equivalent to about 10^{7} dyn/cm^2, and the measured values for erythrocytes are in the region of 3×10^{7} dyn/cm^2 (14). The values obtained for artificial lipid bilayers, recently summarized by White (15), are 1 to 2 orders of magnitude too low to account for the observed membrane elasticities.

In all of the measurements of elasticity and yield of protein films, the properties vary with the degree of packing of the molecules in the array. Furthermore, the changes in the two properties appear to be related, in that the yield does not develop until the film has reached a close packed arrangement (of about 1 mg/m^2) and its maximum elasticity.

The rheological parameters of protein films discussed above, i.e., elasticity, yield, the ionic dependence of yield, the variation of all of the measurements with surface packing, are of the right magnitude or have the right

TABLE 2. *Relative yields of an ovalbumin monolayer in contact with various cations*

Ion	Relative yield (at same monolayer concn.)
Ba^{++}	8.6
Ca^{++}	6.4
Mg^{++}	5.2
K^{+}	3.1
Na^{+}	1.2
Li^{+}	1.0

qualitative dependence to account for the parallel properties of natural membranes. Additional measurements on the effects of continuous small-amplitude distortions on the properties of protein monolayers (13) may also shed light on the reported changes of membrane properties with age. Obviously, further measurements are required on protein films, and especially on proteins derived from membranes. But in the absence of these measurements, the experiments with model systems suggest that we can reconcile the biochemical and rheological paradigms of the membrane.

MEMBRANE FLUIDITY

When discussing membrane rheology, we should consider the term "fluidity," which has been used with increasing frequency in the last few years, but in a very loose way. Fluidity has been used to describe at least three different types of motion: (i) the freedom of movement of a group in a molecule (i.e., "wobble"), (ii) the diffusivity, D, of a molecule in an array, and (iii) the flow of molecules under a pressure gradient giving a fluidity, η^{-1}, of the array of molecules. These are not at all interchangeable, and attempts to draw general conclusions about "membrane fluidity" from one measurement may lead to erroneous conclusions. Some calculations on monolayer arrays may serve to illustrate the great differences between D and η^{-1}.

The physical properties of lipid monolayers at the air/water interface can be understood in terms of the relatively small number of molecules involved in a monolayer process and the consequent large fluctuations in monolayer density that result (16,17). On this basis, it has been possible to formulate the transport properties (diffusion, viscous flow, thermal conduction) and the permeability of monolayers. These ideas were able to predict the permeability of a 1-eicosanol monolayer to water (18) and the self-diffusion coefficient (17) in a stearic acid monolayer. [We obtained $D \sim 10^{-8}$ cm²/sec, that agrees with the measured value of 1.8×10^{-8} cm²/sec in a monolayer of hydrated lecithin. The predicted frequency of diffusion of a molecule between the two planes of a bilayer for the same stearic acid monolayer is about 10^{-5} flips/sec (16), compared to the observed value of 2×10^{-5}/sec for lecithin.] The agreement that exists between the calculation and the measurement of a monolayer property suggests that the theoretical approach based on fluctuations in surface density should be useful for understanding the properties of lipid layers in membrane systems.

When we calculate the value of η from the same theoretical model, we find that it is quite different (by orders of magnitude) from the measured values. This is due to the influence of the aqueous subphase during monolayer flow. The water exerts a considerable drag on the monolayer, and the measurement always includes the effect of the subphase (19). It may even be that the effective surface viscosity is unrelated to monolayer viscosity and determined only by the interaction of the polar group with the water. The actual mechanism of monolayer flow is quite different from the mechanism of diffusion, and it appears that one cannot learn about fluidity (η^{-1}) from measurements of D.

Another way of making the same point is in terms of the Stokes-Einstein equation for spherical molecules in aqueous solutions. [At constant temperature $D\eta = f$ (molecular radius).] This equation was derived for large spheres in a sea of points, which is very different from the situation of a long cylindrical lipid molecule in an array of molecules of the same size and shape. The equation does not apply to a lipid molecule in a monolayer or in a membrane, and one would not expect D to be proportional to η^{-1}.

THE EFFECTS OF PROTEINS ON TRANSPORT

By introducing more protein into our model of the natural membrane, we have been able to account for many of the observed rheological properties. However, the presence of protein in our picture of the membrane is bound to have other effects, and we must begin to take these into account in our thinking. The most obvious effects are on the transport functions and we shall now consider these.

In model studies proteins offer negligibly small resistances to the passage of small molecules, like water (20). Therefore, there has been a tendency to consider the protein in the membrane as having no influence on passive transport, while the lipid is the permeability barrier. Of course, proteins can alter the structure of the lipid bilayer and have an indirect effect on transport, but they can also directly influence the passive transport of ions, in addition to being involved in specialized transport mechanisms (e.g., the Na-K ATPase).

Proteins have the capacity to bind ions, and relatively thick layers of proteins can act like an additional compartment at the membrane surface in ion transport systems. [We have recently shown that such bound ions can be released by ion exchange processes and can behave as a reservoir of ions (*unpublished*)]. The presence of a protein at an interface can also change the net orientation of the water molecules and the surface potential, ΔV (21). The effect of the protein on the surface potential is asymmetric due to the ΔV arising from the unperturbed water structure at the surface (22), and there is a larger ΔV when a positively charged protein molecule adsorbs, as opposed to a negative one.

Proteins can also influence the transport of ions directly because of their charge. In earlier work on charged surface films, it was possible to show that the permeability of a monolayer to an ion varies sharply with the surface charge density (23), and we have been able to demonstrate the same effect with a bilayer system (24). Presumably protein films have similar effects when they are charged.

MEMBRANES AND DISEASES

In reviewing our ideas about natural membranes and the ability of experiments with simple model systems to contribute to our understanding of the more complicated real systems, we have been able to integrate two previously parallel paradigms. We have also seen that the various paradigms of the membrane are so interrelated that even though we have focused on the rheological properties due to the membrane proteins, we should not forget that the proteins can also influence the transport properties. It is the interrelationship between the various properties of membranes that brings us back to the problem of diseases and how to characterize them. Departures from the normal state are apt to affect many properties and this often makes it difficult to understand the nature of the problem. (For example, a defect in a membrane protein may lead to changes in passive ion transport and consequent osmotic swelling, while at the same time it may decrease the strength of the membrane. The net effect of the defect would be cell membrane lysis, but the mechanism would involve a complicated interplay of transport and rheological properties.) From this it appears that the understanding of disease processes must be based on a broad foundation of fundamental knowledge of the normal system, and that the diseased state should be examined from many different experimental points of view.

ACKNOWLEDGMENT

This study was supported in part by Research Grant HD 06908 from the U.S. Public Health Service.

REFERENCES

1. Ashkenazi, I. E., and Hartman, H. (1975) private communication.
2. Danielli, J. F., and Davson, H. (1935): *J. Cell. Comp. Physiol.* 5:495.
3. Bretscher, M. S. (1973): *Science,* 181:622.
4. Blank, M., and Britten, J. S. (1975): *Biorheology,* 12:271.
5. Malcolm, B. R. (1968): *Proc. Soc. Lond.* (*Ser. A*), 305:363.
6. Loeb, G. I., and Baier, R. E. (1968): *J. Colloid Interface Sci.,* 27:38.
7. Augenstine, L. F., Chiron, C. A., and Nims, L. F. (1958): *J. Phys. Chem.,* 62:1231.
8. Fourt, L. (1939): *J. Phys. Chem.,* 43:887.
9. Tachibana, T., Inokuchi, K., and Inokuchi, T. (1957): *Biochim. Biophys. Acta,* 27:174.
10. Blank, M. (1969): *J. Colloid Interface Sci.,* 29:205.
11. Blank, M., and Britten, J. S. (1970): *J. Colloid Interface Sci.,* 32:62.
12. Blank, M., Lee, B. B., and Britten, J. S. (1973): *J. Colloid Interface Sci.,* 43:539.
13. Blank, M., Lucassen, J., and van den Tempel, M. (1970). *J. Colloid Interface Sci.,* 33:94.
14. Skalak, R., Tozeren, A., Zarda, R. P., and Chien, S. (1973): *Biophys. J.,* 13:245.
15. White, S. H. (1974): *Biophys. J.,* 14:155.
16. Blank, M., and Britten, J. S. (1965): *J. Colloid Sci.,* 20:789.
17. Blank, M., and Britten, J. S. (1970): In: *Physical Principles of Biological Membranes,* edited by F. Snell et al., p. 143. Gordon and Breach, New York.

18. Bockman, D. O. (1969): *Ind. Eng. Chem. Fundamentals,* 8:77.
19. Joly, M. (1964): *Recent Prog. Surface Sci.* 1:1.
20. Blank, M., and Mussellwhite, P. R. (1968): *J. Colloid Interface Sci.,* 27:188.
21. Blank, M., Lee, B. B., and Britten, J. S. (1975): *J. Colloid Interface Sci.,* 50:215.
22. Blank, M., and Ottewill, R. H. (1964): *J. Phys. Chem.,* 68:2206.
23. Miller, I. R., and Blank, M. (1968): *J. Colloid Interface Sci.,* 26:34.
24. Sweeney, G. D., and Blank, M. (1973): *J. Colloid Interface Sci.,* 42:410.

Membranes and Disease, edited by L. Bolis, J. F. Hoffman, and A. Leaf. Raven Press, New York, © 1976.

Hereditary Stomatocytosis: A Disease of Cell Water Regulation

James S. Wiley

Hematology–Oncology Section, Department of Medicine, Hospital of the University of Pennsylvania, Philadelphia, Pennsylvania 19174

Hereditary stomatocytosis is perhaps the prototype disease of cell volume regulation since red cell destruction seems to result directly from the abnormality in water content. Much is already known about water and ion movements in human erythrocytes, and it is well established that the monovalent cation content of the red cell determines its water content and cell volume (1,2). Stomatocytes show marked alteration in both content and unidirectional fluxes of Na^+ and K^+, and it seems important to establish that these changes account for the abnormal water content.

Two varieties of hereditary stomatocytosis have been recognized, one with cell overhydration and stomatocytes on blood smear (Fig. 1A) and one showing cell dehydration with only target cells on smear (Fig. 1B). In both varieties there is a shortened red cell life-span, although hemolysis may range from severe to mild within families and may be even absent in affected members of some families. The cation and water content of red cells from two representative patients are shown in Table 1. Patient R.Y. was one of the first reported with this condition (3) and shows reversal of the usual cell K^+/Na^+ ratio having "high Na^+, low K^+" red cells (Table 1). The sum of red cell monovalent cations ($Na^+ + K^+$) in R.Y. was 125 μeq/ml cells, which is substantially above the normal Na^+ plus K^+ of 107 ± 5 μeq/ml cells. Red cells from R.Y. show a corresponding overhydration (697 mg water/g wet wt RBC) compared with normal (658 ± 6 mg/g), and this change is also reflected by the reciprocal decrease in mean cell hemoglobin concentration. The dehydrated variety is illustrated by patient N.S. who has the typical cation changes of this condition (4). Cell Na^+ is slightly increased (by 7 μeq/ml cells), whereas cell K^+ is substantially reduced below normal (by 20 μeq/ml cells), so that the total sum of Na^+ plus K^+ is 95 μeq/ml cells or 12% below normal. Red cell water in N.S. is low (622 mg/g wet wt RBC), and again there is a reciprocal increase in mean corpuscular hemoglobin concentration. Total monovalent cations, Na^+ plus K^+, therefore change in the same direction as cell water content, which is consistent with a defect in cation permeability being responsible for the pathogenesis of this disease.

The different morphologies in the two varieties (stomatocytes and target

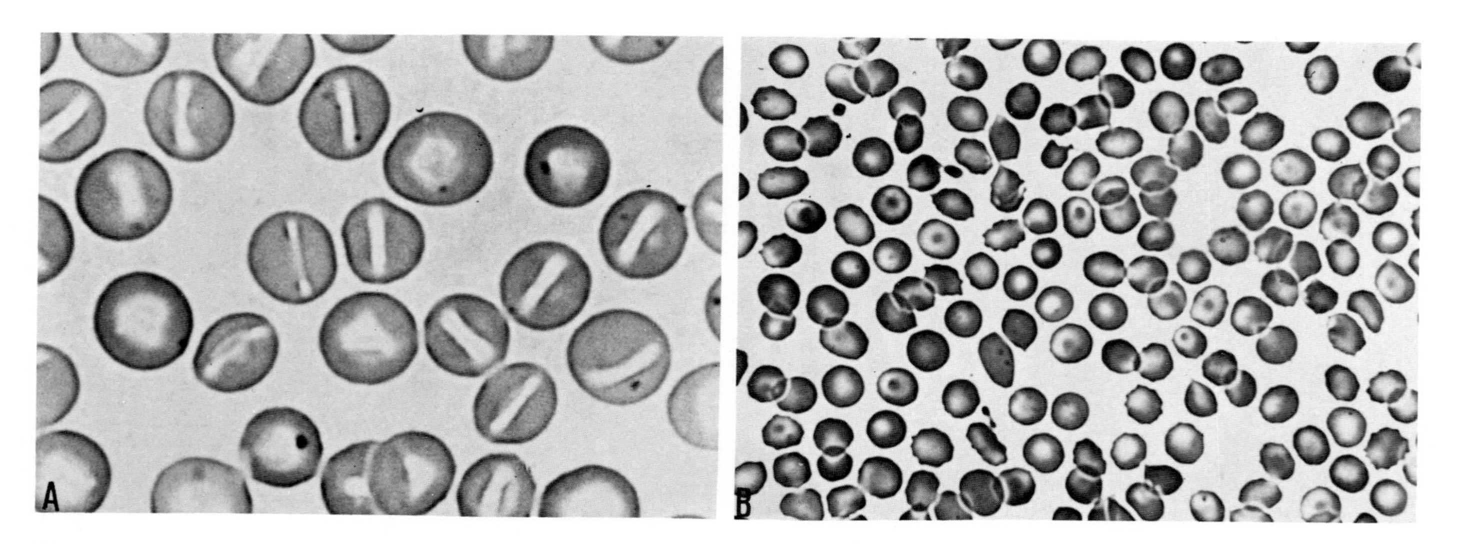

FIG. 1. A: Peripheral blood smear, stained by Wright-Giemsa, from patient R.Y. showing stomatocytes, i.e., cells with slitlike central pallor. Reticulocyte count 8%. **B:** Blood smear from patient N.S. showing target cells. Reticulocyte count, 6.5%.

TABLE 1. *Cation and water content in two varieties of hereditary stomatocytosis*

	μeq/ml cells			Cell water (mg/g wet wt)	MCHC (g/dl)
	Na^+	K^+	$Na^+ + K^+$		
Stomatocytes (R.Y.)	87.0	38	125	697	27
Targets (N.S.)	15.1	79.5	95	622	37
Normals	7.9	99	107	658	33
±SD	1.5	5.2	5	6	2

cells) may superficially suggest two separate syndromes, linked only in the sense that each represents a defect in cell water regulation. In practice there are many similarities between the two varieties, both morphologically and biochemically. Strong evidence for considering the two together is the conversion of stomatocytes to target cells caused by doubling the tonicity of the surrounding plasma. This maneuver induces an outward water movement and dehydration of the stomatocytic cells, and it is little surprise that such an *in vitro* procedure reproduces the characteristic target morphology of the dehydrated cell variant. Thus, both stomatocytes and target cells in this disease represent red cells with excessive surface area (vide infra), but with different water content. Another similarity between the overhydrated and underhydrated varieties comes from the results of lipid analyses performed in collaboration with Dr. Richard Cooper. Lipid in the mature red cell is localized exclusively in the membrane (5) and osmotic fragility studies have demonstrated a direct relation between the lipid content and the surface area of the red cell in at least two disease states (4,6). Patients with hereditary stomatocytosis have increased red cell lipids on a per cell basis and the increment ranges from +0% to +38% (Fig. 2). Moreover, the increment in

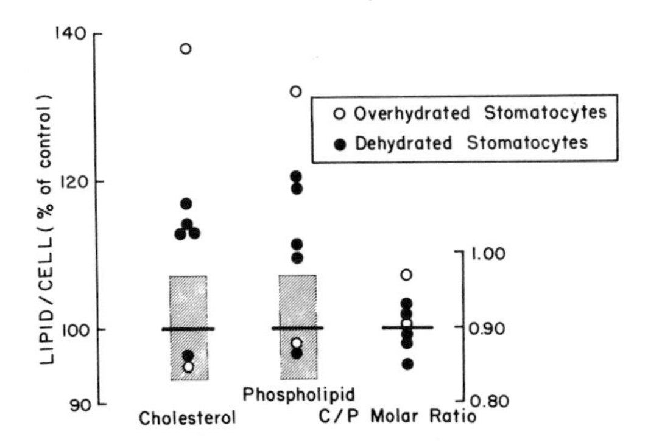

FIG. 2. Red cell lipids in hereditary stomatocytosis. Three patients were splenectomized and their lipids were compared with a group of postsplenectomy controls. The other patients retained their spleens and lipids were compared with the mean for normal subjects. Shaded areas show ±SD of normal subjects.

phospholipid and cholesterol is the same for each patient, so that the cholesterol:phospholipid molar ratio does not deviate from the normal value of 0.9. Thus, red cell surface area is normal or increased in this disorder. Another important result is that both the overhydrated and the dehydrated variety of hereditary stomatocytosis show the same defect, i.e., a variable increase in red cell lipid, which supports the inclusion of the two varieties within the one syndrome.

Isotopic fluxes of both Na^+ and K^+-ions are elevated in hereditary stomatocytosis anywhere between 3- and 25-fold normal and our laboratory routinely employs a Na^+ influx measurement for the diagnosis of hereditary stomatocytosis. Sodium efflux and K^+ influx are also increased in about the same proportion to the Na^+ influx. Na^+ efflux and K^+ influx are, however, uphill (i.e., against the electrochemical potential gradient) and both fluxes are mediated by the active cation pump which is an (Na^+ + K^+)-stimulated ATPase localized in the membrane (7). Cardiac glycosides such as ouabain bind to and inhibit the activity of the active cation pump with an affinity which is so high that the number of ouabain-binding sites may be easily quantitated (8). Normal cells possess 336 ± 14 sites/cell, measured with a ^{3}H-ouabain binding technique. Hereditary stomatocytes have increased numbers of ouabain-binding sites (i.e., pumps) per cell, as might be predicted from the increase in active cation fluxes. This increase in pumps, however, is closely connected with the increase in passive Na^+ leak, as shown by the positive correlation between Na^+ influx and numbers of ouabain-binding sites in Fig. 3. Two patients with overhydrated cells and

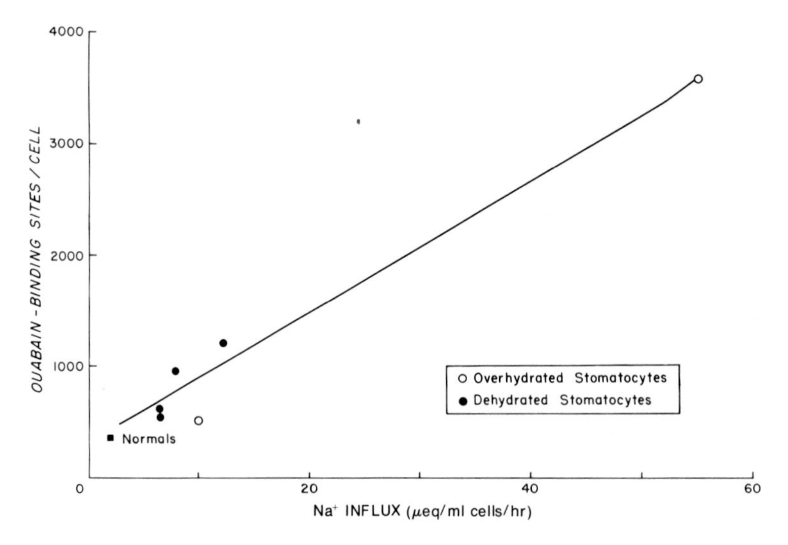

FIG. 3. Correlation of Na^+ influx and numbers of ouabain-binding sites per cell (i.e., cation pumps) in six patients with hereditary stomatocytosis.

four patients with dehydrated cells were studied and both varieties showed the same coordinated increase in pumps and leak. The increase in Na^+ leak and pump was far greater than expected from the increment in surface area revealed by the lipid analyses. Sodium influx and ouabain-binding sites were 25- and 12-fold normal in patient R.Y., but his lipids were only increased by +35%. Likewise patient N.S. had a 6- and 3.5-fold increase in leak and pump but only a +12% increment in cell lipids.

Many hypotheses may be advanced to account for these findings, although only two seem consistent with all the data. The first assumes some defect in the process of membrane degradation which occurs during the maturation of the reticulocyte. Such a defect could lead to a selective sparing of those parts of the membrane containing pumps and leak elements, while overall reduction in membrane mass (as shown by the lipid data) would be only slightly affected. The second hypothesis assumes that the underlying mutation leads to a variable increase in the passive Na^+ leak. Inward movement of Na^+ would occur in the developing erythroblast and increase the cell Na^+ concentration, which, in turn, would induce a compensatory synthesis of more cation pump molecules. Such a compensatory synthesis of more cation pump molecules is well documented in a remaining kidney following unilateral nephrectomy (9), as well as in HeLa cells which have been exposed to low concentrations of ouabain, causing partial but irreversible inhibition of the cation pumps (10). An increased cell Na^+ in the erythroblast might also interfere with membrane lipid turnover in such a way as to increase the lipid content of these cells.

Little is known about the mechanisms responsible for the passive transport of Na^+ and K^+ ions, although it is clear that both a facilitated and a simple leak pathway are present in the human erythrocyte (11). The close correlation observed between increased Na^+ leak and numbers of cation pumps observed in hereditary stomatocytosis may offer a clue which relates these two variables, although it does not indicate if the relationship is at the level of structural similarity or topographical proximity. It seems certain, however, that accurate studies of mutations affecting cation transport and volume regulation will enlarge our understanding of these normal physiological mechanisms.

ACKNOWLEDGMENT

This work was supported by a grant (HL-16201) from the National Institutes of Health.

REFERENCES

1. Tosteson, D. C., and Hoffman, J. F. (1960): *J. Gen. Physiol.*, 44:169.
2. Beilin, L. J., Knight, G. J., Munro-Faure, A. D., and Anderson, J. (1966): *J. Clin. Invest.*, 45:1817.

3. Zarkowsky, H. S., Oski, F. A., Shaafi, R., Shohet, S. B., and Nathan, D. G. (1968): *N. Engl. J. Med.,* 278:573.
4. Wiley, J. S., Ellory, J. C., Shuman, M. A., Shaller, C. C., and Cooper, R. A. (1975): *Blood* 46:337.
5. Dodge, J. T., Mitchell, C., and Hanahan, D. J. (1963): *Arch. Biochem. Biophys.,* 100:119.
6. Cooper, R. A., and Jandl, J. H. (1968): *J. Clin. Invest.,* 47:809.
7. Glynn, I. M., and Karlish, S. J. D. (1975): *Ann. Rev. Physiol.,* 37:13.
8. Hoffman, J. F. (1969): *J. Gen. Physiol.,* 54:343s.
9. Katz, A. I., and Epstein, F. H. (1967): *J. Clin. Invest.,* 46:1999.
10. Boardman, L., Huett, M., Lamb, J. F., Newton, J. P., and Polson, J. M. (1974): *J. Physiol. (Lond.),* 241:771.
11. Wiley, J. S., and Cooper, R. A. (1974): *J. Clin. Invest.,* 53:745.

Membranes and Disease, edited by L. Bolis, J. F. Hoffman, and A. Leaf. Raven Press, New York, © 1976.

Studies on the Cation Transport in High Sodium and Low Potassium Red Cells in Hereditary Hemolytic Anemia Associated with Stomatocytosis

Werner Schröter and Klaus Ungefehr

Department of Pediatrics, University of Göttingen, and Max-Planck Institute of Biophysics, Department of Cell Physiology, Frankfurt/Main, West Germany

Clinical and laboratory studies have demonstrated that the syndrome of hereditary hemolytic anemia seems not to be a homogeneous entity (1–4). Abnormal red cell Na^+ and K^+ concentrations were found in most cases. An increased passive cation permeability and, in some cases, also an increased active cation pumping rate of the red cells were prominent findings. It is generally accepted that the basic defect of these cells is located within the red cell membrane. However, the true nature of the membrane defect is not yet clearly defined. In this chapter, kinetic studies on the red cell cation transport of one distinct family with hereditary hemolytic anemia associated with stomatocytosis are reported.

Table 1 gives the hematologic data of the patient, a 7-year-old girl, who was first seen at the Department of Pediatrics, University of Hamburg, in 1971. A moderate, macrocytic hypochromic hemolytic anemia was improved after splenectomy. Hydrocytosis of the red cells and a nearly normal autohemolysis in the presence of glucose were characteristic findings in this patient. In the dried blood smear, about 40% of the red cells were stomatocytes. No enzyme defect or abnormal hemoglobin could be estimated. Her father, mother, and brother were hematologically normal. The following studies on the red cells of the patient were performed after splenectomy.

The intracellular Na^+ concentration was increased to 95–103 meq/liter cells ($n = 10.0 \pm 3.0$ meq/liter cells) and the K^+ concentration was decreased to 16–22 meq/liter cells ($n = 100.0 \pm 6.0$ meq/liter cells). The cation concentrations in the red cells of the mentioned family members were within the normal range. The cholesterol and phospholipid contents of the red cells of the patient were slightly increased to a degree consistent with the young red cell population and with erythrocyte populations of splenectomized individuals. No significant deviations of the normal distribution of the phospholipid fractions could be detected. The glucose utilization of the patient's erythrocytes was much more increased than in comparably reticulocyte-rich

TABLE 1. *Hematologic data of a patient with hereditary hemolytic anemia*

	Normal	Patient G.W.	
		Presplenectomy	Postsplenectomy
Hemoglobin (g/100 ml)	11 –15	8	10
Packed cell volume (%)	35 –43	27	35
MCV (μ^3)	84 –88	108	115
MCH (pg)	28 –32	33	33
MCHC (%)	33 –35	30	28
Cell water (%)	64 –68	76	77
Reticulocytes (%)	0.5– 1.5	35	11
Autohemolysis (%)			
Without glucose	0.2– 4.0	52	32
With glucose	0.1– 0.6	5	2

MCV, mean corpuscular volume; MCH, mean corpuscular hemoglobin; MCHC, mean corpuscular hemoglobin concentration.

blood. In the presence of 10^{-4} M ouabain glucose, consumption was reduced markedly, suggesting that a predominant part of glucose was utilized for cation transport. ATP, found at high levels, declined rapidly in the absence of glucose. It remained constantly for a few hours in the presence of 10^{-4} M ouabain.

In order to study the effect of intracellular Na^+ concentrations on active transport red cells with varying intracellular Na^+, concentrations were prepared using the method of Garrahan and Rega (5) in the modification of Sachs (6). The original cells of the patient and cells with varied Na^+ concentrations were incubated in a medium containing different K^+ concentrations. The *K^+ dependent $^{22}Na^+$ outflux* followed a Michaelis–Menten-like kinetic. The Na^+ outflux was 40 times higher in the patient's erythrocytes at high internal Na^+ concentrations than in normal erythrocytes. It was also markedly increased to 45 meq/liter cells-hr at a reduced Na^+ concentration of 22.5 meq/liter cells. The Na^+ outflux in the absence of K^+ is equivalent to Na^+/Na^+ exchange (7). This parameter was increased in the patient to 40 meq/liter cells-hr instead of 1.3 meq/liter cells-hr in normal erythrocytes. Thus, an active Na^+ pump rate of 70.2 meq/liter cells-hr was calculated. In the erythrocytes of both the father and the mother active $^{22}Na^+$ outflux was increased threefold. It showed normal kinetics.

The effect of different intracellular Na^+ concentrations on $^{22}Na^+$ outflux was studied at optimal extracellular K^+ concentrations of 20 meq/liter cells. $^{22}Na^+$ outflux strongly depended on the internal Na^+ concentration. The sigmoid kinetic curve indicates that more than one Na^+ ion must interact with the transport system before transport occurs (6). The abnormal cells did not differ in their kinetic properties from normal red cells.

The *ouabain-sensitive $^{22}Na^+$ outflux* was measured in the presence of 10^{-4} *M* ouabain at 0 meq/liter external K^+. The erythrocytes of the father showed a slightly increased ouabain-sensitive $^{22}Na^+$ outflux of 5.2 meq/liter cells-hr (n = 1.8 meq/liter cells-hr), whereas that of the patient was enormously increased to 87.8 meq/liter cells-hr. Normal red cells with altered Na^+ concentrations of 23.1 and 92.9 meq/liter cells showed an ouabain-sensitive $^{22}Na^+$ outflux of 5.7 and 9.9 meq/liter cells-hr, respectively. On the other hand, the ouabain-sensitive $^{22}Na^+$ outflux was reduced at internal Na^+ concentrations of 22.1 meq/liter cells only to 38.7 meq/liter cells-hr in the red cells of the patient.

The *$^{42}K^+$ influx* and the *ouabain-sensitive ^{42}K influx* of the erythrocytes of the patient were only slightly increased to 7.3 and 5.3 meq/liter cells-hr and they were normal in the erythrocytes of the father (n = 1.2 and 1.0 meq/liter cells-hr).

The *passive cation permeability* was grossly disturbed in the red cells of the patient. The $^{22}Na^+$ influx was 128 meq/liter cells-hr (n = 2.6 meq/liter cells-hr) and the $^{42}K^+$ outflux was 8.3 meq/liter cells-hr (n = 1.7 meq/liter cells-hr). In the father only the $^{22}Na^+$ influx was increased to 8.5 meq/liter cells-hr. The $^{42}K^+$ outflux was normal.

In summary, the erythrocytes of the patient are characterized by an extremely increased Na^+ leak and a very high active Na^+ outflux. The net Na^+ influx amounts to + 17.8 meq/liter cells-hr. In the erythrocytes of the father the active and passive Na^+ fluxes are only slightly increased. The net flux is +1.1 meq/liter cells-hr. The deviations of the K^+ fluxes are not so impressive as are those of the Na^+ fluxes. In the patient, they are increased about sixfold; the net flux was −1.0 meq/liter cells-hr. In the father, the K^+ fluxes were normal. In a first experiment with the erythrocytes of the mother, values were similar to those obtained with the erythrocytes of the father.

Concerning the properties of the cation pump of the red cells of the patient, some preliminary conclusions can be drawn: The Na^+ ions are pumped at a very high level. It is about 10 times higher than expected from the internal Na^+ concentration. The kinetics of the Na^+ transport seem to be normal both in internal Na^+ concentration and external K^+ concentration. An increased number of Na^+ pump sites per cell is suggested. The K^+ pump rate, on the other hand, is lower than expected from the internal Na^+ concentration. Thus, a dissociation between the Na^+ and K^+ pump rates seems to be possible in the abnormal cells. The coupling ratio of active Na^+:K^+ movement, which was 1.6 in our experiments with normal red cells, was increased to about 10 in the red cells of the patient. This ratio was calculated in consideration of a Na^+/Na^+ exchange of 40 meq/liter cells-hr.

When compared with the known cases of hereditary hemolytic anemia associated with stomatocytosis, our patient showed the signs of hydrated red cells (low MCHC, high cell volume, high cell water). The cation concentrations and the fluxes were similar to the case published by Zarkowsky et al. (1), but there were some striking differences: cell K^+ was much lower and the K^+ influx was smaller in our case than in that of Zarkowsky et al. In contrast, ATP was stable in the presence of ouabain, and autohemolysis was corrected nearly completely in the presence of glucose in the patient reported here. The studies on the erythrocytes of the patient and his parents suggest an autosomal recessive inheritance of the membrane defect.

ACKNOWLEDGMENTS

These studies were performed with support of the Deutsche Forschungsgemeinschaft, Bonn–Bad Godesberg. We wish to thank Dr. H. Passow for his generous help throughout the studies reported here and for helpful criticism.

REFERENCES

1. Zarkowsky, H. S., Oski, F. A., Shaafi, R., Shohet, S. B., and Nathan, D. G. (1968): *N. Engl. J. Med.,* 278:573.
2. Oski, F. A., Naiman, J. L., Blum, S. F., Zarkowsky, H. S., Whaun, J., Shohet, S. B., Green, A., and Nathan, D. G. (1969): *N. Engl. J. Med.,* 280:909.
3. Miller, D. R., Rickles, F. R., Lichtman, M. A., La Celle, P. L., Bates, J., and Weed, R. I. (1971): *Blood,* 38:184.
4. Wiley, J. S., Ellory, J. C., Shuman, M. A., Shaller, C. C., and Cooper, R. A. (1975): *Blood (in press).*
5. Garrahan, P. J., and Rega, A. F. (1967): *J. Physiol. (Lond.),* 193:459.
6. Sachs, J. R. (1970): *J. Gen. Physiol.,* 56:322.
7. Garrahan, P. J., and Glynn, I. M. (1967): *J. Physiol. (Lond.),* 192:159.

Membranes and Disease, edited by L. Bolis, J. F. Hoffman, and A. Leaf. Raven Press, New York, © 1976.

Factors Modifying the Action of Internal Sodium and External Potassium on Ouabain Binding to Red Cell Membranes

H. Harm Bodemann

Medizinische Universitätsklinik, Freiburg im Breisgau, West Germany

One of the basic mechanisms of the action of cardiac glycosides such as ouabain is the inhibition of the active transport of Na and K and of the membrane-bound (Na + K)-dependent ATPase activity in a variety of types of cells and tissues. In red cells these cardiotonic steroids act with high specificity and in rather low concentrations by binding at the outer surface of the membrane (1). Using tritiated ouabain, the number of binding sites can be determined: ~275 per red cell membrane (2).

The results presented here concern the effects of Na and K on the kinetics of binding of tritiated ouabain to the red cell membrane and are interpreted by the side specificity of their actions. Further comments are directed to some modifications of these effects by Ca ions and by different concentrations of ATP. The following data were obtained in collaboration with Joseph F. Hoffman at Yale University and are part of a study which has been presented in more detail elsewhere (3*a,b,c*).

Using reconstituted human red cell ghosts (4), first the side-dependent effects of internal and external Na and K on the kinetics of binding of [^{3}H]ouabain to the red cell membrane, in the presence of incorporated Mg-ATP, were studied. The main findings were that the effects of intracellular Na and extracellular K are coupled in the sense that together they both decrease the velocity of ATP-promoted ouabain binding as well as activate Na–K transport (3*a*). Therefore, $(Na)_i$ may be an important parameter with respect to the clinical significance of the antagonism between extracellular potassium and ouabain.

This information about the sidedness of action of Na and K on the ouabain binding rate in presence of Mg-ATP allows interpretation of the effects of Na and K on the rate of ouabain binding to porous red cell membranes (experimental details given in the legends). For illustration, one instance is chosen where Na in exchange for choline is added to a reaction medium which contains constant concentrations of K, Mg-ATP, and [^{3}H]ouabain. The result of this type experiment is contained in Table 1 and shows that the velocity of ouabain binding was decreased markedly by the addition of

TABLE 1. *The effect of Na on the rate of ATP-promoted ouabain binding to ghosts*

Cation concn. (mM)			Molecules ouabain bound per ghost
Na	Choline	K	
0	30	6	122,124
30	0	6	19,21

Hemoglobin-free, frozen-thawed ghosts were incubated for 30 min at 37°C in a medium which contained either 30 mM choline or 30 mM NaCl together with 6 mM KCl, 1.0 mM Tris-ATP, 5 mM Tris Cl, 4.0 mM $MgCl_2$ and 5×10^{-8} M [^{3}H]ouabain. The final pH was 7.4. The experiment and analyses were carried out in accordance with the methods described previously (3*b*). The results presented are representative of the results obtained in two other similar experiments. The measure of the rate of ouabain binding is taken as the number of molecules of ouabain bound per ghost after 30 min exposure to [^{3}H]ouabain.

Na. In this case Na has become effective at the inner surface of the porous membrane by means of its affinity constant and by allowing coupling of its own action with that of K, which has a relatively high affinity at the outer surface of the membrane (3*a*).

In general, there is now agreement that the positive inotropic effect of ouabain on heart muscle is mediated by an increase of the intracellular Ca concentration, although the mechanism by which this increase in the presence of digitalis-like compounds occurs is not quite understood. We were interested in this regard to see if there was any direct action of Ca on the kinetics of ouabain binding to red cell membranes. When intact erythrocytes were used, it was found that extracellular Ca does not have any effect on ouabain binding. It is difficult to study sidedness as far as the action of intracellular Ca is concerned. Ca cannot be incorporated into resealed ghosts in sufficient amounts because of the Ca pump, which readily extrudes Ca from the inside of the red cell (5), and because resealing does not take place in the presence of ionized Ca in concentrations above 0.1 mM (4). Therefore hemoglobin-free frozen-thawed red cell membranes were used for further studies of Ca effects, assuming that any effect of Ca observed will be due to an effect on the inside of the porous membrane.

It has already been shown that Ca ions cannot substitute for Mg ions, which are an essential requirement for ouabain binding to red cell membranes (6). For this reason, ouabain binding was carried out in the presence of Mg-ATP. The effect of Ca was then studied in the presence of different Na and K concentrations. Table 2 shows the result of a typical experiment.

TABLE 2. *The effect of Ca on the rate of ATP promoted ouabain binding to ghosts*

Incubation medium (mM)						
Choline Cl	NaCl	KCl	$[^3H]$ouabain (M)	$CaCl_2$ (mM)	Molecules ouabain bound per ghost	Fraction of binding
46	0	0	3.7×10^{-8}	0	149,151	—
				0.6	67,71	0.46
40	0	6	1.0×10^{-7}	0	88,88	—
				0.6	51,55	0.60
6	40	0	2.5×10^{-8}	0	138,140	—
				0.6	108,111	0.79
0	40	6	2.5×10^{-7}	0	78,78	—
				0.6	10,12	0.14

Hemoglobin-free, frozen-thawed ghosts were incubated for 15 min at 37°C in a medium which, in addition to the constituents shown, contained 1 mM Tris-ATP, 0.25 mM EDTA, and 2 mM $MgCl_2$. The final pH was 7.4. Ca, when present, was added as $CaCl_2$. The concentration of $[^3H]$ouabain was varied in an attempt to keep the ouabain binding rate in the presence of the different concentrations of Na and K approximately comparable. The results presented in this table are duplicate analyses and are representative of similar results obtained in three other experiments of the same design.

As far as the specified concentrations of Mg, ATP, and Ca are concerned, this experiment was performed under conditions under which the effect of Ca was most pronounced. The concentration of $[^3H]$ouabain in the reaction medium was varied in order to obtain approximately comparable ouabain binding rates in the absence of Ca. Isotonicity and ionic strength were maintained with choline Cl. In presence of Na alone the rate of ouabain binding is high as indicated by the relatively low concentration of $[^3H]$ ouabain used in the reaction medium. There is a slight decrease of the binding rate in presence of Ca, but this effect is small. This observation agrees with the results of Hoffman, who found under similar experimental conditions only a small, if significant, effect of Ca (6). Slightly more pronounced are the Ca effects in the presence of KCl or of choline Cl alone. However, when both Na plus K are present in the reaction medium, the rate of ouabain binding is decreased markedly by the addition of Ca to the medium. Ouabain binding under these conditions is already low because of the coupled action of Na and K described above. Therefore, a relatively high concentration of $[^3H]$ouabain was applied to obtain substantial binding in absence of Ca.

The meaning of these results is not quite clear. Possibly a Ca-binding site becomes accessible to Ca at the inner surface of the membrane, provided that a suitable enzyme configuration is induced by Na (inside) and K (outside). The effects of Na, K, and Ca observed may reflect one mechanism of the regulation of ouabain binding *in vivo* by intracellular Ca and Na and by extracellular K, because an inhibition of the Na,K-ATPase by ouabain will be followed by an increase of these three elements and therefore would interfere with further ouabain binding.

Ouabain binding to red cell membranes requires besides Mg ions also the presence of a suitable substrate like ATP, although this requirement is not restricted to ATP and can be met by some other triphosphonucleotides (6) as well as by ADP and inorganic phosphate. Compared to large amounts, small amounts of ATP down to less than 10 μM are equally effective in the presence of Na and in absence of K when ouabain binding to leaky frozen-thawed red cell membranes is measured (6).

In additional experiments we have determined rates of ouabain binding to porous red cell membranes in the presence of K as a function of the ATP concentration present, since evidence has been presented previously that K alters the affinity of the enzyme for ATP (7). Figure 1 shows the result of such an experiment and contains rates of ouabain binding in the presence of different concentrations of K and ATP. It turns out that the lower the ATP concentration, the more effective K becomes in slowing ouabain binding rate. Therefore, it can be supposed, as mentioned above, that K displaces ATP from the enzyme and therefore markedly decreases the rate of ouabain binding when the ATP concentration is diminished 100-fold. Further support for this assumption comes from experiments in which ouabain binding was measured in the presence of UTP instead of ATP. UTP has previously been shown to be just as effective as ATP in promoting oua-

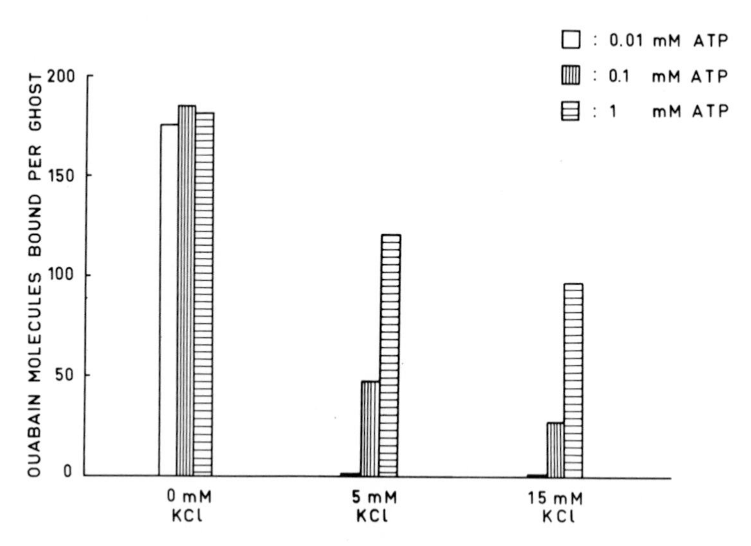

FIG. 1. The effect of K on the rate of ATP-promoted ouabain binding to ghosts. The experimental procedure used was the same as that described in the footnotes to Tables 1 and 2, except that the ghosts were incubated in a medium which contained the indicated concentrations of K, where the quantity NaCl + KCl was 40 mM together with 1.0, 0.1, or 0.01 mM Na_2ATP, 1.25 mM $MgCl_2$, 0.25 mM EDTA, 10 mM Tris Cl (final pH 7.5), and 3.4×10^{-7} M [^{3}H]ouabain. The results presented are typical of the results obtained in several similar experiments. The measure of the rate of ouabain binding is taken as the number of molecules of ouabain bound per ghost after 30 min incubation at 37°C.

bain binding in the absence of K (6), although it does not support active transport of Na and K and is not split by the (Na + K)-dependent ATPase of the red cell membrane (8). Competition experiments between UTP and ATP indicated that in the red cell the affinity for UTP to the enzyme is more than 100 times less than for ATP (9), a result similar to that obtained for a microsomal preparation of stomach Na,K-ATPase (10). Figure 2 contains a comparison of the K effects on the rates of ouabain binding which was promoted in the presence of identical concentrations of ATP or UTP. This experiment was performed in the absence of Na in the reaction medium. In the presence of 5 mM KCl, UTP promoted ouabain binding is almost completely inhibited. This effect of K on ouabain binding in the presence of 0.66 mM UTP resembles the effect of K when binding was studied in presence of 10 μM ATP (Fig. 1).

Using UTP at about 0.5 mM it was also possible to study the side-specific actions of internal Na and external K on ouabain binding to intact red cell ghosts. But it was not possible to incorporate adequately small concentrations of ATP into resealed ghosts which are comparable to the UTP concentrations used with regard to the K effect described above because of

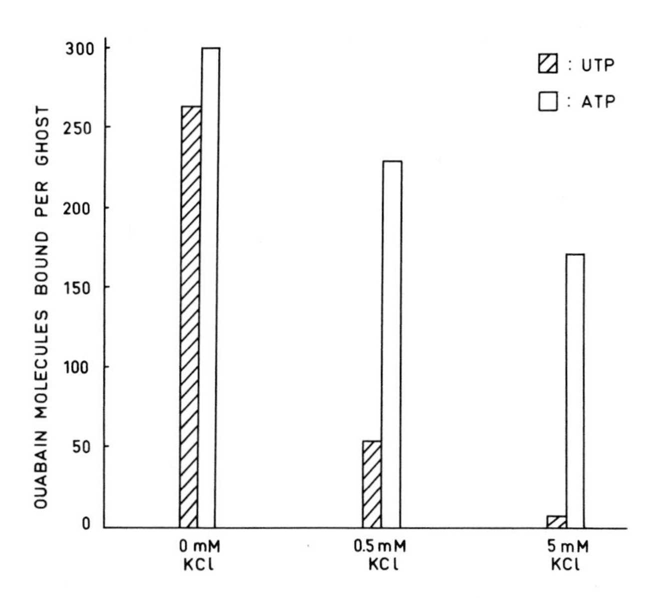

FIG. 2. Comparison of the effect of K on the rate of ATP- and UTP-promoted ouabain binding to ghosts. The experimental procedure used was the same as that described in the footnotes to Tables 1 and 2, except that the ghosts were incubated in a medium which contained the indicated concentrations of K, 0.5 mM EDTA, 3 mM $MgCl_2$, 10 mM Tris Cl (final pH 7.4), and either 0.66 mM Na_3UTP or 0.66 mM Na_2ATP + 0.66 mM NaCl and 1×10^{-7} M [^{3}H]ouabain. Choline Cl was also present in the medium such that the quantity, KCl + choline Cl, was kept constant at 35 mM. Incubation was carried out at 37°C for 30 min. The datum points in the figure are the average of duplicate determinations. The results presented are representative of two other similar experiments.

breakdown during the preparation of the ghosts, and because of adenylic kinase activity. From these experiments, it appears that the K effect on UTP-promoted ouabain binding to porous ghosts may be due mainly to external K.

In summary, ouabain binding seems to be strongly inhibited by K, probably external K, when ATP is diminished. This lack of binding may be already of considerable significance at about 100 μM ATP and may contribute to digitalis ineffectivity in heart failure due to lack of metabolic energy.

ACKNOWLEDGMENTS

This work was supported by Deutsche Forschungsgemeinschaft Fellowships Bo 425/1, 425/2, National Institutes of Health Grants HE09906 and AM05644 and National Science Foundation Grant GB18924.

REFERENCES

1. Hoffman, J. F. (1966): *Am. J. Med.*, 41:666.
2. Ingram, C. J. (1970): Ph.D. Dissertation, Yale University, New Haven, Conn.
3. *a,b,c* Bodemann, H. H. and J. F. Hoffman: In Press. *J. Gen. Physiol.*
4. Bodemann, H. H., and Passow, H. (1972): *J. Membr. Biol.*, 8:1.
5. Schatzmann, H. J. (1966): *Experientia*, 22:364.
6. Hoffman, J. F. (1969): *J. Gen. Physiol.*, 54:343s.
7. Hegyvary, C., and Post, R. L. (1971): *J. Biol. Chem.*, 243:5234.
8. Hoffman, J. F. (1962): *Circulation*, 26:1201.
9. Hoffman, J. F. (1975): Unpublished observations.
10. Siegel, G. J., and Goodwin, B. (1972): *J. Biol. Chem.*, 247:3630.

Membranes and Disease, edited by L. Bolis, J. F. Hoffman, and A. Leaf. Raven Press, New York, © 1976.

Potassium and Calcium Permeability Changes in Normal and Pathological Red Cells

G. Gárdos, I. Szász, and S. R. Hollán

National Institute of Haematology and Blood Transfusion, Budapest, Hungary

Orringer and Parker (1), in their review on ion and water movements in red blood cells, present a table summarizing the conditions and agents that induce a selective K^+ leak. Let us complete their list with some of our recent results and, at the same time, divide the agents into two groups according to their Ca^{2+} dependence (Table 1). The additions in the left column are:

1. Not only propranolol, but also other compounds: pronethalol and tetracaine, elicit selective K^+ transport, concentration and pH range being critical in every case (10)
2. Ca^{2+}-ionophore, A-23187, is also a very potent inducer of the phenomenon (11,12).

In the right column, it is noteworthy that, in addition to valinomycin, other antibiotics, such as nigericin and gramicidin, have practically the same effect under certain conditions (e.g., in choline chloride medium). Hence, the term "K^+-ionophores" was introduced instead of valinomycin.

Orringer and Parker (17) demonstrated that acetylphenylhydrazine induces a specific K^+-leak in human red cells, even in the presence of EDTA. Let us explain the reason why we placed acetylphenylhydrazine, in spite of the latter fact, between the two columns.

We checked Ca^{2+} influx into the red cells, and it increased in all cases listed in the left column (triosereductone was not tested). Acetylphenylhydrazine treatment was found to increase Ca^{2+} influx as well. The increase in Ca^{2+} influx, however, was smaller than that induced by the presence of Heinz bodies formed from unstable hemoglobins and HbSS (Fig. 1). Ca^{2+} influx proved to enhance K^+-leak in all the above cases including acetylphenylhydrazine-treated cells. This means that Ca^{2+} effect is superimposed on the Heinz body effect. According to Riordan and Passow (16) this is not the case with lead ions. In order to express this difference, we attributed an intermediary position to acetylphenylhydrazine.

Let us inform you at the same time about the detailed studies conducted in our Institute on the membrane lipid constitution of red cells containing Heinz bodies (18). *In vitro* acetylphenylhydrazine treatment caused no

TABLE 1. *Calcium-dependent and calcium-independent selective increase of K^+ permeability in human red cells*

Induction of selective K^+ leak	
Ca^{2+}-dependent	Ca^{2+}-independent
Metabolic depletion	K^+ ionophores (13,14)
Substrate deprivation (2)	Lead and other heavy metal salts (15,16)
IA + purine nucleosides (3,4)	
NaF (4–6)	
Triosereductone (7)	
Propranolol, pronethalol, tetracaine (8–10)	
Ca^{2+} ionophore (11,12)	
Acetylphenylhydrazine (17)	Acetylphenylhydrazine (17)

significant change in membrane lipids. But *in vivo* acetylphenylhydrazine treatment in the rabbit, as well as Heinz body anemias in a great number of patients studied, caused significant membrane lipid alteration: an increase in cholesterol, fatty acid esters, and phospholipids. Within the latter category lecithin increased, whereas sphingomyelin decreased. The changes were highly significant in splenectomized patients and were not due to the pres-

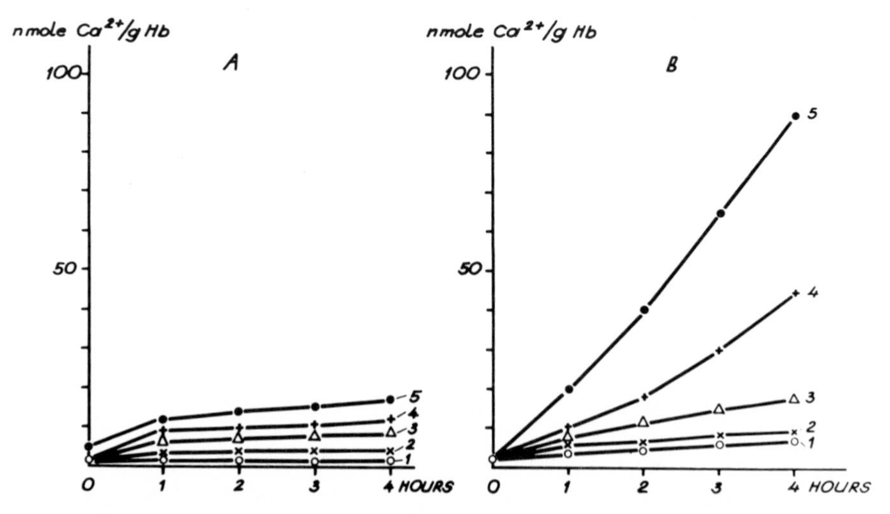

FIG. 1. $^{45}Ca^{2+}$ penetration into the red cells from patients with different types of hemolytic anemia and into normal human red cells treated with acetylphenylhydrazine *in vitro.* **A:** ATP-containing red cells; **B:** ATP-depleted red cells. **1:** Control; **2:** Paroxysmal nocturnal hemoglobinuria (PNH); **3:** Acetylphenylhydrazine-treated normal cells; **4:** HbSS; **5:** Hb Santa Ana. Temperature: 37°C.

ence of young red cells. In these "chronic" cases transport changes were also more definite than in the case of acetylphenylhydrazine treatment. These findings might indicate that, in addition to the disulfide bridge, formation between the Heinz bodies and red cell membrane secondary lipid changes may contribute to the altered membrane permeability.

REFERENCES

1. Orringer, E. P., and Parker, J. C. (1973): In *Progress in Hematology,* Vol. 8, edited by E. B. Brown, p. 1. Grune and Stratton, New York.
2. Kregenow, F. M., and Hoffman, J. F. (1972): *J. Gen. Physiol.,* 60:406.
3. Gárdos, G. (1956): *Acta Physiol. Acad. Sci. Hung.,* 10:185.
4. Gárdos, G. (1958): *Biochim. Biophys. Acta,* 30:653.
5. Wilbrandt, W. (1940): *Pfluegers Arch.,* 243:519.
6. Lindemann, B., and Passow, H. (1960): *Pfluegers Arch.,* 271:497.
7. Passow, H., and Vielhauer, E. (1966): *Pfluegers Arch.,* 288:1.
8. Ekman, A., Manninen, V., and Salminen, S. (1969): *Acta Physiol. Scand.,* 75:333.
9. Manninen, V. (1970): *Acta Physiol. Scand.* [*Suppl.*], 80:355.
10. Szász, I., and Gárdos, G. (1974): *FEBS Lett.,* 44:213.
11. Reed, P. W. (1973): *Fed. Proc.,* 32:635 (Abstr.).
12. Sarkadi, B., Szász, I., and Gárdos, G. (1975): *J. Membr. Biol.* (*in press*).
13. Tosteson, D. C., Cook, P., Andreoli, T. E., and Tieffenberg, M. (1967): *J. Gen. Physiol.,* 50:2513.
14. Gárdos, G. (1972): *Haematologia,* 6:237.
15. Joyce, C. R. B., Moore, H., and Weatherall, M. (1954): *Br. J. Pharmacol.,* 9:463.
16. Riordan, J. R., and Passow, H. (1971): *Biochim. Biophys. Acta,* 249:601.
17. Orringer, E. P., and Parker, J. C. (1973): *Clin. Res.,* 21:94.
18. Hollán, S. R., Hasitz, M., and Breuer, J. H. (1975): *FEBS Proc.,* 35:59.

Membranes and Disease, edited by L. Bolis, J. F. Hoffman, and A. Leaf. Raven Press, New York, © 1976.

The Mechanism of Action of Lithium Ion in Affective Disorders: A New Hypothesis

A. K. Sen,* R. Murthy,* H. C. Stancer,† A. G. Awad,† D. D. Godse,† and P. Grof*,†

**Department of Pharmacology, Faculty of Medicine, University of Toronto, Toronto, Ontario, † Department of Neurochemistry, Clarke Institute of Psychiatry, Toronto, Ontario, and *† Lithium Clinic, Hamilton Psychiatry Hospital, Hamilton, Ontario*

> This time it looks like lithium is here to stay in medicine and may well be a crucial element in unlocking the door to our understanding of affective disorders.
>
> (Nathan S. Kline)

The etiology of affective disorders is unknown. Experimental evidence suggests however, that biochemical changes in the brain may be the most important factors in the etiology of the disease process. Cade's (1) original observations of the effectiveness of lithium in treating mania have been amply confirmed by numerous investigators (for references see refs. 2 and 3). To elucidate the pharmacological mechanism of action of lithium, investigations have been centered around two major areas—biogenic amines and electrolytes. From these fields of research, several theories have been advanced to provide a biological basis for the development of affective disorders. One of the most widely discussed current ideas is that the metabolism of biogenic amines may be disturbed in these conditions (4–9). In its simplest form, the hypothesis states that functional deficit or excess of neurotransmitter amines at central synapses is causally related to depression and mania, respectively. The amines that have been implicated in the process are norepinephrine (NE), dopamine (DA), 5-hydroxytryptamine (5-HT), and acetylcholine (Ach) (for references see ref. 9). One or more of these amines may be responsible for the modulation of mood. The findings, however, remain controversial. Thus, the question is still open and further experimentation is necessary to arrive at a definitive answer. However, it is reasonable to conclude that the levels of these physiologically active amines in the synaptic cleft regions of the brain are intimately related to the mood changes in laboratory animals and to affective disorders in humans.

It is obvious from the above discussion that synthesis, storage, release, metabolism, and uptake of these amines are important factors modulating the concentrations of these amines at their active sites under both normal and pathological conditions. The uptake of these amines in brain tissues is known to be active, i.e., they are accumulated against their chemical gradients. It is known to be Na^+- and K^+-dependent and has been shown to be ouabain-sensitive (10–12). The energy required for this process is postulated to be derived from the inward-directed Na^+ concentration gradient across the cell membrane, which, in turn, is maintained by the membrane-bound transport enzyme Na^+/K^+-ATPase. It has further been demonstrated that the intracellular sodium ion concentration increases in affective disorder patients and reverts to normal with the recovery of the mood. Thus, it has been postulated that, in manic-depressive conditions, the fault in the transport of these amines may be due primarily to a fault in the membrane transport (7,13).

In this chapter, we present some of our findings on sodium and potassium ion transport in the red blood cells (RBC) from control subjects and bipolar type manic-depressive patients before and after lithium treatment. We also present some results derived from animal experiments on brain catecholamine (CA) turnover and Na/K-ATPase in rat brain using lithium ion as a pharmacological tool.

II. METHODS

A. Human Experiments

1. Control Group

Volunteers in the age group of 20–50 years were recruited for this study. A detailed clinical and laboratory screening were conducted, and those with abnormal findings on their personal or family history or clinical and laboratory findings were excluded.

Screening and a 1-week placebo period were used to collect base-line data and to develop regular intake habit. Following that, the participants were placed on lithium carbonate for a period of 7 weeks. The dosage was individually tailored to maintain plasma levels exceeding 0.6 meq lithium/liter. The follow-up data were collected 4 weeks after discontinuation of lithium.

On each volunteer, venous blood samples for electrolyte studies in plasma and RBC were taken once a week, thus providing three base-line values, seven determinations on lithium treatment and at least one in the drug-free follow-up.

2. Patient Group

Prior to admission to the Clinical Investigation Unit of the Clarke Institute of Psychiatry in Toronto, patients were interviewed and the diagnosis of bipolar affective disorder was established independently by two psychiatrists. On admission, patients were taken off all medications except for occasional chloral hydrate at bedtime and were kept so during the entire study period except for lithium carbonate treatment. Although the patients were free to move in the unit, they were not involved in any strenuous physical activity. All patients were maintained on normal hospital diet with no more than three cups of coffee a day.

Mood ratings were conducted twice daily, morning and evenings, by a psychiatrist and by the patients themselves using a visual analogue scale (14).

Venous blood samples were collected in heparin as necessary. Serial 24-hr urine collections were made throughout the study. The urine samples were refrigerated until the collection was complete. An aliquot of each 24-hr pool was frozen and stood at −70°C for future assay of cyclic 3′5′-AMP (cyclic AMP). The assay was carried out in triplicate by Gilman's simplified protein binding assay method (15).

The intracellular Na^+ and K^+ concentrations in the RBC and the active Na^+/K^+ transport across the RBC membrane were measured by the methods described by Sen and Post (16).

B. Animal Experiments

Male Wistar rats weighing between 150 and 200 g were housed in individual cages and were given free access to food and water. Following 3–5 day acclimatization, they were treated with 4 meq LiCl/kg-day in divided doses of 2 meq/kg by intraperitoneal route. The control group received equivalent amounts of NaCl. The treatment was for either 5 or 10 days.

The rats were sacrificed by cervical dislocation 2 hr after the last injection in the morning. The brains were removed immediately and used either for Na/K-ATPase preparation, or for estimation of catecholamines.

The heavy microsomal Na/K-ATPase was prepared and assayed according to the method described by Post and Sen (17). The turnover rates of NE and DA were estimated by using the method of Brodie et al. (18). DL-α-methyl-*p*-tyrosine methyl ester (α-MPT) was used as the tyrosine hydroxylase inhibitor. α-MPT (250 mg/kg) was injected intraperitoneally, and the rats were sacrificed at 0, 2, 3, 4, or 6 hr after the injection. The brains were quickly removed and used for NE and DA estimation by the method described by Shellenberger and Gordon (19).

III. RESULTS

As the transport characteristics for sodium ions and potassium ions in the RBC and nerve cells are qualitatively similar (20), we felt there might be parallel changes in the CNS neurons and RBCs during the disease process. Furthermore, blood was the only readily available tissue in which the transport characteristics could easily be studied.

A. Intracellular K^+ to Na^+ Ratio as an Indicator of Na/K-ATPase Activity in RBC Membrane

The intracellular concentration of Na^+ and K^+ is maintained by the transport enzyme Na/K-ATPase, located in the cell membrane. We assumed that measurement of intracellular concentrations of these ions should reflect the enzyme activity. Since the efflux of Na^+ and the influx of K^+ are coupled, the intracellular ratio of these ions was taken as a more reliable index than absolute concentrations. The results obtained from normal subjects are shown in Fig. 1. The intracellular ion concentrations were measured within 10 min of withdrawal of the blood and the Na/K-ATPase activity was measured in the ghost preparations as described previously (16). As can be seen, the intracellular ratio of these ions correlated with the Na/K-ATPase activity. The correlation coefficient, was $r = 0.586$ ($p < 0.05$). It should be pointed out that such correlation could not be obtained in postmenopausal women. The results indicate that measurement of this ratio is an acceptable indicator of Na/K-ATPase activity.

B. Effect of Lithium Carbonate Therapy in Normal Volunteers on Intracellular K^+:Na^+ Ratio

Representative results are shown in Fig. 2. As mentioned under methods, the volunteers were put on placebo for 1 or 2 weeks before the lithium treatment. This allowed us to establish a base line or control value for each

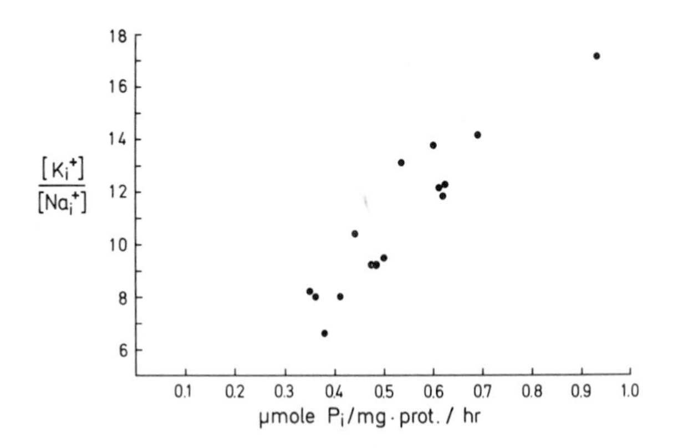

FIG. 1. K_i/Na_i ratio vs red blood cell membrane Na/K-ATPase activity. Venous blood was obtained in heparin. Ten volumes of isotonic sodium chloride was added per volume of blood and spun at $2,000 \times g \times 5$ min. Supernatant along with the buffy coat was discarded. An aliquot of approximately 0.1 ml of red blood cells was washed three times with isotonic MgCl solution by alternate respension and centrifugation. The packed cells were then hemolyzed with deionized distilled water. Sodium and potassium was measured as described by Sen and Post (16). The ATPase was prepared and assayed as described in the methods.

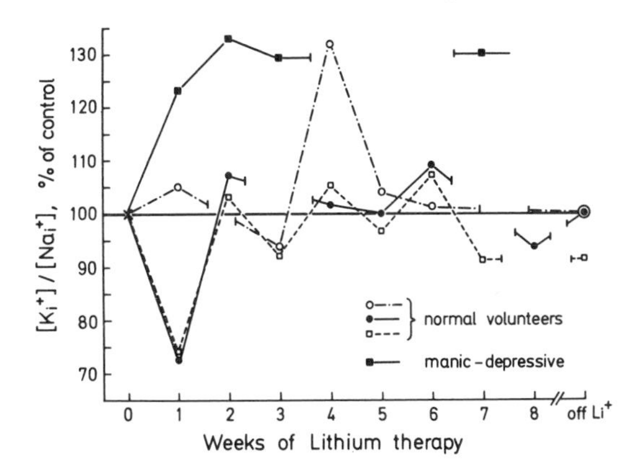

FIG. 2. K_i/Na_i ratio in normal volunteers before and after lithium therapy. The procedure for measuring the intracellular concentration of these ions was described in Fig. 1. The base line of 100% for each individual was established during the placebo treatment period.

individual and this was represented as 100% in Fig. 2. The individual results shown are typical of the whole group. Within the first 2 weeks there was a fall in the K_i/Na_i ratio in 5 of the 6 controls; this was followed by a recovery phase with an overshoot of 130% in one of the volunteers. During the later part of the treatment, the values appeared to be stabilized about the pretreatment level. The values were back to initial levels in 5 subjects out of 6 within 4 weeks of lithium withdrawal. In contrast, the manic-depressive patient showed a significant rise in the ratio following treatment, which remained elevated during the whole course of treatment. The initial fall in ratio in the controls was probably due to the known inhibitory effect of Li^+ on both the K^+ site and the Na^+ site of the ATPase (21). The rise could have been due to direct stimulation of the enzyme activity by lithium ion, the mechanism for which is unknown, or new enzyme synthesis in these cells, or formation of new cells. Some evidence has been found to confirm the third possibility. The results are shown in Fig. 3. As can be seen, the reticulocyte counts increased during the course of the treatment, suggesting that lithium has stimulated the erythropoietic mechanism.

C. Effect of Lithium Carbonate Treatment on Active Transport of Sodium and Potassium Ion Across Red Blood Cell Membranes from Manic–Depressive Patients

To follow the changes in transport properties further, we examined the active transport of sodium and potassium in the patient's blood before and after treatment and compared the transport rates during the manic or depressed phase of the patient versus his normal state arrived at after treatment. The results are shown in Table 1. Two of the three patients examined

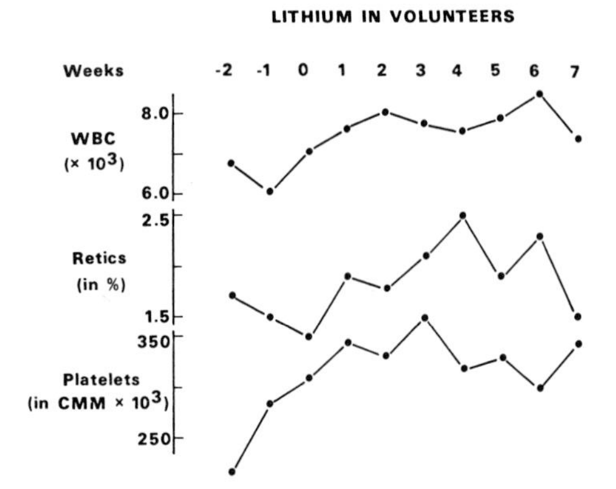

FIG. 3. Effect of lithium ion on white blood cells, reticulocytes, and platelet counts in volunteers.

(#1 and #3) showed significant increase in transport rate for sodium and potassium ions. This paralleled an improvement in mood. The third patient did not show any change in transport parameters even though he became normal following lithium treatment. However, it may be pointed out that before treatment, the patient was depressed but did not show as much inhibition of transport rate as the other two patients. One should take into account that the number of patients examined is insufficient to arrive at any definitive conclusion.

TABLE 1. *Effect of lithium carbonate treatment on active transport of Na and K across red blood cell membranes from manic-depressive patients*

			mmoles cation transported per 5 mmoles Hb per hr	
Patient	Treatment	Mood	Na^+	K^+
1	Before	Manic	0.48	0.39
	After	Normal	3.22	2.60
2	Before	Depressed	1.96	1.64
	After	Normal	1.93	1.64
3	Before	Depressed	0.61	0.53
	After	Normal	3.01	1.62

Washed red blood cells were stored in potassium-free sodium storage solution containing NaCl (150 mM) as described by Sen and Post (16). The cells were stored at 0–4°C for sufficient length of time to increase the intracellular sodium concentration between 35 and 40 mM/5 mM hemoglobin. The cells were then incubated at 37°C in a solution containing both Na (140 mM) and K (10 mM) plus or minus ouabain 2.5×10^{-4} M. The cells were incubated for different lengths of time. The intracellular concentration of sodium and potassium ions was measured as described in the legend of Fig. 1. The net active transport of both ions was calculated by taking the difference between the control and the ouabain-containing tube.

D. Effect of Lithium Ion Treatment of Intracellular K_i/Na_i Ratio in Bipolar Manic Depressive Patients

Since measurement of active transport of sodium and potassium ions is a time consuming process, we decided to look into the K_i/Na_i ratio which is, as shown in Fig. 1, indicative of active transport of these ions. We decided

TABLE 2. *Effect of Li on intracellular* $[K_i^+]/[Na_i^+]$ *ratio in bipolar manic–depressive patients*

Patient	Day	Treatment	Mood ratings	$[K_i^+]/[Na_i^+]$
A	1	–	D-1	12.7
	3	–	N	12.2
	4	Lithium		
	10	Lithium	N	14.2
	12	Lithium	N	13.7
	15	Lithium	N	13.8
	18	Lithium	N	14.4
B	4	–	M-1	4.0
	8	Lithium	M-1	–
	10	Lithium	N	5.1
	13	Lithium	N	5.4
	18	Lithium	N	5.0
	20	Lithium	N	5.1
	25	Lithium	N	5.0
C	4	–	D-1	11.0
	6	–	D-1	10.0
	7	–	D-1	11.0
	11	Lithium	D-1	13.2
	15	Lithium	D-1	15.8
	18	Lithium	N	15.0
D	1	–	D-2	13.3
	5	–	D-1	12.6
	7	–	D-1	12.7
	8	Lithium	D-1	–
	12	Lithium	D-1	12.9
	14	Lithium	D-2	11.8
	18	Lithium	N	12.8
E	2	–	M-1	4.8
	5	Lithium	N	6.7
	13	Lithium	N	7.5
F	1	–	M-1	10.6
	5	Lithium	M-1	10.7
	8	Lithium	D-1	9.4
	12	Lithium	D-1	9.4
	19	Lithium	D-1	8.6
	22	Lithium	D-1	8.0
	26	Lithium	D-1	8.0

Blood was collected in the morning and the intracellular concentration of sodium and potassium ions was measured as described in the legend of Fig. 1. The diagnosis and the mood ratings of the patients were made as described under Methods. Mood ratings were given a maximum score of 2 on either side of normal mood. N = Normal; D = depression; M = mania.

to measure this parameter only, in a longitudinal fashion in six patients and compare it with mood changes. The results are shown in Table 2. Of the six patients examined, five responded to treatment; four of these patients have increased K_i/Na_i ratios (Patients A, B, C, E) as their moods became normal. Even though patient D became normal, his K_i/Na_i ratio did not increase. Patient F, on the other hand, did not become normal and his ratio decreased as his mood changed from mania to depression. Once again the number of patients examined was not sufficient to make a general conclusion. It appears, however, from these experiments that measurement of these ratios could be of value in predicting clinical improvement in patients under lithium treatment. Our results on sodium pump activity and those on K_i/Na_i ratio are in agreement with those of Naylor et al. and Hokin-Neaverson et al. (22,23). We are, however, in disagreement with Mendels and Frazer (24) who showed an increase in intracellular sodium content following lithium treatment. The reason for this discrepancy is unclear.

E. Animal Experiments

Results described so far indicate that lithium ion can stimulate active transport of sodium and potassium ion in peripheral cells. Since measurements of the active transport process could not be done on human brains because of nonavailability of tissue, we had to resort to investigating ATPase activity and amine turnover rates in the rat brain.

F. Effect of Lithium Treatment of Dopamine and Norepinephrine Turnover in Rats

In these experiments, the turnover was measured in the whole brain by inhibiting the synthesis of these monoamines by treating the rats with α-MPT, a tyrosine hydroxylase inhibitor. The results are shown in Table 3.

TABLE 3. *Effect of lithium treatment on dopamine and norepinephrine turnover in rats*

	Turnover rate (μg/g-hr)			
	DA		NE	
Days of treatment	Saline	Li^+	Saline	Li^+
5	0.061	0.036	0.032	0.027
10	0.105	0.071	0.067	0.041

Following LiCl treatment of 4 meq/kg-day in two divided doses, the rats were treated with α-methyl-*p*-tyrosine as described under Methods. The rats were sacrificed at 0, 2, 3, 4, and 6 hr postinjection. The whole brain dopamine and norepinephrine content were estimated. The turnover rate was calculated from the slope.

The dopamine turnover was significantly decreased both by 5 and 10 days treatment. The norepinephrine turnover rate, on the other hand, was decreased only after 10 days treatment and was statistically only marginally significant. This observation is in agreement with Corrodi et al. (25). In chronically lithium-treated rats Friedman and Gershon (26) also found a decrease in dopamine synthesis in striatal slices of brain.

G. Effect of Lithium Treatment on Rat Brain Na/K-ATPase

The Na/K-ATPase activity was measured in different brain regions and compared with the control group. Results are shown in Table 4. Five days treatment did not elicit any significant difference in Na/K-ATPase activity in any of the regions tested, although a tendency toward increased enzyme activity could be seen. With 10 days lithium ion treatment there was a significant increase in only the cerebral hemisphere. The increased activity, though significant, was small. This finding is in essential agreement with those observed by Gutman et al. (27). As can be seen, there was no parallelism between the ATPase activity and the turnover rates of dopamine or norepinephrine. This finding was rather unexpected because the turnover of these amines should be highly dependent on the ionic gradient maintained by the Na/K-ATPase enzyme activity. It occurred to us that there could be some other factor(s) available in an *in vivo* situation for the modulation of the activity of Na/K-ATPase. If this assumption is true, an *in vitro* assay will not allow us to demonstrate a change in the enzyme activity in the absence of such a modulator. The results described in the next paragraph indicate the existence of such a modulatory effect on this enzyme by a cAMP-dependent protein.

TABLE 4. *Effect of lithium treatment on rat brain ATPase*

	$Na^+ + K^+$-ATPase activity (μmoles Pi per mg Protein per hr)			
	5 days treatment		10 days treatment	
	Saline	Lithium	Saline	Lithium
Cerebral Hemisphere	25.9 ± 2.7	30.0 ± 1.6	20.5 ± 3.0	$25.3^{a} \pm 2.9$
Cerebellum	26.8 ± 2.9	28.0 ± 3.6	40.5 ± 5.6	36.1 ± 4.6
Medulla	29.5 ± 2.0	31.3 ± 2.8	32.9 ± 3.4	34.2 ± 2.1

Rats were treated with LiCl as described in Methods for 5 or 10 days. The heavy microsomal fraction from different brain areas was isolated and the Na/K-ATPase assays were performed according to Post and Sen (17).

[a] Significance at <0.05 by paired t-test.

H. Effect of cyclic AMP-Dependent Protein on Rat Brain Na/K-ATPase

The results are shown in Table 5. The enzyme activity was tested in the presence of 0.05 mM cAMP, or in the presence of protein fractions isolated from the cytosol fraction of the brain or in the presence of both cAMP and the protein fractions. Cyclic AMP alone did not have any effect on the enzyme activity. The protein fractions alone did not exhibit any effect on the enzyme activity. However, a combination of cAMP and K_2 fraction consistently showed an inhibition of about 20%. Thus, the finding supports the idea that in an *in vivo* situation the enzyme activity could be modulated by cAMP and that in isolated enzyme preparations the effect of lithium treatment may not be observed.

TABLE 5. *Effect of cyclic AMP and soluble protein fractions on rat brain Na/K-ATPase*

Condition	Na/K-ATPase activity (%)
Control	100[a]
Cyclic AMP	99
K_1	100
K_1 + cyclic AMP	95
K_2	98.5
K_2 + cyclic AMP	82
K_3	100
K_3 + cyclic AMP	98

Na/K-ATPase from rat cerebellum was prepared and assayed as described in Methods. Following initial homogenization of the tissue the cytosol fraction was separated by centrifugation at 200,000 × *g* × 1 hr. Solid ammonium sulfate was added to the supernatant fraction to 20% saturation. The supernatant was centrifuged at 27,000 × *g* × 30 min, the supernatant decanted and the precipitated protein was labeled K_1. The remaining supernatant was further saturated to 40% and procedure repeated. The precipitated protein was labeled K_2. The remaining supernatant was labeled K_3. The precipitated proteins were redissolved in buffer containing glycylglycine 10 mM and imidazole 10 mM at pH 7.0, and were dialyzed for 48 hr with two changes against the same buffer. Final concentration of these proteins was 0.25–0.50 mg/ml. The final assay mixture when necessary contained 0.025–0.05 mg/ml of the fractionated soluble protein and 0.05 μmoles of cyclic AMP.

[a] Specific activity = 150 μmoles $P_i \cdot mg^{-1}$ protein hr^{-1}.

DISCUSSION

Based on numerous published observations, the possible sites of action of lithium on central synapse are diagrammatically presented in Fig. 4. Our data indicate that in bipolar affective disorder patients there is a deficiency in "sodium pump" activity in both manic and depressed phases of the disease and that lithium therapy tended to correct this deficiency. In agreement with this observation, Naylor et al. (22) reported an increase in erythrocyte Na/K-ATPase activity in patients following lithium treatment. In animal studies also, lithium has been shown to lower brain sodium concentration (28,29). The presence of a genetic factor in affective disorder has been suggested (30). It is therefore not unreasonable to extrapolate these observations to the central nervous system. Thus, an increment in intracellular sodium ion concentration in affective disorder patients and its decrease concomitant with return to normal mood, following lithium treatment, has been reported (7,13).

The inhibitory effect of lithium ions on the Na/K-ATPase system *in vitro* is well documented (21). It competes for the K site on the outer surface and Na site on the inner surface of the membrane. Yet it appears that *in vivo* lithium has a stimulatory effect on the "sodium pump" mechanism. In our experiments, the rat brain Na/K-ATPase isolated following lithium treatment showed only a slight increase in activity after 10 days and therefore cannot explain the observations. It appears that failure to see any effect of

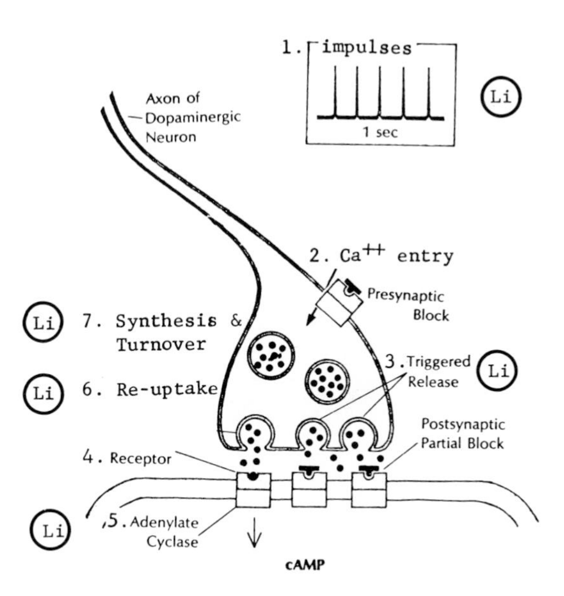

FIG. 4. Diagrammatic presentation of possible sites of lithium action. [Adapted from Seeman (34).]

lithium on this enzyme was probably because lithium does not have a direct stimulatory effect on the system, but modulates the enzyme activity through a cAMP-dependent protein. The evidence for the existence of such a protein has been presented. Furthermore, lithium ion has been shown to inhibit in a dose-dependent manner, the neurotransmitter-dependent adenylate cyclase system in the brain cortex from rat and rabbit (31,32). Even though the diagram shows only a postsynaptic adenylate cyclase, a presynaptic localization of this enzyme system was reported by De Robertis, Rodriguez de Loreo Arraiz, Alberici, Butcher, and Sutherland (33).

Although our findings are of a preliminary nature, we suggest the following mechanism for the observed effects of lithium. A direct inhibitory effect of lithium on the Na/K-ATPase system would account for the observed initial fall in K_i/Na_i ratio in normal subjects, illustrated in Fig. 2. The delayed return of ratio to normal would then result from lithium's inhibition of the adenylate cyclase system, leading to fall in the intracellular level of cAMP. This, in turn, would inactivate the cAMP-dependent inhibitory protein, resulting in activation of the Na/K-ATPase and return of the K_i/Na_i ratio to normal. In the affective disorder patients, the ATPase system is already inhibited and pretreatment K_i/Na_i values are low. Therefore, only the elevation is seen after initiation of lithium therapy.

The normalization of mood by lithium therapy would, on this basis, be the result of normalization of turnover of biogenic amines as a result of restoration of the normal intracellular electrolyte concentrations and transport rates.

Although this scheme is quite speculative, it has the merit that all of the steps can be experimentally tested. This may lead to better attempts at formulation of explanatory or predictive hypotheses.

ACKNOWLEDGMENTS

This project was supported by Ontario Mental Health Foundation grant #527–74B and Medical Research Council, Canada grant #MT2485. We wish to thank Mrs. Joan Sax for literature survey and expert technical help. Technical assistance was also provided by Mrs. S. Chandramowli, Mr. M. Finley and Mrs. R. Funk.

REFERENCES

1. Cade, J. F. J. (1949): Lithium salts in the treatment of psychotic excitement. *Med. J. Aust.*, 36:349–352.
2. Goodwin, F. K., and Ebert, M. H. (1973): Lithium in Mania: Clinical trials and controlled studies. In: *Lithium, Its Role in Psychiatric Research and Treatment,* edited by S. Gershon and B. Shopsin, pp. 237–252. Plenum Press, New York.
3. Mendels, J. (1973): Lithium and depression. In: *Lithium, Its Role in Psychiatric Research and Treatment,* edited by S. Gershon and B. Shopsin, pp. 253–267. Plenum Press, New York, London.

4. Bunney, W. E., Jr., and Davis, J. M. (1965): Norepinephrine in depressive reactions. *Arch. Gen. Psychiat.,* 13:483–494.
5. Schildkraut, J. J. (1965): The catecholamine hypothesis of affective disorders. A review of supporting evidence. *Am. J. Psychiat.,* 122:509–522.
6. Schildkraut, J. J. (1970): *Neuropsychopharmacology and the affective disorders.* 111 pages. Little, Brown, Boston, Mass.
7. Coppen, A. J. (1969): The biochemistry of affective disorder. *Br. J. Psychiat.,* 113:1237–1264.
8. Davis, J. M. (1970): Theories of biological etiology of affective disorders. *Int. Rev. Neurobiol.,* 12:145–175.
9. Baldessarini, R. J. (1975): An overview of the basis for amine hypothesis in affective illness. In: *The Psychobiology of Depression,* edited by J. Mendels, pp. 69–83. Spectrum Publications, New York.
10. Bogdanski, D. F., Tissari, A., and Brodie, B. B. (1968): Role of sodium, potassium, ouabain and reserpine in uptake, storage and metabolism of biogenic amines in synaptosomes. *Life Sci.,* 7:419–428.
11. Colburn, R. W., Goodwin, F. K., Murphy, D. L., Bunney, W. E., Jr., and Davis, J. M. (1968): Quantitative studies of norepinephrine uptake. *Biochem. Pharmacol.,* 17:957–964.
12. Sugrue, M. F., and Shore, P. A. (1970): The mode of sodium dependence of the adrenergic neurone amine carrier. Evidence for a second sodium dependent, optically specific and reserpine sensitive system. *J. Pharmacol. Exp. Ther.,* 170:239–245.
13. Davis, J. M., Janowsky, D. S., and El-Yousef, K. (1973): Pharmacology—the biology of lithium. In: *Lithium, Its Role in Psychiatric Research and Treatment,* edited by S. Gershon and B. Shopsin, pp. 167–188. Plenum Press, New York, London.
14. Aitkin, R. C. B., and Zeally, A. K. (1970): Measurement of moods. *Br. J. Hosp. Med.,* 4:215–224.
15. Gilman, A. G. (1970): A protein binding assay for adenosine 3′:5′-cyclic monophosphate. *Proc. Natl. Acad. Sci. USA,* 67:305–312.
16. Sen, A. K., and Post, R. L. (1964): Stoichiometry and localization of adenosine triphosphate-dependent sodium and potassium transport in the erythrocyte. *J. Biol. Chem.,* 239:345–352.
17. Post, R. L., and Sen, A. K. (1967): Sodium and potassium-stimulated ATPase. In: *Methods of Enzymology,* edited by S. P. Colowick and N. D. Kaplan, Vol. 10, pp. 762–768. Academic Press, New York.
18. Brodie, B. B., Costa, E., Dlabec, A., Neff, N. H., and Smookler, H. H. (1966): Application of steady state kinetics to the estimation of synthesis rate and turnover time of tissue catecholamines. *J. Pharmacol. Exp. Ther.,* 154:493–498.
19. Shellenberger, M. K., and Gordon, J. H. (1971): A rapid, simplified procedure for simultaneous assay of norepinephrine, dopamine and 5-hydroxytryptamine from discrete brain areas. *Anal. Biochem.,* 39:356–372.
20. Dahl, S. L., and Hokin, L. E. (1974): The sodium-potassium adenosinetriphosphatase. *Annu. Rev. Biochem.,* 43:327–356.
21. Skou, J. C. (1965): Enzymatic basis for active transport of Na and K across cell membrane. *Phys. Rev.,* 45:596–617.
22. Naylor, G. J., Dick, D. A. T., Dick, E. G., and Moody, J. P. (1974): Lithium therapy and erythrocyte membrane cation carrier. *Psychopharmacologia,* 37:81–86.
23. Hokin-Neaverson, M., Spiegel, D. A., and Lewis, W. C. (1974): Deficiency of erythrocyte sodium pump activity in bipolar manic-depressive psychosis. *Life Sci.,* 15:1739–1748.
24. Mendels, J., and Frazer, A. (1974): Alterations in cell membrane activity in depression. *Am. J. Psychiatry,* 131:1240–1246.
25. Corrodi, H., Fuxe, K., and Schou, M. (1969): The effect of prolonged lithium administration on cerebral monoamine neurones in the rat. *Life Sci.,* 8:643–651.
26. Friedman, E., and S., Gershon (1973): Effect of lithium on brain dopamine. *Nature,* 243:520–521.
27. Gutman, Y., Hochman, S., and Strachman, D. (1973): Effect of lithium treatment on microsomal ATPase activity in several tissues. *Int. J. Biochem.,* 4:315–318.
28. King, L. J., Carl, J. Z., Archer, E. G., and Castellanet, M. (1969): Effect of lithium on brain energy reserves and cations *in vivo. J. Pharmacol. Exp. Ther.,* 168:163–170.

29. Baer, L., Kassir, A., and Fieve, R. R. (1970): Lithium-induced changes in electrolyte balance and tissue electrolyte concentration. *Psychopharmacologia,* 17:216–224.
30. Tsuang, M. T. (1975): Genetics of affective disorder. In: *The Psychobiology of Depression,* edited by J. Mendels, pp. 85–100. Spectrum Publications, Inc., New York.
31. Dousa, T., and O., Hechter (1970): Lithium and brain adenyl cyclase. *Lancet,* 1:834–835.
32. Forn, J., and Valdecasas, F. G., (1971): Effects of lithium on brain adenyl cyclase activity. *Biochem. Pharmacol.,* 20:2773–2779.
33. DeRobertis, E., Rodriguez de Loreo Arraiz, C., Alberici, M., Butcher, R. W., and Sutherland, E. W. (1967): Subcellular distribution of adenyl cyclase and cyclic phosphodiesterase in rat brain tissue. *J. Biol. Chem.,* 242:3487–3493.
34. Seeman, P. (1974): The actions of nervous system drugs on cell membranes. *Hosp. Pract.,* 9:93–101.

Membranes and Disease, edited by L. Bolis, J. F. Hoffman, and A. Leaf. Raven Press, New York, © 1976.

Genetic Regulation of Membrane Synthesis in Chloroplasts as Studied with Lethal Gene Mutants

Diter von Wettstein

Department of Physiology, Carlsberg Laboratory, DK-2500 Copenhagen, and Institute of Genetics, University of Copenhagen, Copenhagen, Denmark

A considerable number of diseases in higher plants can be considered as membrane diseases. For instance, the ugly brown stripes of the eyespot disease on sugarcane leaves result from an interaction between the toxin released by the fungus *Helminthosporium sacchari* and the cell membranes —the plasmalemmas—of the mesophyll cells (1). The toxin is an α-galactoside named helminthosporoside. When sprayed on sugarcane leaves it causes the brown stripes by binding to a specific protein in the plasmalemma, the helminthosporoside-binding protein (2). The binding of the toxin to the membrane protein apparently changes the conformation of the plasma membrane and thereby activates the K-Mg-ATPase in the membrane. This upsets the ion balance of the cells with ensuing effects on photosynthesis and respiration eventually leading to cell death (3). Mutants of sugarcane that have been selected for resistance toward the eyespot disease contain in their plasma membrane a protein which has lost the capacity to bind the toxin because of an altered amino acid composition (1). This example demonstrates the important role played by components of the plant's plasma membrane in the defense against attack from pathogens.

A sensitive indicator for plant diseases is provided by the elaborate internal membrane system of the chloroplast. These photosynthetic membranes contain all the chlorophyll pigment and constitute the dominating membrane component of the mature leaf. Chlorophyll deficiency is associated with diseases ranging from nutritional or environmental stress to inborn errors of metabolism. To the extent that this has been investigated, such chlorophyll deficiency is always associated with either restricted and abnormal development of chloroplast membranes or their breakdown.

We have studied in barley diseases of the internal chloroplast membranes caused by induced mutations in nuclear genes (4,5). These lethal or conditional lethal mutants can form one well-developed seedling leaf, using the endosperm as a complete growth medium and are recognized by chlorophyll deficiencies as well as by chloroplast membrane aberrations visible in electron micrographs. More than 100 genes have been identified, which, upon mutation, give rise to defects in the organization or development of the photosynthetic membranes (6). The inner membrane system of mature chloroplasts consists of flattened membrane sacks—thylakoids—in parallel arrangement. In the columns of the grana about twice as many thylakoids are present as in the intergrana regions and the neighboring thylakoids are closely stacked, their membranes being appressed at a distance of a few nanometers. The chlorophyll in the membranes harvests the light. The absorbed energy is used via the electron transport chains of the membranes to split water and to produce reducing power in the form of adenosine triphosphate and reduced pyridine nucleotides. These compounds drive the carbon reduction cycle located in the soluble stroma surrounding the membrane system.

As an example of a drastically altered organization of the thylakoids, mutants in the *xantha-a* gene of barley may be mentioned. The thylakoids in the plastids of these mutants are frequently arranged in a honeycomb-like configuration instead of the parallel arrangement, and disorganized aggregates of lipids are deposited together with membrane fragments (4). This particular membrane syndrome is associated with an hyperactivity of acetate incorporation into chloroplast-specific lipids (7). Chloroplasts with grana-rich or grana-poor membrane systems can be obtained in other mutants. These are used to probe into the importance of membrane organization and composition for part reactions of photosynthesis (8).

The collection of mutants in barley has been particularly useful in the study of the formation of chloroplast membranes. For synchronized organelle development the grain is germinated and the primary leaf grown for 7 days in darkness. The leaf cells then contain a uniform population of etioplasts, which, upon continuous illumination, differentiate in 24 hr into mature chloroplasts. The etioplasts form a certain amount of internal membranes, the majority of which are organized into one or several crystalline prolamellar bodies. As shown by mutants blocked in chlorophyll synthesis, this particular membrane configuration is only attained in the presence of protochlorophyllide, the precursor to chlorophyllide (4). The protochlorophyllide is bound to a holochrome protein with an apparent molecular weight of 63,000 daltons and located in the membranes of the prolamellar body (9). When a certain amount of protochlorophyllide has been formed, the porphyrin pathway is shut down leaving only a little turnover synthesis. Evidence summarized earlier (10, 11) is strong that this takes place by repression of one of the enzymes synthesizing δ-aminolevulinate, whereas the enzymes converting δ-aminolevulinate into protochlorophyllide continue to be present and active in the etioplasts.

In the differentiation of chloroplasts from etioplasts the following events take place upon illumination. The protochlorophyllide is photoreduced to chlorophyllide within a time scale of milliseconds. This elicits within minutes a conformational change of the holochrome molecules in the prolamellar body membranes (12) and/or a dissociation of the chlorophyllide holochrome into a pigment protein complex and a photoenzyme (13). The crystalline configuration of the prolamellar bodies is lost and the latter are dispersed in the first half hour into sheets of primary lamellar layers. Simultaneously chlorophyllide is esterified with phytol (14) and the reaction centers for photosystems I and II are established (15). The small amount of chlorophyll formation together with the reorganization of the internal etioplast membranes suffices to organize the photosynthetic units within 30 min after onset of illumination.

Conversion of protochlorophyllide to chlorophyllide also induces δ-aminolevulinate synthesis and thereby reestablishes the porphyrin pathway for further synthesis of chlorophyll (cf. ref. 11). During the first 8 hr of light-dependent greening and grana formation, five or five groups of thylakoid polypeptides are synthesized and inserted into the etioplast membranes (16). It appears that one high molecular weight polypeptide is inserted into the intergrana membranes, whereas three polypeptides of lower molecular weight are characteristic for the thylakoid areas that can stack into grana. After 8 hr of greening, additional synthesis of all chloroplast membrane components take place.

Nuclear genes maintain a tight control over the processes just described as is evidenced by the fact that for most of them one or several interfering mutations have been found. Mutations in four genes, *tigrina-d, tigrina-b, tigrina-n,* and *tigrina-o,* result in defective regulation of chlorophyll synthesis. These *tigrina* mutant seedlings accumulate in their etioplasts 1.5 to 15 times the wild-type amount of protochlorophyllide in the dark (11,17,18). The synthesis of δ-aminolevulinic acid from glutamic acid (19) is not halted in the dark grown mutants, and they can be distinguished from the dark grown, yellow wild-type seedlings by their green color. We consider, therefore, these genes in barley to be regulatory genes for chlorophyll synthesis. Mutant *tigrina-d*^{12} is of special interest. In alternating light and dark cycles the primary leaves are tiger-striped with green and yellow-brown bands. The necrotic bands result from photodynamic destruction of cells that have accumulated large amounts of protochlorophyllide in the dark cycle and this damage leads to the death of the seedlings. Regulation of chlorophyll synthesis by repression of δ-aminolevulinic acid synthesis should be dispensible in continuous light. In agreement with this expectation, mutant *tigrina-d*^{12} is fully viable in continuous light and can be grown to maturity under this condition. This indicates that the product of the *tigrina-d* gene functions exclusively in the repression of δ-aminolevulinate and thereby protochlorophyllide formation in the dark.

Mutations that block steps in chlorophyll biosynthesis can be recognized by feeding dark grown mutant seedlings with δ-aminolevulinic acid. In the wild-type seedlings, this precursor is converted all the way to protochlorophyllide (20), whereas in mutants with a complete or partial block in chlorophyll synthesis porphyrins accumulate (4,5,11,21). In this way mutants in five *xantha* genes and one *albina* mutant have been identified which accumulate either protoporphyrin IX, Mg-protoporphyrins, or, in one case, uroporphyrinogen. We designated these genes as structural genes in contrast to the *tigrina* genes and want to stress that mutations in these genes do not interfere with the regulation of the pathway, since the block first becomes apparent when the mutants are provided with a metabolite beyond the point of regulation.

Double mutants homozygous for a mutant allele of a structural gene and a mutant allele of a regulatory gene are expected to accumulate constitutively the porphyrin which serves as the substrate for the step controlled by the structural gene. This is borne out by double mutant combinations, such as *tigrina-d*^{12} with *xantha-l*^{35} which accumulates constitutively Mg-protoporphyrin and its monomethylester or *tigrina-o*^{34} with *xantha-f*^{10} which accumulates constitutively protoporphyrin IX (11). We have now analyzed the possible combinations among mutant alleles of the above-mentioned six structural and four regulatory genes with regard to constitutive accumulation of porphyrins (Avivi, Kahn, and von Wettstein, *unpublished*). In all combinations involving mutant *xanthal-l*^{35}, which places a block in the pathway between Mg-protoporphyrin and protochlorophyllide, the expected constitutive accumulation of Mg-protoporphyrins is found. Mutants blocked between protoporphyrin IX and Mg-protoporphyrin in combination with mutants in the regulatory genes *tigrina-b* and *tigrina-o* synthesize constitutively protoporphyrin IX, but not in quantities which would be expected. In double mutants containing *tigrina-d*^{12} and mutant alleles blocking the conversion of protoporphyrin IX to Mg-protoporphyrin, accumulation of protoporphyrin was not detectable. These results can be explained with the assumption that protoporphyrin IX exerts feedback inhibition on one of the enzymes involved in the conversion of glutamate to δ-aminolevulinate, whereas Mg-protoporphyrins or protochlorophyllide do not seem capable of this interaction.

The mutation in the gene *tigrina-d* thus gives constitutive synthesis of an enzyme forming δ-aminolevulinate, but the enzyme is inhibited in double mutants in which the uncontrolled synthesis must lead to the accumulation of porphyrins with the capacity to exert feedback inhibition. On the other hand, double mutants involving *tigrina-b* or *trigrina-o* mutants accumulate some protoporphyrin IX, which suggests that the defective regulation in these mutants involves the feedback inhibition mechanism rather than repression. Further analysis of the regulatory mechanisms for chlorophyll synthesis and the role of the *tigrina* genes requires the isolation and characterization of the two or three newly recognized enzymes (19) catalyzing the synthesis of δ-aminolevulinate in the chlorophyll pathway.

The conditional lethal mutant *tigrina-d*^{12} offers interesting possibilities for isolating new structural mutants in the porphyrin pathway. Since repression of the protochlorophyllide pathway does not take place in the dark, new mutants accumulating protochlorophyllide precursors can be screened for spectrophotometrically. We have treated homozygous *tigrina-d*^{12} grains with the potent mutagen sodium azide and grown the M_1 generation in continuous light in the Stockholm phytotron. Screening of the M_2 generation seedlings has so far revealed two new, nonleaky Mg-protoporphyrin-accumulating mutants recognizable by the characteristic Mg-protoporphyrin spectrum of the leaf *in vivo*. We are also looking for mutants which accumulate larger amounts of the holochrome protein to see what effects an uncontrolled accumulation of a membrane protein might have. Of the large amounts of protochlorophyllide accumulated in the regulatory *tigrina* mutants, only wild-type amounts can be photoreduced to chlorophyllide (17,18,22). This is due to the limiting amounts of holochrome protein in the prolamellar body membranes to which the protochlorophyllyde has to be bound specifically for photoconversion. By repeated use of the same holochrome molecule, inactive protochlorophyllide molecules of the mutant can be converted successively by light flashes interspaced with short dark periods. In the latter the newly formed chlorophyllide molecule is removed from the conversion site in the holochrome and replaced by an inactive protochlorophyllide molecule which now becomes convertible. Obviously, the genetic defect leading to uncontrolled synthesis of the chromophore does not abolish the regulation for the amount of the holochrome protein made. A regulatory mutation resulting in increased amounts of the holochrome protein should be detectable in the presence of the *tigrina-d*^{12} allele by conversion of a large amount of protochlorophyllide with a single saturating light flash.

In a sense the opposite situation to the unregulated protochlorophyllide synthesis in *tigrina-d*^{12} is encountered in the barley mutant *albina-f*^{17}. It has the characteristics of a stringent mutant for protochlorophyllide synthesis (23). Whereas in the wild-type a small amount of inactive protochlorophyllide, i.e., protochlorophyllide not bound in the conversion site of the holochrome, is always made before the pathway is closed, this mutant lacks inactive protochlorophyllide. This mutation stops the synthesis of the chromophore when available holochrome conversion sites have been filled. Upon illumination of the mutant, the protochlorophyllide is photoreduced, but the ensuing events are blocked. The prolamellar body membranes remain undispersed, the *in vivo* spectral blue shift of chlorophyllide normally occurring simultaneously with the membrane dispersal is absent as well as phytolation (14). Further chlorophyll synthesis and membrane differentiation is not possible without the proper *albina-f* gene product. However, by supplying δ-aminolevulinate to the mutant, inactive protochlorophyllide is produced and thereupon the initial events following protochlorophyllide reduction become possible as evidenced by the occurrence of a spectral blue shift and the reuse of the photoconversion sites in the holochrome of the mutant's fed leaves. Esterification of the chlorophyllide with phytol remains impossible. The mutant thereby reveals that this reaction is not a prerequisite for the reuse of the photoconversion sites and that some inactive protochlorophyllide may be necessary for the release of chlorophyllide from the conversion site (23).

Differentiation of the membrane system of the chloroplast during the first 8 hr of greening in the light involves the preferential insertion of five polypeptides (16). Polyacrylamide gel electrophoresis after dissolution of the membranes in phenol–acetic acid–urea and removal of all pigments showed these proteins to have apparent molecular weights of 100,000, 63,000, 41,000, 39,000, and 34,000 daltons. Mutants in the genes *xantha-b, xantha-c,* and *xantha-d* were previously found to assemble most of their thylakoids into grana, thereby forming giant grana and a few single intergrana thylakoids in their plastids. These mutants were now found to be strongly inhibited in the synthesis of the 100,000-dalton membrane protein, but not in the synthesis and insertion of the other light-induced polypeptides (24). Conversely, mutants which are deficient in grana, such as *viridis-k*^{23}, appear to lack the 39,000-dalton polypeptide. These results suggest that the differentiation of the thylakoids into grana and intergrana (stroma) membrane areas is accompanied by the insertion of specific proteins. The 39,000-dalton protein is inserted probably together with the 41,000- and 34,000-dalton polypeptides into membrane areas that have the capacity to stack the thylakoids into grana. The unstacked intergrana thylakoids contain the 100,000-dalton polypeptides and we may therefore assign an antistacking function to the latter.

The membrane proteins discussed are major constituents of the chloroplast membranes and their presence or absence can be detected by staining gels with suitable dyes. There are other minor protein components that require other techniques to be recognized. An instructive example is provided by acetyl-CoA carboxylase, a key enzyme complex in fatty acid biosynthesis. The carboxylation of acetyl-CoA to malonyl-CoA is catalyzed by the three polypeptides of the complex: the biotin carboxylase, the biotin carboxyl carrier protein and the transcarboxylase. In the barley embryo the complex is soluble and has a molecular weight of 610,000 daltons. Gel electrophoresis assigns apparent molecular weights of 41,000, 32,000, and 21,000 daltons to the three polypeptides. The 21,000-dalton polypeptide contains one biotin molecule covalently linked to a lysyl residue and thereby identifies it as the carboxyl carrier protein (25). In the chloroplasts of the leaves a biotin carboxyl carrier protein with the same molecular weight is contained in the thylakoid membrane (26). It can be specifically labeled with ^{14}C-biotin and detected as a radioactive band after electrophoresis of the membrane proteins on gels. Isolation of the labeled biotin protein from the gels is possible and its function in the acetyl-CoA carboxylase reaction was ascertained. The other two components of the carboxylase complex appear as soluble proteins in the chloroplast stroma. The biotin carboxyl carrier protein illustrates that enzymes present in soluble form in one organ or organelle become membrane bound in other organs or organelles. It remains to be determined whether the same or different genes code for such enzymes with variable location.

In concluding this survey of our studies on the genetic control of chloroplast membrane formation in a higher plant, I would like to point out that genetic defects have provided a very helpful tool in recognizing some of the complex processes involved in membrane biosynthesis of this organelle. Whereas the structural and functional characterization of mutants has told us which mutants to use in pursuing the analysis of chlorophyll biosynthesis and membrane differentiation, a molecular characterization of the involved enzymes and other membrane components remains a necessary prerequisite in order to determine the precise nature of the genetic regulation of chloroplast membrane biogenesis.

ACKNOWLEDGMENT

The support by grant GM10819 from the U.S. Public Health Service, National Institutes of Health is gratefully acknowledged.

REFERENCES

1. Strobel, G. A. (1973): *Proc. Natl. Acad. Sci. USA,* 70:1693.
2. Strobel, G. A. (1973): *J. Biol. Chem.,* 248:1321.
3. Strobel, G. A. (1975): *Sci. Am.,* 232:81.
4. von Wettstein, D., Henningsen, K. W., Boynton, J. E., Kannangara, G. C., and Nielsen, O. F. (1971): In: *Autonomy and Biogenesis of Mitochondria and Chloroplasts,* edited by N. K. Boardman, A. W. Linnane, and R. M. Smillie, p. 205, North-Holland, Amsterdam.
5. Henningsen, K. W., Boynton, J. E., von Wettstein, D., and Boardman, N. K. (1973): In: *The Biochemistry of Gene Expression in Higher Organisms,* edited by J. K. Pollak and J. W. Lee, p. 457. Australian and New Zealand Book Co., Sydney.
6. von Wettstein, D., and Kristiansen, K. (1973): *Barley Genet. Newslett.,* 3:113.
7. Appelqvist, L. A., Boynton, J. E., Henningsen, K. W., Stumpf, P. K., and von Wettstein, D. (1968): *J. Lipid Res.,* 9:513.
8. Smillie, R. M., Nielsen, N. C., Henningsen, K. W., and von Wettstein, D. (1974): *Proc. 3rd Int. Congr. Photosynthesis* (M. Avron ed.) p. 1841. Elsevier Publ. Comp. Amsterdam.
9. Henningsen, K. W., and Kahn, A. (1971): *Plant Physiol.,* 47:685.
10. von Wettstein, D. (1974): *Biochem. Soc. Trans.,* 2:176.
11. von Wettstein, D., Kahn, A., Nielsen, O. F., and Gough, S. (1974): *Science,* 184:800.
12. Foster, R. J., Gibbons, G. C., Gough, S., Henningsen, K. W., Kahn, A., Nielsen, O. F., and von Wettstein, D. (1971): *Proc. 1st Eur. Biophys. Congr.,* 4:137.
13. Henningsen, K. W., Thorne, S. W., and Boardman, N. K. (1974): *Plant Physiol.,* 53:419.
14. Henningsen, K. W., and Thorne, S. W. (1974): Physiol. Plant. 30:82.
15. Henningsen, K. W., and Boardman, N. K. (1973): *Plant Physiol.,* 51:1117.
16. Nielsen, N. C. (1975): *Eur. J. Biochem.,* 50:611.
17. Nielsen, O. F. (1974): *Arch. Biochem. Biophys.,* 160:430.
18. Nielsen, O. F. (1974): *Hereditas,* 76:269.
19. Beale, S. I., Gough, S. P., and Granick, S. (1975): *Proc. Natl. Acad. Sci. USA,* 72:2719.

20. Granick, S. (1959): *Plant Physiol.,* 34:18.
21. Gough, S. (1972): *Biochim. Biophys. Acta,* 286:36.
22. Nielsen, O. F. (1973): *FEBS Lett.,* 38:75.
23. Nielsen, O. F. (1975): *Biochem. Physiol. Pflanzen,* 167:195.
24. Nielsen, N. C., Henningsen, K. W., and Smillie, R. M. (1974): *Proc. 3rd Int. Congr. Photosynthesis,* edited by M. Avron, p. 1603. Elsevier, Amsterdam.
25. Brock, K., and Kannangara, C. G. (1975): *Compt. Rend. Trav. Lab. Carlsberg* (*in press*).
26. Kannangara, C. G., and Jensen, C. J. (1975): *Eur. J. Biochem.,* 54:25.

Membranes and Disease, edited by L. Bolis, J. F. Hoffman, and A. Leaf. Raven Press, New York, © 1976.

Lysosomotropic Drugs: Biological and Therapeutic Aspects

P. Tulkens* and A. Trouet

International Institute of Cellular and Molecular Pathology, and Université Catholique de Louvain, B-1200 Brussels, Belgium

Tremendous efforts have been spent these last years in attempting to understand the physiology and pathology of lysosomes (1–3). However, little use has been made of this knowledge to design drugs acting through or in lysosomes, although these organelles display very remarkable properties which should make them extremely attractive to the pharmacologist. Indeed, they are the unique intracellular organelles readily accessible to exogenous substances, thanks to the physiological route of endocytosis; they contain a collection of acid hydrolases susceptible, by their cooperative action, to deal with the vast majority of the biological compounds; they are the most acid compartment of the cell, and are therefore expected to concentrate a variety of basic molecules. As a consequence, many pharmacological compounds spontaneously accumulate or are conveyed through lysosomes; further, many others can be converted into "lysosomotropic" drugs, and their action may be, in that way, dramatically modified and possibly improved. These facts have been little considered in recent pharmacology, although definite proof of their value in therapeutics is already available. The aim of this chapter is to illustrate some aspects of *lysosomotropism* and to discuss their implications in cell chemotherapy.

ROUTE OF ENTRY IN LYSOSOMES

Molecules of the extracellular milieu can access lysosomes through three different mechanisms: permeation, endocytosis, and piggyback endocytosis (Fig. 1).

The first one characterizes substances of low molecular weight susceptible to permeate through biological membranes and to be trapped in lysosomes. Most of them are weak organic bases, and therefore trapping by protonation has been proposed as the major mechanism for their accumulation in the acid milieu of lysosomes (Fig. 2). Organic bases diffuse indeed much faster across membranes under their unprotonated than under their protonated

* Chargé de Recherches of the Belgian Fonds National de la Recherche Scientifique.

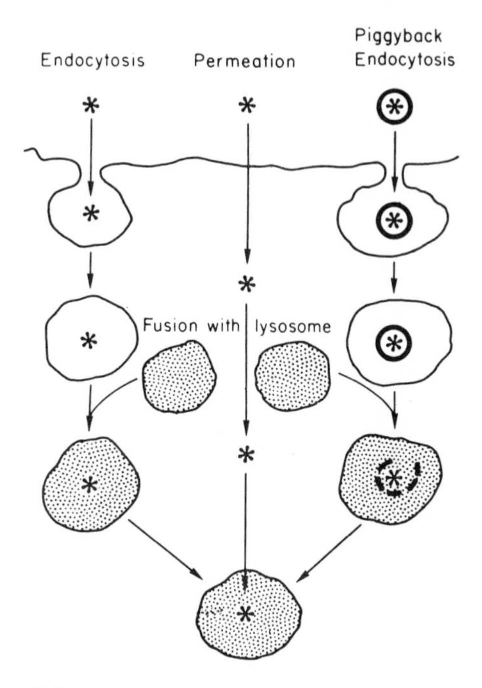

FIG. 1. Mechanisms of entry in lysosomes.

form. At equilibrium, the concentration of the uncharged form of the molecules will be uniform in all cell compartments, but the concentration of the charged form will be influenced by the local pH (14). Accordingly, weak bases will accumulate in the most acid compartment of the cell, namely, lysosomes the pH of which lies between 4 and 5. The factors influencing rate and level of intralysosomal accumulation of weak bases have been analyzed and extensively discussed by de Duve et al. (4), who showed that concentration ratios of 1,000 can be expected, on this basis, between lysosomes and the extracellular fluid. In order to sustain storage of large quantities of base, protons should, however, be furnished to replace those lost by protonation. This could be achieved by a proton pump located in

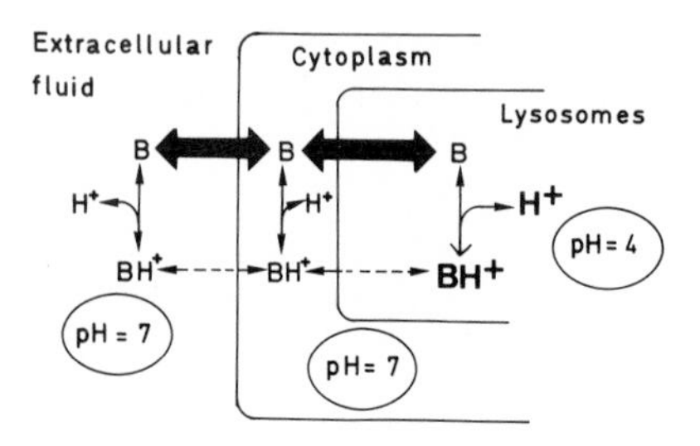

FIG. 2. Mechanism of accumulation of organic bases in lysosomes. The accumulation of BH^+ in lysosomes relies on the acid pH prevailing therein and on the faster permeation rate of the uncharged (*thick arrow*) than the charged (*broken arrow*) form of the base through membranes.

the lysosome membrane (5), or could result from a very high buffering capacity of the lysosomes constituents (D. J. Rijngoud and J. Tager, *submitted for publication*).

The second mechanism is endocytosis. It consists in the enclosure of some part of the extracellular medium in vacuoles derived from plasma membrane. The content of these vacuoles will normally be delivered in lysosomes, where accumulation will occur. The main characteristics of endocytosis and their implications has been discussed by Jacques (6), but little is known about the exact mechanism of the membrane fusions involved in that process. There is theoretically no compound which could not be accumulated in lysosomes through endocytosis. However, substances able to diffuse through lysosomal membrane and without affinity for lysosomal constituents will not remain stored in lysosomes. On the other hand, endocytosis has been shown very selective regarding the nature of the molecule (7) and may vary enormously from one type of the cell to the other.

The third mechanism derives directly from endocytosis. It concerns molecules, which for a reason or another gain no access to lysosomes, but which can be converted into an endocytozable complex, usually by linkage to a macromolecule. This mode of entry has been referred as *piggyback endocytosis* (4). It may also concern substances which diffuse across membranes, but which are restricted to the endocytosis route by coupling to an appropriate carrier.

Each of these three routes will now be illustrated with examples taken from our recent experimental work and concerned with compounds of pharmacological interest.

AGENTS ENTERING BY PERMEATION

We gained interest in this aspect of lysosomotropism along a study on the intracellular accumulation and localization of antibiotics. It is known that bacteria which multiply or sojourn inside the living cells are resistant to many antibiotics, even if they are highly sensitive to them *in vitro* (8,9). It was therefore concluded that antibiotics did not penetrate cells in sufficient amount to exert their effect. Development of antibiotics susceptible to being actively taken up in cells by endocytosis was a reasonable goal to pursue and efforts along that line were made soon after the development of antibiotics (10,11), even before lysosomes were reckoned, but this lead was unfortunately not followed up. We came to this problem when it became clear that many intracellular microorganisms stay and develop in vacuoles closely related to lysosomes.

Stimulated by the work of Bonventre and Imhoff (12) who detected accumulation of ^{3}H-dihydrostreptomycin in cultured macrophages and BHK cells, we observed that cultured fibroblasts, if maintained in presence of streptomycin long enough, accumulate the drug at a level about two to three times that of the culture fluid. Surprisingly, the accumulation was a direct function of the extracellular concentration up to a concentration of 3 mg/ml. Cells incubated with streptomycin for 4 days were fractionated by differential and isopycnic centrifugation, according to Tulkens et al. (13) (Fig. 3). In all instances, the distribution patterns of streptomycin followed closely that of the lysosomal hydrolases acid phosphatase, *N*-acetyl-β-glucosaminidase and cathepsin D and dissociated from all the other detected cell components (mitochondria:cytochrome oxidase; peroxysomes:catalase; plasma membrane:5′-nucleotidase; cell sap:neutral pyrophosphatase; bound and free ribosomes:RNA) (Fig. 3). It is therefore highly probable that intracellular streptomycin is really associated with lysosomes *in vivo*. Its concentration can be calculated on the basis of the morphometric measurements of the lysosomal volume of fibroblasts and was found about 100 times higher in lysosomes than in the culture fluid. The association of streptomycin to lysosomes was not only observed in cultured fibroblasts. Indeed, strepto-

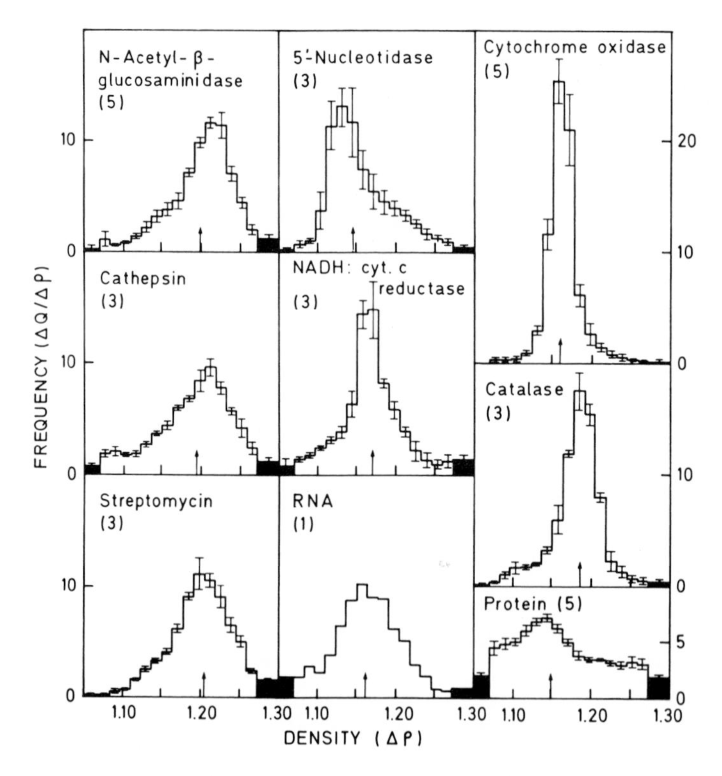

FIG. 3. Isopycnic centrifugation of cytoplasmic extracts of fibroblasts cultivated 4 days in presence of 100 μg/ml streptomycin. Abscissa is the density scale of the sucrose gradients; ordinate is the frequency of constituents in each subfraction; numbers in parentheses refer to the number of independent experiments. Details of the experimental procedures and mode of representation are given in ref. 13.

mycin injected in the bloodstream of rat and accumulated by the liver also displays a distribution pattern very similar to that of lysosomal acid phosphatase and largely different from microsomal glucose 6-phosphatase (Fig. 4).

Accumulation of streptomycin or dihydrostreptomycin in fibroblasts, macrophages, and BHK cells is a slow process, and a plateau value is obtained only after 4–5 days of incubation. Of particular interest are the facts that no difference in the *rate* of uptake of streptomycin is observed between phagocytozing and nonphagocytozing cells (12), and that this rate is directly proportional to the external concentration of the antibiotic. Fibroblasts loaded with streptomycin and transferred in fresh medium lose the antibiotic according to a first-order kinetic and with a half-life of about 1.6 day. This figure is too high to be accounted for by exocytosis. It is noteworthy that the net rates of accumulation and loss are dose-dependent and similar to each other.

All these data suggest that streptomycin does not enter lysosomes by endocytosis. Other mechanisms of accumulation should therefore be considered. Streptomycin is an organic base and its lysosomotropism could be explained on that basis. However, besides one amino group of pK ~7.6 on its N-methylglucosamine moiety, streptomycin displays two strong guani-

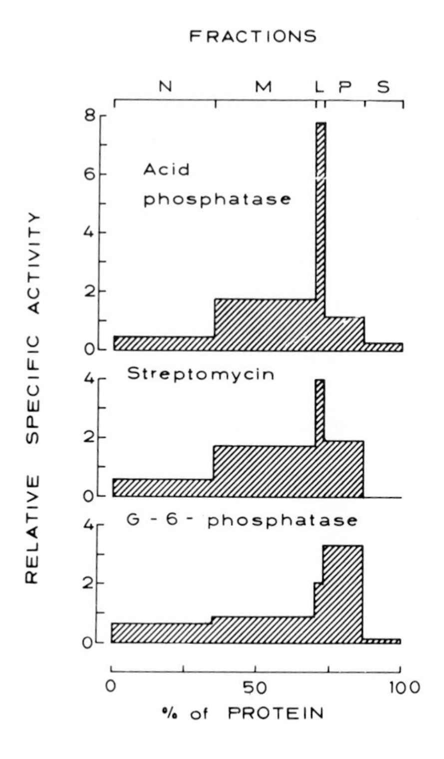

FIG. 4. Fractionation of rat liver by differential centrifugation according to de Duve et al. (42). Animals were injected intravenously 1 hr before sacrifice with 20 mg of streptomycin. (Unpublished results of A. Trouet.)

dinium groups on the streptidin part. Accordingly, the concentration of the uncharged form should be very low and accumulation should process rather slowly; the concentration ratio achieved in lysosomes at the steady state will also be lower than expected, as the relative impermeability of the membranes toward the charged form might be partially defeated by its high concentration (4). This is exactly what we found. Further evidences favoring this mechanism of lysosomal accumulation will be presented in a forthcoming paper (15).

Other antibiotics of the aminoglycoside group were similarly studied. All those which were tested (neomycin, kanamycin, gentamycin) showed the same properties, although the level at which they accumulate in lysosomes, compared to the external milieu, varied from 150- to 300-fold (16). Similarly, two antibiotics of the anthracycline group, daunorubicin and adriamycin, endowed with antitumoral properties, were shown to localize largely in lysosomes (17). For these latter drugs, however, significant amounts (two-thirds of the intracellular drug for adriamycin, one-third for daunorubicin) are found in nuclei, as they tightly bind to DNA, exerting in that way their antitumoral action (18). The drug localized in lysosomes was concentrated about 1,000-fold therein and was accumulated at a very extensive rate, with a half-accumulation time of about 90 min. These data are in accordance with those predicted by the permeation model for drugs displaying a pK_a around 8.0, like that of these two drugs (4).

Other basic compounds were shown on the same grounds to accumulate selectively or partially into lysosomes by a trapping phenomenon. Among them are chloroquine (18), neutral red (19), and orange acridine (20).

In view of these findings, it seems very surprising that intracellular bacteria escape the action of basic antibiotics if they are to be found within the same lysosomal vacuoles. Two explanations may be put forward to account for this paradox.

Basic antibiotics, driven to lysosomes by a gradient of pH, would have little tendency to reach thereafter the cytoplasm of bacteria, which is undoubtly more alkaline than the lysosome milieu. Intralysosomal aminoglycosides will therefore be ineffective in the control of the intralysosomal bacteria. Such a phenomenon is difficult to demonstrate, but evidences were found very early that antibiotics of this class are markedly less active at acid than alkaline pH (21). Although quantitative determinations are somewhat difficult to obtain, we could demonstrate about a 10-fold decrease of the antibiotic power of streptomycin and gentamicin for a drop on one unit of pH, which is the value expected if diffusion through bacterial membrane is the rate-limiting factor in the antibiotic activity of these drugs. Whatever the exact mechanism, all aminoglycosides, and most of the basic antibiotics, act poorly at acid pHs and will therefore be ineffective against microorganisms that develop in acidic compartments of the cell, such as lysosomes. A number of antibiotics display optimum activity at acid pHs;

however, most are weak acids which will be excluded from lysosomes by the same force which drives in the basic drugs. Investigations should therefore be directed toward antibiotics which could avoid these difficulties.

Another mechanism should also be considered. As pointed out by Armstrong and d'Arcy Hart (22) and Jones and Hirsch (23), virulent strains of some microorganisms are characterized by their ability to prevent fusion of the phagosome in which they are enclosed with the lysosomal vacuoles. In that way they are protected against lysosomal hydrolases and the other compounds accumulated in lysosomes by any route. What would be needed in that case is a "phagosomotropic" drug, something much more difficult to achieve.

In essence, it can be said that chemotherapy of the intracellular infection is undertaken with weapons either ineffective in the environment they will have to act or endowed with a wrong tropism. A field is therefore open for the development of new forms of antibiotherapy based on the knowledge of the subcellular localization of both drugs and bacteria and of the physico-chemical conditions prevailing at their sites of accumulation.

Besides being inactive, basic antibiotics can prove very toxic through their lysosomal accumulation. Kosek et al. (24) reported dramatic alterations

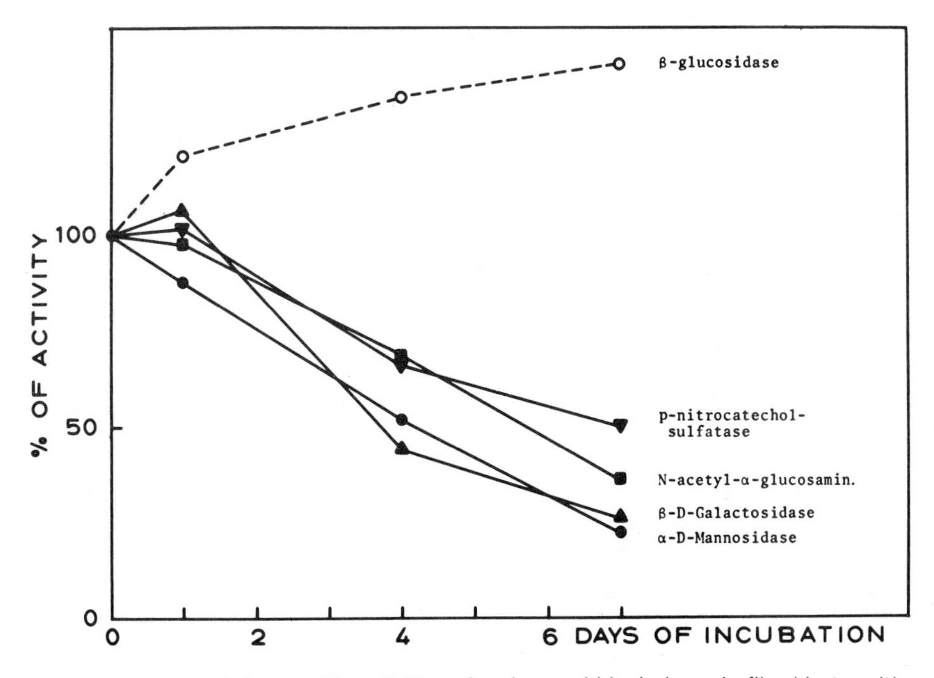

FIG. 5. Variation of the specific activities of various acid hydrolases in fibroblasts cultivated in presence of 3 mg/ml streptomycin. During the course of the experiment, streptomycin-treated cells kept on multiplying, although at a slower rate than control cells. (Unpublished results of F. van Hoof and P. Tulkens.)

of the morphology of rat kidney lysosomes following administration of gentamicin. Our findings provide a clue for this pathology. If aminoglycosides accumulate in lysosomes, they could interfere with their functions. In fibroblasts incubated with high concentrations of streptomycin, we observed a very significant fall of activity of several acid hydrolases (Fig. 5). The part of the cell volume occupied by lysosomes was increased about 20-fold and they showed manifest signs of overloading like those observed after immunological impairment of their enzymes (25,26) or in mucolipidosis type II, a disease characterized by a genetic deficiency of several acid hydrolases (27). Simultaneously, the endocytotic pathway and the digestive capacities of lysosomes were severely hindered. Such side effects of lysosomotropic antibiotics might well be rewarding to be considered in the general assessment of the toxicity of these drugs. This is particularly important as dysfunction of lysosomes may prove very dramatic in view of the multiple functions of these organelles (28).

AGENTS ENTERING BY ENDOCYTOSIS

As mentioned before, any macromolecular compound could, in principle, be delivered to lysosomes through endocytosis. Of particular interest would be the replacement therapy in those lysosome storage diseases characterized by the absence of one acid hydrolase (29,30). We would like here to present some data illustrating the selectivity of endocytosis at the cellular level and show what properties of the molecule may render it more or less endocytozable and direct it toward different targets in the cell.

Stimulation of endocytosis by polyions has been very largely documented, although with some controversy regarding the effects. For instance, Ryser (31) found that basic polymers stimulated uptake of albumin, whereas Cohn and Parks (32) observed this effect only with polyglutamate and other polyanions. The subcellular localization of these inducers, as well as the albumin itself was not studied, and only *"cell association"* was recorded. Prompted by these studies, G. Noël, in our laboratory, examined the cell accumulation of three types of polymers:polylysine, acetyl-labeled IgG, and fluorescein-labeled IgG. The levels of cellular accumulation were found strikingly different, polylysine reaching the highest value, and acetyl-IgG the lowest. But when the subcellular localization of these substances was investigated, the differences in cellular accumulation were shown to bear no relationship with lysosomal accumulation. By fluorescence microscopy, polylysine was found to be localized exclusively at the cell surface, no fluorescence being detected inside the cytoplasm; conversely, fluorescein-labeled IgG were accumulated exclusively in granules, which were shown thereafter by isopycnic centrifugation to be lysosomes; acetyl ^{3}H-IgG were shown to localize largely at the cell surface, after short exposure, and to move thereafter to lysosomes, but at a much lower rate than fluorescein-IgG.

The charge and nature of the molecule influences thus dramatically both its accumulation and localization inside fibroblasts. As judged from the conflicting results of the literature in this area, it is highly probable that general rules will be difficult to establish, but it should be of great interest to know at least some of them if chemotherapy through endocytosis is looked for in a given cell system.

Selectivity in subcellular localization could also be gained by the use of molecules susceptible to recognize and bind selectively to some constituent of the vacuolar system. For this purpose antibodies were raised against purified plasma membranes and soluble constituents of lysosomes. Selectivity of the antibodies was obtained by immunoadsorption on the heterologous cell components and purification on immobilized antigens. It was found by Y. J. Schneider that all these antisera were accumulated by fibroblasts at a level severalfold higher than control IgG; all of them bound primarily to plasma membrane, but their further behavior differed dramatically. Anti-plasma membrane antibodies remained for long periods at the cell surface, whereas antilysosomes antibodies were soon transferred to lysosomes where they accumulated (Fig. 6). The implications of these

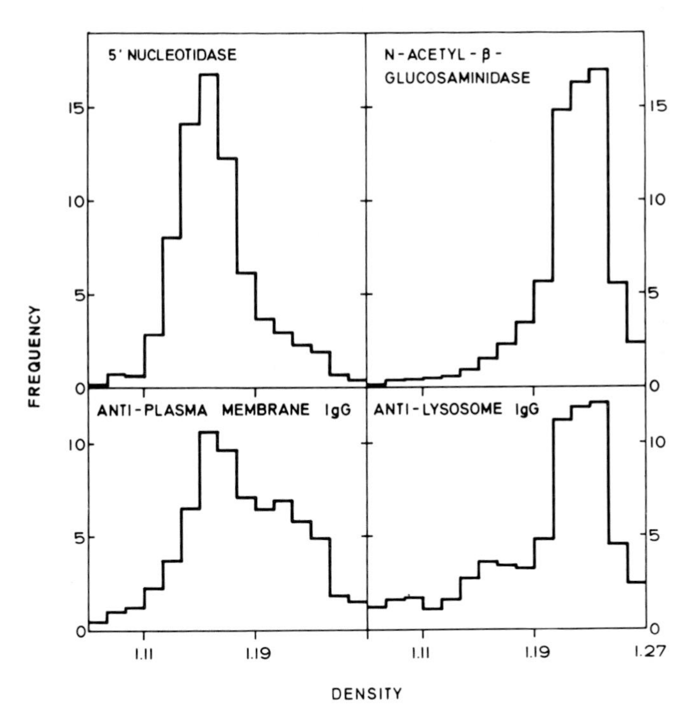

FIG. 6. Isopycnic centrifugation of cytoplasmic extracts of rat fibroblasts incubated 30 hr in presence of ^{3}H-acetyl-labeled IgG directed against liver plasma membrane **(left)** or soluble constituents of liver lysosomes **(right).** Details on the experimental procedures and mode of representation are given in ref. 13.

findings in the general mechanism of endocytosis has been recently presented elsewhere (33,34), and we think that they may lead to the development of drugs that would selectively reach specified parts of the vacuolar apparatus, from the cell surface up to the lysosomes.

Besides selectivity of the intracellular target, selectivity between different cells will certainly be a field open to promising experiments. The work of Morell and co-workers (7), who showed that desialation of glycoproteins increase their uptake by hepatocytes, gives hints for the development of such tissue specific molecules.

AGENTS ENTERING BY PIGGYBACK ENDOCYTOSIS

As mentioned earlier, theoretically any compound could be directed to lysosomes if coupled to a substance carrier, which will normally be accumulated in these organelles. For practical reasons, the carrier has always been, up to now, a macromolecule and such complexes enter thus lysosomes by endocytosis.

Some work has already been done in the design of lysosomotropic complexes by coupling drugs to polymers or encapsulating them in lipid microspheres (liposomes). We would like here to describe results obtained in our laboratory with the two antitumoral drugs daunorubicin and adriamycin.

These drugs are already spontaneously lysosomotropic to a certain extent (17); they reach lysosomes, and nuclear DNA as well, by permeation through pericellular membrane. Thanks to their high affinity for high molecular weight DNA, they may form with this macromolecule tight complexes, the size of which will prevent entry of the drug in cells by permeation, restricting it to the sole endocytosis route. DNA-complexed daunorubicin or adriamycin loose their antibacterial properties, but these are fully restored if DNA is hydrolyzed by lysosomal DNAse (35). Accordingly, complexed drugs, reaching lysosomes by endocytosis, would be digested therein and free drug released. As we have seen that diffusion of basic drugs through lysosome membrane is a reversible process, governed by concentration ratios, free daunorubicin or adriamycin will partially leave lysosomes and finally reach the nucleus, where they will exert their cytostatic properties. As shown in Fig. 7, the intracellular distribution of the drugs at equilibrium, and their cytotoxic action should be identical, whether given free or complexed. This is effectively observed on cultured leukemic cells as illustrated at Fig. 8, the complexed drugs acting somewhat slower, as could be expected, since they have to follow the endocytosis route and be dissociated from the carrier before they could act. Further observations, made by G. Nöel on cultured fibroblasts, showed that the model was in essence correct, i.e., that free permeant drugs were effectively transformed in endocytozed compounds, the fate of which was as predicted.

What advantages could be gained by such a transformation? Very potent

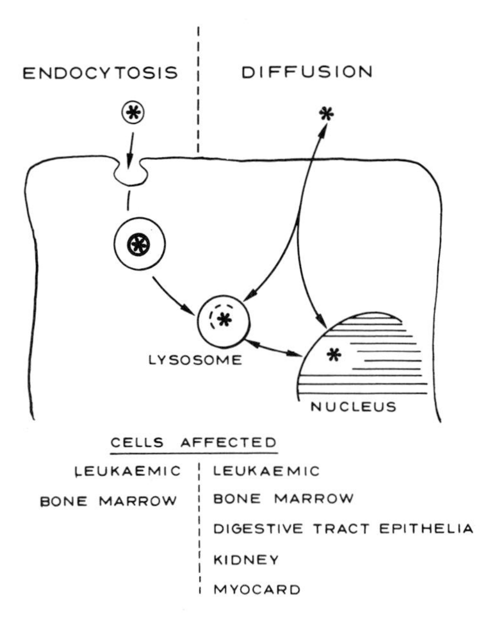

FIG. 7. Conversion of a lysosomotropic *and* nucleotropic drug from a permeant to an endocytozable form, by linkage to a macromolecule. If the endocytosed drug is released free from its carrier in lysosomes (by acid or enzymic action), its final subcellular distribution will be identical to that of the permeant drug. The *cellular* concentration will, however, depend on the endocytotic activity of cells. Such a transformation may confer to the drug a new basis for tissue selectivity; examples are given for antitumoral drugs daunorubicin or adriamycin injected intravenously.

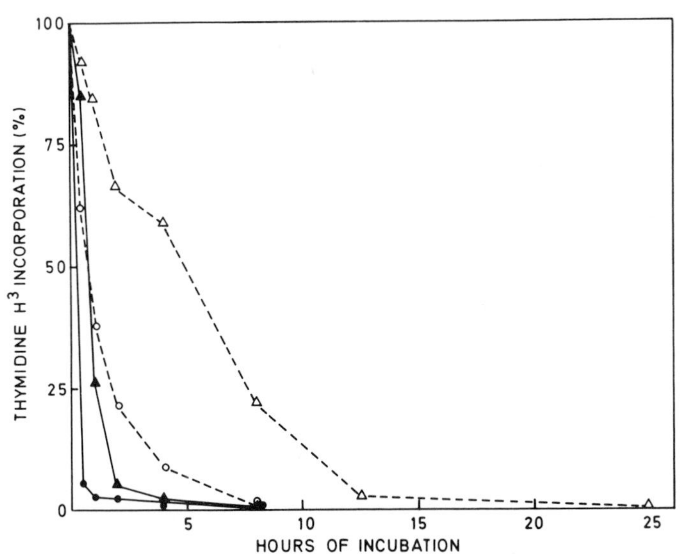

FIG. 8. ^{3}H-thymidine incorporation (1 hr incubation) by leukemic L1210 cells cultivated *in vitro* in presence of free daunorubicin (left solid line), free adriamycin (right solid line), DNA-complexed daunorubicin (left broken line), DNA-complexed adriamycin (right broken line). The molar concentration of the drugs was identical in all experiments (1.7×10^{-6} M). (Unpublished results of A. Zenebergh and A. Trouet.)

antileukemic drugs, adriamycin and daunorubicin show distinct toxic effects toward all highly dividing cells, chiefly normal bone marrow cells and digestive epithelia. They display also a poorly understood but very grave toxicity toward the myocard. It was reasoned that a drug converted into an exclusive endocytozable compound, would loose its toxicity toward these tissues which have little or no endocytotic activity. This is effectively the case for the myocard and at a lesser extent for the digestive epithelia, which take little or no endocytozable material from their vascular pole. It was also hoped that leukemic cells would endocytoze more than normal bone marrow cells. Conversely, those cells with high endocytotic activities, like macrophages and Küpfer cells, would not be harmed, as their very low mitotic index should make them insensitive to these cytotoxic agents. Experiments along this line were very rewarding. For instance, it could be demonstrated that mice tolerated about twice more complexed adriamycin than free drug without toxicity. The antileukemic effects at these doses were, of course, distinctly better, so that the therapeutic index was largely increased (Fig. 9). Details of these experiments and others with daunorubicin have been already reported (35,37). Accordingly, very careful, at first,

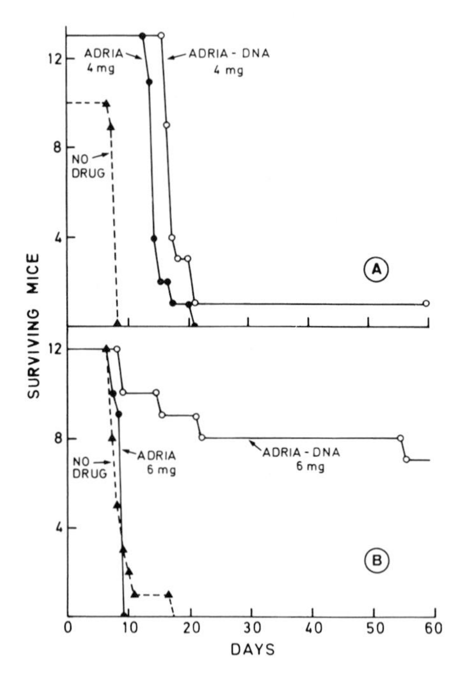

FIG. 9. Survival of DBA_2 mice injected intravenously with 10^4 L-1210 leukemic cells at day 0. Treatment was given by the same route on days 1 to 5, at adriamycin dosages indicated. The overlapping of the survival curves of the nontreated animals and the animals treated with 6 mg/kg-day free adriamycin is accidental, but illustrates the toxicity of such dose of drug. (From ref. 36).

and thereafter systematic treatments of human leukemic patients was undertaken with DNA-complexed adriamycin and daunomycin. Results have been published elsewhere (38–40). They largely confirmed the hopes gained from our previous experiments and several patients have been treated at doses which would have caused severe toxic effects, had the drug not been complexed.

For these attempts, DNA was chosen as carrier because it spontaneously forms a tight complex with the drugs mentioned, without chemical engineering. It is, however, by no means certain that other carriers, such as proteins, immunoglobulins, or other polymers would not have much more attractive properties, especially regarding the cell selectivity they might confer to the complex. Experiments are in progress along this line (41). No doubt that they will prove to be of dramatic interest for these substances which are given selectivity through endocytosis, and still more for those for which piggyback endocytosis may constitute the sole mechanism for entering the cell.

CONCLUSIONS

In this chapter, we presented briefly the concept of *lysosomotropic drugs* and we illustrated it with examples taken from different fields of pharmacology. Lysosomes may constitute a target for some drugs; they may also be only a tool through which drugs are delivered more specifically or more efficiently to a given cell type. The knowledge gained in physiopathology of lysosomes should enable us to design these drugs in a reasoned fashion.

ACKNOWLEDGMENTS

The collaboration of Mrs. Deprez-De Campeneere and Dr. F. Van Hoof was greatly appreciated. This work was supported by the Belgian Fonds National de la Recherche Scientifique Médicale, Fonds Cancérologique de la Caisse Générale d'Epargne et de Retraite and Fonds de la Recherche Fondamentale Collective.

REFERENCES

1. Wattiaux, R. (1969): In: *Handbook of Molecular Cytology,* edited by A. Lima-De-Faria, p. 1159. North-Holland, Amsterdam.
2. Dingle, J. T., and Fell, H. B., editors (1969): *Lysosomes in Biology and Pathology,* Vols. 1 and 2; Dingle, J. T. editor (1973): *Ibid.,* Vol. 3. North-Holland, Amsterdam.
3. Hers, H. G., and van Hoof, F., editors (1973): *Lysosomes and Storage Diseases,* p. 666. Academic Press, New York.
4. de Duve, C., de Barsy, T., Poole, B., Trouet, A., Tulkens, P., and van Hoof, F. (1974): *Biochem. Pharmacol.,* 23:2495.
5. Lie, S. O., and Shofield, B., (1973): *Biochem. Pharmacol.,* 22:3109.
6. Jacques, P. (1969): In: *Lysosomes in Biology and Pathology,* edited by J. T. Dingle and H. B. Fell, Vol. 2, p. 395. North-Holland, Amsterdam.

7. Morell, A. G., Gregoriadis, G., Scheinberg, I. H., Hickman, J., and Ashwell, G. (1971): *J. Biol. Chem.,* 246:1461.
8. Solberg, C. O. (1972): *Acta Med. Scand.,* 191:383.
9. Mackaness, G. B. (1960): *J. Exp. Med.,* 112:35.
10. Markham, N. P., Wells, A. Q., Heatley, N. G., and Florey, H. W., (1951): *Br. J. Exp. Pathol.,* 32:353.
11. Nelson, W. E., Forgacs, J., and Kucera, J. L. (1947): *Proc. Soc. Exp. Biol. Med.,* 64:20.
12. Bonventre, P. F., and Imhoff, J. G. (1970): *Infect. Immun.,* 2:89.
13. Tulkens, P., Beaufay, H., and Trouet, A. (1974): *J. Cell Biol.,* 63:383.
14. Waddel, W. J., and Bates, R. G. (1969): *Physiol. Rev.,* 49:285.
15. Tulkens, P., and Trouet, A. (*In preparation.*)
16. Tulkens, P., and Trouet, A. (1974): *Arch. Int. Phys. Biochem.,* 82:1018.
17. Noel, G., Trouet, A., Zenebergh, A., and Tulkens, P. (1975): In: *Adriamycin Review: EORTC International Symposium,* edited by M. Staquet, H. Tagnon, Y. Kenis, G. Bonadonna, S. K. Carter, G. Sokal, A. Trouet, M. Ghione, C. Praga, L. Lenoz, and O. S. Karim, p. 99. European Press Medikon, Ghent, Belgium.
18. Wibo, M., and Poole, B. (1974): *J. Cell Biol.,* 63:430.
19. Allison, A. C., and Young, M. R. (1964): *Life Sci.,* 3:1407.
20. Canonico, P. G., and Bird, J. W. C. (1969): *J. Cell Biol.,* 43:367.
21. Abraham, E. P., and Duthie, E. S. (1946): *Lancet,* 250:455.
22. Armstrong, J. A., and d'Arcy Hart, P. (1971): *J. Exp. Med.,* 134:713.
23. Jones, T. C., and Hirsch, J. G. (1972): *J. Exp. Med.,* 136:1173.
24. Kosek, J. C., Mazze, R. I., and Cousins, M. J. (1974): *Lab. Invert.,* 30:48.
25. Tulkens, P., Trouet, A., and van Hoof, F. (1970): *Nature,* 228:1282.
26. Dingle, J. T., Poole, A. R., Lazarus, G. S., and Barrett, A. J. (1973): *J. Exp. Med.,* 137: 1124.
27. van Hoof, F. (1973): In: *Lysosomes and Storage Diseases,* edited by H. G. Hers and F. van Hoof, p. 217. Academic Press, New York.
28. de Duve, C., and Wattiaux, R. (1966): *Ann. Rev. Physiol.,* 28:435.
29. Desnick, R. J., Bernlohr, R. W., and Krivit, W. (1973): *Birth Defects,* 9:236.
30. Fratantoni, J. C., Hall, C. W., and Neufeld, E. F. (1968): *Science,* 162:570.
31. Ryser, H. J. P. (1968): *Science,* 159:390.
32. Cohn, Z. A., and Parks, E. (1967): *J. Exp. Med.,* 125:213.
33. Tulkens, P., Schneider, Y. J., and Trouet, A. (1975): In: *Proceedings of the 2nd International Symposium of Intracellular Catabolism, Ljubljana, Yugoslavia.* (*in press.*)
34. Schneider, Y. J., Trouet, A., and Tulkens, P. (1975): *Arch. Int. Phys. Biochim.* (*in press.*)
35. Trouet, A., Deprez-de Campeneere, D., and de Duve, C. (1972): *Nature* [*New Biol.*], 239:110.
36. Trouet, A., Deprez-de Campeneere, D., Zenebergh, A., and Sokal, G. (1974): In: *Activation of Macrophages,* edited by W. H. Wagner and H. Hahn, p. 330. Excerpta Medica, Amsterdam.
37. Trouet, A., Deprez-de Campeneere, D., de Smedt-Malengreaux, M., and Atassi, G. (1974): *Eur. J. Cancer,* 10:405.
38. Cornu, G., Michaux, J. L., Sokal, G., and Trouet, A. (1974): *Eur. J. Cancer,* 10:405.
39. Michaux, J. L., Cornu, G., Sokal, G., and Trouet, A. (1975): In: *Adriamycin Review: EORTC International Symposium,* edited by M. Staquet, H. Tagnon, Y. Kenis, G. Bonadonna, S. K. Carter, G. Sokal, A. Trouet, M. Ghione, C. Praga, L. Lenoz, and O. S. Karim, p. 216. European Press Medikon, Ghent, Belgium.
40. Lie, S. O., Lie, K. K., and Langslet, A. (1975): In: *Adriamycin Review: EORTC International Symposium,* edited by M. Staquet, H. Tagnon, Y. Kenis, G. Bonadonna, S. K. Carter, G. Sokal, A. Trouet, M. Ghione, C. Praga, L. Lenoz, and O. S. Karim, p. 226. European Press Medikon, Ghent, Belgium.
41. Levy, R., Hurwitz, E., Maron, R., Arnon, R., and Sela, M. (1975): *Cancer Res.,* 35:1182.
42. de Duve, C., Pressman, B. C., Gianetto, R., Wattiaux, R., and Appelmans, F. (1955): *Biochem. J.,* 60:604.

Membranes and Disease, edited by L. Bolis, J. F. Hoffman, and A. Leaf. Raven Press, New York, © 1976.

The Responses of Biological Structure to Stress: The Case of the Endoplasmic Reticulum

Philip Siekevitz

Rockefeller University, New York, New York 10021

The theme of this chapter is that membranes are so constituted as to resist stress, to resist drastic changes in their immediate environment. They do this by being constructed not like a rigid rod, which cannot be broken until a certain amount of stress is brought to bear upon them. For it is becoming clear that membrane structure is not really as rigid as once thought (cf. ref. 1), perhaps not as fluid as the original fluid-mosaic model would make it out to be (2), but nevertheless, a structure within which a good deal of movement is allowed its constituent components (cf. ref. 3). These movements cannot only occur in the plane of the membrane (3), for the structure is so constituted that membrane components can efficiently "hop in and out" of the structure itself (1). Because of these properties, changes—and sometimes drastic changes—can occur in a membrane without the destruction of the membrane structure. I will point out selective instances of this with regard to the endoplasmic reticulum (ER) membranes, but I think I can generalize from these to membranes in general. I have mentioned some examples of this property in a previous article (1), and actually two reviews have already appeared on changes in ER membranes after various conditions of stress (4,5). In the present review, I will try to give examples of alterations in the function of the ER, whether due to genetics, or dietary or various other environmental conditions, and try to show what the alterations mean in terms of ER membrane structure.

I think the earliest example of a ER genetic defect was shown by Ganschow and Paigen (6) in a mouse mutant whose liver lacked the ER form of β-glucoronidase, although it did contain the lysosomal form of the enzyme. Later work (7) showed that the four proteins making up the liver enzyme complex were missing in the mutant ER, and instead a precursor form was present, the same precursor which could be induced in the normal mouse. Since genetic evidence indicated a malfunction in a structural gene, and since the lysosomal enzyme consisted of only one protein, the conclusion was that in the mutant a regulatory factor was missing which was required to anchor the enzyme in the ER membranes (7). Although no morphological evidence for structural abnormality was included in the papers, we can infer that the ER membranes in the mutant mice exist as such, even in the absence of several proteins. A more fully worked out morphological description was given in a series of papers (8–13) concerning another mouse mutant having greatly decreased glucose 6-phosphatase, serine dehydratase, and tyrosine aminotransferase activities. Whereas the genetics of the system are still somewhat obscure, the available evidence does indicate that the mutation does not deal with the structural genes for these enzymes. Since one of the enzymes is an ER membrane protein, glucose 6-phosphatase, changes in lipid or protein composition (by gels and by immunological methods) were looked for, and were not found (13). The membranes of the ER in the mutant are apparently normal, except that a marked vesiculation had taken place in the cisternae of the ER (11). The tentative conclusion was that all of the proteins of the ER were present in the membrane, and that except for the phosphatase, all the other tested proteins showed normal activity; however, at least this one, the phosphatase, was so altered that its function was markedly decreased, and the presence of this altered protein somehow resulted, not in a visible alteration (by EM observation) of membrane structure, but in a large expansion of the space enclosed by these membranes. This dilation and vesiculation of the ER system has been noted before (cf. ref. 5) and is remarked upon as being a seemingly common form of response of the ER to cellular injury (cf. ref. 5 and below for other instances).

Another obvious environmental stress condition is the diet, particularly starvation, and the 1969 review (4) cites numerous instances of the effects of various dietary regimes on the form and function of the ER. Starvation, and also a protein-deficient diet, does not have much effect on the liver ER; the membrane seems to be intact, the ribosomes are still there attached, but there is some vacuolization of the cisternae and some disarray of the overall parallel arrangement of the ER membranes (cf. ref. 4). However, there is a changeover in metabolic function during starvation, in that while the concentrations of cytochromes remain unchanged in the membranes (14,15), glucose 6-phosphatase activity increases (14) and the terminal stearyl CoA desaturase system diminishes (15), again indicating a possible rearrangement of proteins in the membrane without altering greatly its structure. Refeeding the animal brings about a shift once again to normal levels (15), and because cycloheximide inhibits this shift, it would appear that a synthesis and insertion of new protein is involved. Indeed, a conservation of the liver ER cytochromes was found during a regime of iron deprivation, as opposed to a decrease in the mitochondrial cytochromes (16).

The great perturbability of the ER membranes is well illustrated by these diet-deficient or diet-change studies. Rats fasted for 2 days showed changes in the fatty acid composition of their ER membranes (17); when the activities of enzymes from the ER membranes were examined, all three possible variations were found, as glucose 6-phosphatase increased twofold in specific activity, stearyl CoA desaturase went down some threefold, while the alkenylglycerylphosphorylcholine hydrolase showed no change (17). Of course, the overall fatty acid composition of microsomal lipids changed (17), and one again has to reflect on the individuality of the responses of the different enzymes within a common structural unit. Again, rats force-fed a protein-deficient diet displayed a large increase in the amounts of lysophosphatides in their microsomal membranes, and at the same time a twofold increase in the ER enzyme, UDP-glucoronyltransferase (18). Since treatment with lysophosphatides of microsomes from normally fed rats caused a similar increase *in vitro,* it appears that a perturbation in the enzyme environment *in vivo* was responsible for the activation. In the same vein, rats kept on an ethanol diet developed an increase in triglycerides in their microsomes (19) and showed increases in specific concentrations of the cytochromes b_5 (19) and P_{450} (19,20) and in specific activities of aniline hydroxylase (19), aminopyrine demethylase (19), and the overall NADPH oxidase (21).

Perhaps the most investigated case of ER pathology is that induced by CCl_4 (cf. ref. 4). The morphological picture that emerges is that of an extensive vacuolization of the ER, with dilated cisternae, and with loss of regular polyribosomal arrays. There is also an alteration in the phospholipid composition (22). However, there is a marked heterogeneity of response with regard to the various ER enzymes; it has been known for a long time (23) that glucose 6-phosphatase is markedly decreased in CCl_4-intoxicated animals, but, while cytochrome P_{450} and the oxidative demethylations associated with it are also decreased (24), the concentration of the other ER cytochrome, b_5, remains unchanged (24). The prevalent view of the mode of action of CCl_4 is not one based on its lipid—solvent properties but one based on its induction of the peroxidation of membrane lipids (25). This attack on the polyunsaturated lipids is particularly accentuated in ER membranes because of the presence of these lipids (cf. ref. 26) and of the enzyme system predominantly involved, the so-called NADPH-cyt c reductase (27–31). The interesting point is that even after extensive change in the ER membrane lipids (26,32–34), with differential effects on several ER enzymes (34–36), little change was found to occur in the morphological appearance of the membranes (34,37,38); the ribosomes were more detached, the fragmentation was more apparent, but the familiar trilaminar structure was easily observed. The point is that even after *in vivo* and *in vitro* treatments that had marked effects on the membrane lipids, some sort of membrane structure still persists, with even a good deal of enzymatic activity, perhaps of those proteins not dependent on lipids for their function.

Another interesting point can be inferred from studies dealing with cytochrome P_{450}. As this protein is the single largest species in the ER membrane, which, in the normal animal accounts for at least 10% of the protein therein, and because it could be thought of as being a structural protein, decreases in this protein might have profound effects on the membrane. The destruction of the heme of cytochrome P_{450} can be caused by allylisopropylacetamide (AIA) administration (39–41), by CCl_4 (40), by CS_2 (42), and by certain barbiturates (43); this effect seems to be mediated by a lipid peroxidation (44–46). The effect of AIA, at least, seems to be specific, in that cytochrome b_5, the other heme protein in the ER membranes, is not affected (40,42), nor is the NADPH-cyt c reductase (40), whereas the electrophoretic pattern of ER membrane protein shows only a decrease in the cyt P_{450} band (47). Even though AIA injection can decrease the amount of this major protein in the ER membranes by two-thirds, it is clear that such a large perturbation does not lead to major alterations in the ER membrane.

A large number of compounds have been tested for their action on either ER function or morphology. One interesting response to a variety of agents, such as ethionine (48), dimethylnitrosoamine (49), vinblastine (50), protein synthesis inhibitors (51), and various other compounds (cf. ref. 5), is the production of "whorls" of ER membranes; whether this is due to a proliferation of membranes or a rearrangement is not known, although the protein synthesis inhibitor experiments would argue against the former interpretation.

Although there have been a large number of studies on the biochemical effects of various hormones upon ER function, there have been very few studies on the correlation of these effects with membrane morphology. All that we can infer is that these endogenous promoters of regulatory change would have much less effect on the structural integrity of the membrane than the gross attacks of the exogenous agents.

Another type of correlation has been to compare microsomes from normal liver and from hepatomas. Differences were found in phospholipid composition (52), although not confirmed (53), in a decreased amount of phospholipid with no change in cholesterol content (53), in electrophoretic patterns of the membrane proteins, although minor (54), and in changes in enzyme activities (55,56). An interesting point is that the various types of hepatomas had quite dissimilar ranges of enzyme activities when compared among themselves and compared to normal liver (55,56); this took in the glucose 6-and ATP phosphatases, the cytochromes, and the various activities oxidizing NADH and NADPH. Whereas it was stated that marked membrane alteration occurred in one of the tumor lines, the published pictures are inconclusive (56). The main point is that an abnormality in a liver cell giving rise to abnormal growth leads to a pleiomorphic picture with regard to the protein composition of the ER membranes; different kinds of alterations can occur in the structure without altering it as a structure.

Finally, some mention should be made of the stress conditions imposed by differentiation. First, no changes are apparent in regenerating liver microsomes in their enzyme activities (55). Second, ever since the earliest report (57), it has been known that fetal and early neonatal rat liver microsomes are deficient in drug-metabolizing enzymes and their associated cytochromes (also cf. refs. 58–60). Changes also occur in other microsomal enzymes (58) and in lipid composition (58,61). In the differentiation during aging of the chick sarcoplasmic reticulum membranes, marked changes occur in the ATPase and in the Ca^{2+} uptake, with smaller changes in protein electrophoretic patterns and in phospholipid and fatty acid compositions (62). The only change observed in morphology was the marked increase in 75 Å particles on the surface of the membrane, perhaps equivalent to the Ca^{2++} sensitive ATPase (63). The point to be made here, and emphasized also earlier (59), is that despite these changes in occurrence and function of certain proteins and lipids, the membranes remained intact as distinctive structural entities throughout the whole differentiation process.

I think one conclusion can be reached in light of all the references I have cited and which have been cited in the previous reviews (4,5). That is, any disturbance in the ER membrane structure gives rise to a local perturbation only, even what appear to be large-scale destructive influences, such as CCl_4 toxicity, do not lead to the obliteration of the structure. I think we might have inferred all this from a survey of the dynamic structure of the ER membrane (cf. ref. 1), for we learned that the protein and lipid composition of the membrane can vary within rather wide limits. That is, during differentiation and development, the ratios among the various ER enzymes are changing all the time. Specificity with regard to protein still resides in the membrane, but the absolute amounts of these specific proteins, or the relative amounts among them, can vary. What I am stating is that the concentration of any one protein in the membrane can change with absolutely no effect on the concentrations of the other membrane proteins. This being so, any perturbation of the system, be it disease-generated or not, will lead only to a localized disorder of the membrane. Even when an enzyme protein like glucose 6-phosphatase, which may be considered also a structural protein of the ER membrane, is modified, very little effect is seen on the other enzymes of the ER, as mentioned previously in the mutant case. Even when a protein such as cytochrome P_{450}, which again may be considered also a structural protein in that it is the largest single protein species, by far, in the ER membrane, is present in very small or very large amounts in the membrane, only negligible effects are seen on the activities of some of the other membrane proteins, as mentioned earlier in the AIA cases.

Finally, I would guess that the other membranes of the cell behave in the same manner, under any and all conditions of stress. The only exception may be the plasma membrane, in which the number of species of protein is probably less than in the intracellular membranes (cf. reviews 64–69), and thus any effect on any one of these species would be more destructive of the membrane as a whole. In the case, however, of the intracellular membranes we can surmise that specific conditions of stress should not give rise to pleiomorphic results, that alteration of membrane structure and function become localized events, with specific effects. Hence, paradoxically, the somewhat fluid nature of the membrane tends to "seal off" perturbations, for at first glance we would have thought that a perturbation at one point would "roll along" like a wave, along and through the membrane. This might still be the case in certain situations wherein a destructive influence on one protein species could be transferred to another protein species next to it in the membrane, but I would guess this does not happen in too many cases. The altered protein may be lost from the membrane, the gap is filled up, the structure of the membrane remains intact, and all other functions of the membrane are preserved.

REFERENCES

1. Siekevitz, P. (1972): *Ann. Rev. Physiol.* 34:117.
2. Singer, S. J., and Nicolson, G. L. (1972): *Science,* 175:720.
3. Edidin, M. (1974): *Annu. Rev. Biophys. Bioenerget.,* 3:179.
4. Smukler, E. A., and Arcasoy, M. (1969): *Int. Rev. Exp. Pathol.,* 1:305.
5. Goldblatt, P. J. (1972): *Sub-Cell. Biochem.* 1:147.
6. Ganschow, R., and Paigen, K. (1967): *Proc. Natl. Acad. Sci.* USA, 58:938.
7. Swank, R. T., and Paigen, K. (1973): *J. Mol. Biol.,* 77:371.
8. Erickson, R. P., Gluecksohn-Waelsch, S., and Cori, C. F. (1968): *Proc. Natl. Acad. Sci.* USA, 59:437.
9. Gluecksohn-Waelsch, S., and Cori, C. F. (1970): *Biochem. Genet.,* 4:195.
10. Thorndike, J., Trigg, M. J., Stockert, R., Gluecksohn-Waelsch, S. and Cori, C. F. (1973): *Biochem. Genet.,* 9:25.
11. Trigg, M. J., and Gluecksohn-Waelsch, S. (1973): *J. Cell Biol.,* 58:549.
12. Gluecksohn-Waelsch, S., Schiffmann, M. B., Thorndike, J., and Cori, C. F. (1974): *Proc. Natl. Acad. Sci.* USA, 71:825.
13. Erickson, R. P., Siekevitz, P., Jacobs, K., and Gluecksohn-Waelsch, S. (1974): *Biochem. Genet.,* 12:81.
14. Bock, K. W., Fröhling, W., and Remmer, H. (1973): *Biochem. Pharmacol.,* 22:1557.
15. Oshino, N., and Sato, R. (1972): *Arch. Biochem. Biophys.,* 149:369.
16. Dallman, P. R., and Goodman, J. R. (1971): *J. Cell Biol.,* 48:79.
17. Ellingson, J. S., Hill, E. E., and Lands, W. E. M. (1970): *Biochim. Biophys. Acta,* 196:176.
18. Graham, A. B., Woodcock, B. G., and Wood, G. C. (1974): *Biochem. J.,* 137:567.
19. Ariyoshi, T., Takabatake, E., and Remmer, H. (1970): *Life Sci.,* 9:Part II, 361.
20. Ishii, H., Joly, J.-G., and Lieber, C. S. (1973): *Biochim. Biophys. Acta,* 291:411.
21. Lieber, C. S., and De Carli, L. M. (1970): *Science,* 170:76.
22. Comporti, M., Burdino, E., and Ugazio, G. (1971): *Ital. J. Biochem.,* 20:156.
23. Recknagel, R. O., and Lombardi, B. (1961): *J. Biol. Chem.,* 236:564.
24. Smukler, E. A., Arrhenius, E., and Hultin, T. (1967): *Biochem. J.,* 103:55.
25. Ghoshal, A. K., and Recknagel, R. O. (1965): *Life Sci.,* 4:2195.
26. Tappel, A. L. (1973): *Fed. Proc.,* 32:1870.
27. Hochstein, P., and Ernster, L. (1963): *Biochem. Biophys. Res. Commun.,* 12:388.
28. Wills, E. D. (1969): *Biochem. J.,* 113:325.
29. McCary, P. B., Poyer, J. L., Pfeifer, P. M., May, H. E., and Gilliam, J. M. (1971): *Lipids,* 6:297.
30. Bidlack, W. R., Okita, R. T., and Hochstein, P. (1973): *Biochem. Biophys. Res. Commun.,* 53:459.
31. Pederson, T. C., Buegi, J. A., and Aust, S. D. (1973): *J. Biol. Chem.,* 248:7134.
32. May, H. E., and McCay, P. B. (1968): *J. Biol. Chem.,* 243:2288.
33. Bidlack, W. R., and Tappel, A. L. (1973): *Lipids,* 8:177.
34. Högsberg, J., Bergstrand, A., and Jakobsson, S. V. (1973): *Eur. J. Biochem.,* 37:51.

35. Hochstein, P., and Ernster, L. (1964): *Ciba Found. Symp. Cell Injury,* p. 123.
36. Schachter, B. A., Marver, H. S., and Meyer, U. A. (1972): *Biochim. Biophys. Acta,* 279: 221.
37. Tam, B. K., and McCay, P. B. (1971): *J. Biol. Chem.,* 245:2295.
38. Arstila, A. U., Smith, M. A., and Trump, B. F. (1972): *Science,* 175:530.
39. De Matteis, F. (1970): *FEBS Lett.,* 6:343.
40. Levin, W., Jacobson, M., and Kunzman, R. (1972): *Arch. Biochem. Biophys.,* 148:262.
41. Satyanarayana, R. M. R., Malathi, K., and Padmanaban, G. (1974): *Biochem. J.,* 127:553.
42. De Matteis, F. (1973): *Drug. Metab. Dispos.,* 1:267.
43. Levin, W., Jacobson, M., Sernatinger, E., and Kunzman, R. (1973): *Drug Metab. Dispos.,* 1:275.
44. De Matteis, F., and Sparks, R. G. (1973): *FEBS Lett.,* 29:141.
45. Levin, W., Lu, A. Y. H., Jacobson, M., Kunzman, R., Poyer, J. L., and McCay, P. B. (1973): *Arch. Biochem. Biophys.,* 158:842.
46. Schachter, B. A., Marver, H. S., and Meyer, U. A. (1973): *Drug Metab. Dispos.,* 1:286.
47. Siekevitz, P. (1973): *J. Supramol. Struct.,* 1:471.
48. Shinozuka, H., Reid, I. M., Shull, K. H., Liang, H., and Farber, E. (1970): *Lab. Invest.,* 23:253.
49. Emmelot, P., and Benedetti, G. L. (1960): *J. Biophys. Biochem. Cytol.,* 7:393.
50. Krishan, A., Hsu, D., and Hutchins, P. (1968): *J. Cell Biol.,* 39:211.
51. Hwang, K. M., Yang, L. C., Carrico, C. K., Schulz, R. A., Shenkman, J. B., and Sartorelli, A. C. (1974): *J. Cell Biol.,* 62:20.
52. Bergelson, L. O., Dyatlovitskaya, E. V., Torkhovskaya, T. I., Sorokina, I. B., and Gorgova, N. P. (1970): *Biochim. Biophys. Acta,* 210:287.
53. Feo, F., Canuto, R. A., Bertone, G., Garcea, R., and Pani, P. (1973): *FEBS Lett.,* 33:229.
54. Chiarugi, V. P. (1972): *Cancer Res.,* 32:2707.
55. Sugimura, T., Ikeda, K., Hirota, K., Hozumi, M., and Morris, H. P. (1966): *Cancer Res.,* 26:1711.
56. Miyake, Y., Gaylor, J. L., and Morris, H. P. (1974): *J. Biol. Chem.,* 249:1980.
57. Jondorf, W. R., Maickel, R. P., and Brodie, B. B. (1958): *Biochem. Pharmacol.,* 1:352.
58. Dallner, G., Siekevitz, P., and Palade, G. E. (1966): *J. Cell Biol.,* 30:73, 97.
59. Basu, T. K., Dickerson, J. W. T., and Parke, D. V. W. (1971): *Biochem. J.,* 124:19.
60. Welch, R. M., Gommi, B., Alvares, A. P., and Conney, A. H. (1972): *Cancer Res.,* 32:973.
61. Hawcroft, D. M., and Martin, P. A. (1974): *Mech. Ageing Dev.,* 3:121.
62. Boland, R., Martonosi, A., and Tillack, T. W. (1974): *J. Biol. Chem.,* 249:612.
63. Tillack, T. W., Boland, R., and Martonosi, A. (1974): *J. Biol. Chem.,* 249:624.
64. Guidotti, C. (1972): *Annu. Rev. Biochem.,* 41:731.
65. Oseroff, A. R., Rollins, P. W., and Burger, M. M. (1972): *Annu. Rev. Biochem.,* 42:647.
66. Juliano, R. L. (1973): *Biochim. Biophys. Acta,* 300:341.
67. Reynolds, J. A. (1973): *Fed. Proc.,* 32:2034.
68. Steck, T. L. (1974): *J. Cell. Biol.,* 62:1.
69. Singer, S. J., (1974): *Annu. Rev. Biochem.,* 43:805.

Membranes and Disease, edited by L. Bolis, J. F. Hoffman, and A. Leaf. Raven Press, New York, © 1976.

The Interaction of Sulfhydryl Reagents with Microsomal Glucose 6-Phosphatase

Rebecca C. Garland, Carl F. Cori, and Hsia-fei Wang Chang

The Enzyme Research Laboratory, Massachusetts General Hospital, and the Department of Biological Chemistry, Harvard University Medical School, Boston, Massachusetts 02114

Iodoacetamide, *N*-ethylmaleimide, *p*-hydroxymercuribenzoate (p-MB), and $HgCl_2$ were tested as inhibitors of glucose 6-phosphatase. Iodoacetamide had no effect up to a concentration of 2 mM. *N*-ethylmaleimide inhibited only crude microsomal preparations not previously exposed to detergents. The purified M_2 fraction (specific activity 2–4 μmoles P_i/min-mg protein, 30°C) was not inhibited and ^{14}C-labeled *N*-ethylmaleimide was not bound to the protein.

p-MB inhibited all types of preparations, and, in contrast to *N*-ethylmaleimide, the inhibition was not counteracted by detergents. The glucose 6-phosphate hydrolysis with M_2 preparations was inhibited 50% by 5×10^{-5} M p-MB, and the inhibition was completely reversible by dithiothreitol, except when the enzyme was preincubated with p-MB in the absence of substrate. In the latter case p-MB greatly accelerated the temperature-dependent inactivation of glucose 6-phosphatase. Binding studies showed that about 3 μmoles ^{14}C-p-MB were incorporated into 100 mg of M_2 protein over a wide range of concentrations. The lowest concentration of p-MB which caused incorporation was not inhibitory to the enzyme, whereas the highest concentration, which caused 80% inhibition, showed no higher incorporation. Thus, there was no direct correlation between the p-MB that was irreversibly bound to the M_2 protein and enzyme activity.

Kinetically, the p-MB inhibition of the enzyme corresponded to that produced by a reversible inhibitor and this was confirmed by dilution experiments. Several compounds, including some amino acids, antagonized the inhibition by p-MB. The order of effectiveness was EDTA > barbital > tryptophan > histidine > lysine > other amino acids. Double reciprocal plots showed that the K_m for glucose 6-phosphate was increased and the V_{max} decreased in the presence of p-MB. $HgCl_2$ was a more effective inhibitor than p-MB with a K_i of 6×10^{-6} M.

It is concluded that a reaction of p-MB or *N*-ethylmaleimide with sulfhydryls of the enzyme does not play a part in the inhibition. It is suggested that p-MB may interact with one or more amino acid side chains in such a way that enzyme conformation is altered. This is at variance with the conclusion reached by Colilla and Nordlie (1).

ACKNOWLEDGMENT

This work was supported by U.S. Public Health Service Grant AM11448–08 and General Research Support Grant RR05486–12.

REFERENCE

1. Colilla, W., and Nordlie, R. C., (1973): *Biochim. Biophys. Acta,* 309:328.

Membranes and Disease, edited by L. Bolis, J. F. Hoffman, and A. Leaf. Raven Press, New York, © 1976.

Selective Labeling of Normal and Transformed Cell Surface by Trinitrobenzene Sulfonate

Paolo M. Comoglio,* Guido Tarone,** and Marilena Bertini†

**Department of Histology and Embryology, University of Trieste, School of Medicine, Trieste, Italy; **Department of Human Anatomy and †Department of Histology and Embryology, University of Torino, School of Medicine, 10126 Torino, Italy*

Neoplastic transformation of a cell is accompanied by one or more of the following changes: loss of control over growth, loss of the ability to respond correctly to outside stimuli, and differences in surface antigenic composition. In each case, alterations of biological properties mediated by the plasma membrane are responsible. In recent years, therefore, attention has been largely directed to changes in membrane structures accompanying neoplastic transformation. Yet it is becoming increasingly clear that the membrane is the site of very delicate metabolic regulation functions. Such changes would thus be limited; otherwise they would be lethal for the cell. Their study requires suitably sensitive methods of analysis. We have recently developed a technique based on trinitrobenzene sulfonate (TNBS), a small probe that selectively labels molecules exposed on the outer surface of the plasma membrane. Its polarity and net charge (Fig. 1) prevent it from crossing the membrane (1–6). Moreover, the sulfonic reactive group gives firm covalent bonds with the nucleophile groups, mostly free amino groups, exposed on the membrane surface. The reaction takes place in physiological conditions of temperature, pH, and solvent. TNBS has two other advantages. First, being yellow, it is readily determined by light absorption at 348 nm. Second, the bound trinitrophenyl group (TNP) is a very good hapten. This allows the use of specific antibodies directed against the probe for purification of the labeled membrane molecules by affinity chromatography.

CELL SURFACE LABELING BY TNBS

Low passage hamster BHK fibroblasts, supplied by Dr. L. Warren, either control (C13) or transformed by the B4 strain of Rous sarcoma virus (B4), were grown to confluency in Eagle medium supplemented with 10% calf serum. After being carefully washed with Earle's solution, monolayers were reacted at 37°C for 10 min in Earle pH 7.5 with 5 mM TNBS twice recrystallized from 1 M HCl. The reaction was stopped by repeated washings with

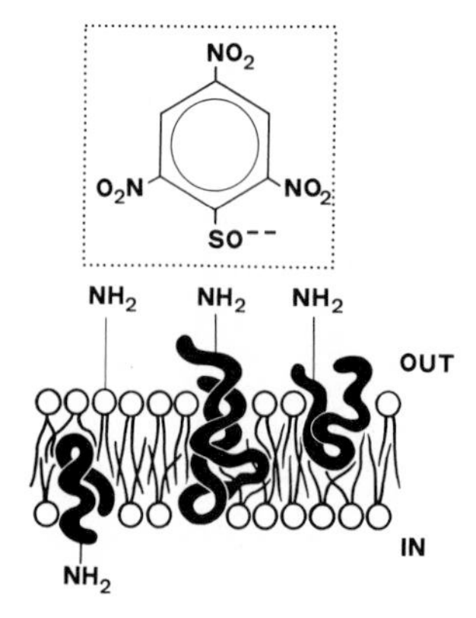

FIG. 1. A model of the plasma membrane showing proteins and phospholipids which expose free amino groups. TNBS binds the amino groups exposed on the outer surface of intact cells.

iced 750 mM glycine, diluted 1:1 (v/v) with Earle's solution. Trypan-blue exclusion tests performed before and after the reaction showed that, under these conditions, cell viability was not affected.

This procedure bound to cells a sufficient number of TNP groups for direct visualization by sandwich immunofluorescence. For this purpose, TNP-labeled cell monolayers either living or fixed by 2×10^{-2} M glutaraldehyde, were incubated for 1 hr with saturating amounts of purified anti TNP rabbit antibodies at 0°C to prevent the antibody mediated cell surface redistribution of TNP-labeled molecules found on living cells. After being carefully washed with medium supplemented with IgG (1 mg/ml) purified from nonimmune sera, monolayers were stained with the fluorescein-isothiocyanate-labeled IgG fraction of a goat antirabbit L chain antiserum diluted (2 mg/ml) with complete culture medium. Controls were performed using cells not labeled by TNBS or rabbit IgG purified from non immune sera, or by inhibiting anti-TNP antibodies from binding by excess TNP-glycin. In both control and transformed fibroblasts the specific fluorescence was observed only at the cell surface (Fig. 2). Prefixation of cells with glutaraldehyde, which makes the plasma membrane fully permeable to fluorescent antibodies, did not alter the staining pattern.

Mild trypsin treatment of monolayers (20 U/ml of crystalline trypsin for 10 min at 37°C) removed as much as 70–80% of the total amount of TNP bound without damage to the cells. TNBS-labeled fibroblasts were then fractionated to investigate the subcellular distribution of TNP groups on the cell after trypsinization. Cells were disrupted by Dounce homogenization in TKM buffer (Tris-HCl pH 7.5 50 mM; KCl 25 mM; $MgCl_2$ 5 mM) 0.25 M sucrose and fractionated in a discontinuous (0.25–2.3 M) sucrose gradient, as described in detail elsewhere (1). Four fractions were recovered, consisting of purified nuclei, mitochondria contaminated with smooth and rough endoplasmic reticulum (ER), plasma membranes contaminated with ER,

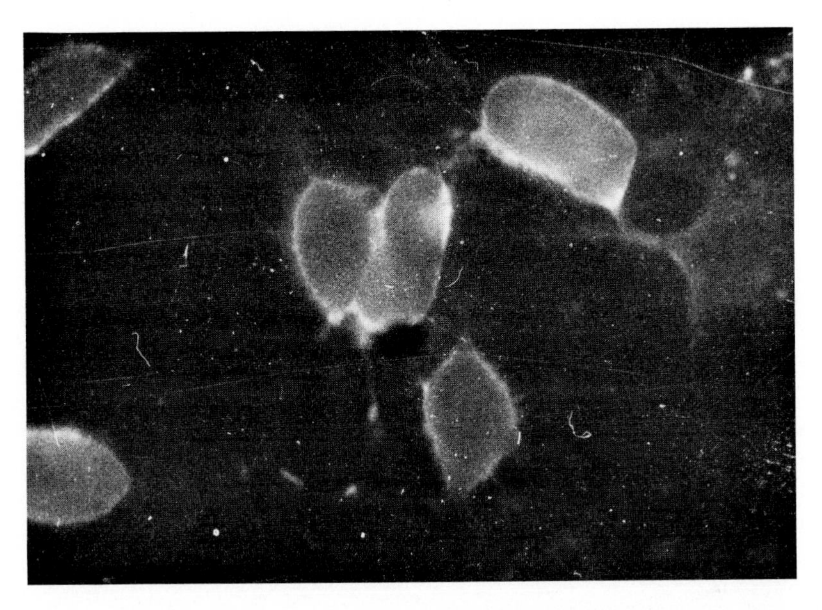

FIG. 2. Visualization of TNP groups bound to BHK fibroblasts (B4). TNP-labeled glutaraldehyde-fixed cells were incubated with purified rabbit anti-TNP antibodies and stained by fluorescent goat anti-rabbit immunoglobulins. The specific fluorescence is exclusively located at the cell surface. [From Comoglio et al. (1).]

and the 150,000 × *g* nonsedimentable cytoplasm fraction. Only the fraction containing plasma membranes showed a significant amount of TNP. This amount added to that released by trypsin from the cell surface accounted for more than 95% of the total probe recovered after fractionation (Table 1). These data, with those obtained from the immunofluorescence experiments, showed that TNBS was bound selectively to molecules exposed on the outer surface of BHK fibroblasts, as previously reported for erythrocytes by us and others (2–5) as well as for 3T3 mouse fibroblasts (6).

TABLE 1. *Subcellular distribution of TNP groups after selective surface labeling with TNBS*

Cell fraction	mmoles × 10^{-12} TNP/cell (C13)	% of total
Trypsinate	2.2	81
1 Plasma membranes + ER	0.4	15
2 (Mitochondria + ER)	0.1	<4
3 (Soluble cytoplasm)	0	0
4 (Nuclei)	0	0

ER, endoplasmic reticulum.

From 70 to 80% of the TNP groups bound to the cell surface after trypsinization were on membrane proteins, the remainder on lipids. This was found by extracting lipids from TNBS-labeled cells by a five-step treatment with chloroform-methanol 1:1 (v/v) according to Weinstein et al. (7) and by measuring the amount of TNP bound to the lipid fraction by light absorption at 337 nm, the absorbtion peak of TNP-phosphatidylethanolamine (4).

PURIFICATION OF PROTEINS EXPOSED ON THE OUTER SURFACE OF THE PLASMA MEMBRANE BY AFFINITY CHROMATOGRAPHY

TNP-labeled surface proteins were purified from unlabeled membrane components by chromatography through columns of insoluble antibodies. Labeled control and transformed cells, carefully rinsed with serum-free medium, were removed from culture flasks with a rubber-tipped spatula. "Crude" membrane fractions were prepared by Dounce homogenization and centrifugation on discontinuous sucrose gradients as described in detail elsewhere (8).

Membranes were solubilized with 1% sodium dodecyl sulfate (SDS), 1% β-mercaptoethanol, and 50 mM ditiothreitol, alkylated by 0.1 M iodoacetic acid for 30 min at 0°C, and ultracentrifuged at $100{,}000 \times g$ for 1 hr. After dialysis against 0.05 M Tris-HCl pH 8.0, 0.05% sodium desoxycholate membrane proteins were labeled by radioactive iodine by the cloramine-T method (9). Iodine-125 was used for proteins solubilized from transformed cells and ^{131}I for those from control cells. Labeled samples were then mixed and, from this point on, processed together.

TNP proteins were purified by affinity chromatography, using purified rabbit antidinitrophenyl (DNP) antibodies covalently linked to the resin Sepharose 4B (Pharmacia) by the cyanogen bromide technique (10). Anti-DNP antibodies cross-react with TNP groups and were preferred to anti-TNP antibodies, as the high affinity of rabbit antibodies for their specific nitrophenylated ligands prevents elution of TNP-protein except under very strong denaturating conditions (2). Each immunoadsorbent column was washed immediately before use with 0.1 M acetic acid and equilibrated with 0.05 M Tris-HCl pH 8.0, 0.25 M NaCl, 0.05% sodium desoxycholate. Solubilized membrane samples were then slowly passed through the column, followed by extensive washing with detergent containing buffer and 14 mM *N*-CBZ-glycine. TNP-labeled molecules were eluted either with 10 mM DNP-glycine (Fig. 3) or by heating Sepharose beads at 100°C for 2 min in SDS-β-mercaptoethanol. TNP-labeled cell surface proteins purified from control and transformed cells were coelectrophoresed in SDS-acrylamide disc gels. The pattern displayed by C13 cells was very similar to that displayed by B4 cells. Only quantitative differences restricted to a few peaks were observed (Fig. 4).

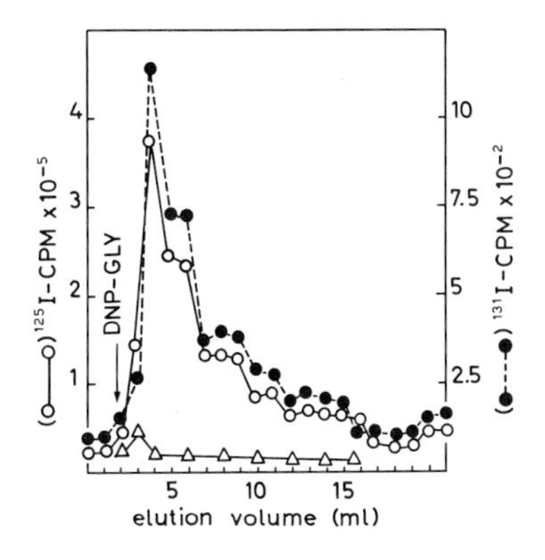

FIG. 3. Purification of TNP-labeled surface membrane proteins by immunoadsorption. SDS-solubilized membrane samples were passed through a 1 × 2 cm column of insoluble anti-DNP antibodies presaturated with unlabeled solubilized hepatocyte membranes. After washing with detergent containing buffer, TNP proteins were eluted by 10 mM DNP-glycine (*arrow*). (●) Proteins solubilized from TNP-labeled C13 cells and from (○) TNP-labeled B4 cells. The elution pattern obtained under the same conditions with proteins solubilized from cells not labeled with TNBS is superimposed (△).

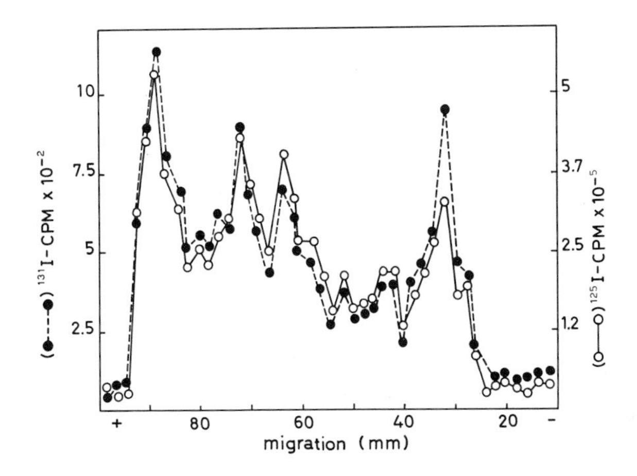

FIG. 4. Electrophoretic profile of TNP-labeled membrane proteins purified by affinity chromatography. 12% acrylamide gels, with a 3% upper spacer, were prepared in SDS according to Maizel (26). Samples were reduced, alkylated as described in the text and heated in 5% SDS–10% β-mercaptoethanol for 3 min at 100°C. The gels were frozen after electrophoresis and cut into 1-mm slices; the amount of ^{125}I and ^{131}I in each slice was then measured by simultaneous counting in a two-channel gamma counter. (●) Control cells; (○) transformed cells.

EXPOSURE OF TNBS BINDING GROUPS ON CONTROL AND TRANSFORMED CELLS

Next, the number of TNBS binding groups exposed on the outer surface of C13 and of B4 cells was measured. Cell monolayers were labeled with 5 mM TNBS for 10 min and detached from glass by mild trypsinization. The number of TNP groups bound was determined by measuring the light absorption at 348 nm of 5 × 10^6 cells dissolved by treatment with 10% SDS, 2 M β-mercaptoethanol at 100°C for 3 min. Blanks were prepared by dis-

TABLE 2. *TNP groups bound to the surface of control and transformed cells*

Cell	mmoles × 10^{-12}/cell		
	Trypsinate	Cell[a]	Total
BHK-C13 (control)	2.3 (± 0.6)	0.8 (± 0.2)	3.3 (± 0.3)
BHK-C13/B4 (transformed)	9.7 (± 1.5)	3.9 (± 1.5)	14.2 (± 3.0)

[a] Number of TNP groups left on the cell after trypsinization.

solving an identical number of unlabeled cells in the same way. The number of TNP groups removed by trypsin from the cell surface was calculated from the light absorption at the same wavelength of the lyophilized trypsin supernatants dissolved in SDS-mercaptoethanol. The molar extinction coefficient of TNP-lysine ($E_M 1^{\text{cm}} = 15{,}400$) was used in calculations. Measurement of TNP groups bound to the lipid fraction was performed as described previously. As can be seen in Table 2, transformed cells bound roughly five times more TNP groups than control cells. Measurement of the surface area by the Follet and Goldman method (11) showed that spread transformed cells were an average of 1.6 times larger than spread control cells. No differences were found between round (mitotic) cells. This means that transformed cells still exposed at least 3.3 times more TNBS-binding groups per surface area unit.

The kinetic curve of TNBS binding reaction showed greater velocity in the case of transformed cells, but no appreciable difference in the velocity constant: the calculated K value was 1×10^{-2}/sec for both cell types (Fig. 5). Therefore, it appears that only quantitative differences exist between TNBS binding radicals exposed on the surface of control and transformed cells. As previously shown (1,2,6), these radicals are predominantly free amino groups (12,13). Since no significant difference could be detected between the amount of TNP bound to lipid fractions from control and trans-

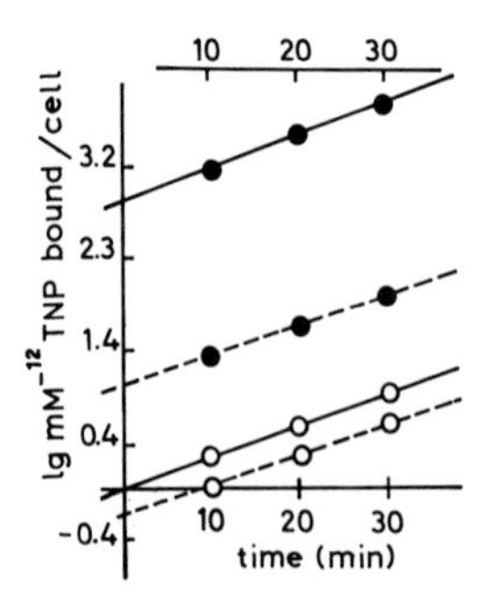

FIG. 5. Kinetic curves of the reaction between TNBS and its binding groups exposed on the outer surface of control (○) or transformed (●) BHK fibroblasts. Solid curves refer to TNP groups released from cells by trypsin; dotted curves to TNP groups left on cells after trypsinization.

formed cells it was concluded that the increase in TNBS reactive radicals on the transformed plasma membrane surface was almost solely dependent on protein differences.

SUMMARY AND CONCLUSIONS

TNBS selectively labeled the amino groups of proteins and lipids exposed on the outer surface of the plasma membrane. The method has the advantage of being preparative as well as analytic, as the labeled membrane molecules can be fairly easily purified by affinity chromatography with insoluble specific antibodies directed against nitrophenylated ligands. When applied to the BHK system it showed that neoplastic transformation is accompanied by a fivefold increase in TNBS binding groups exposed on the outer cell surface. This increase cannot simply be attributed to variations in the surface area of transformed cells. It is not even shared proportionally by membrane proteins and lipids. No differences, in fact, were observed between the amount of TNP bound to phospholipids extracted from control and transformed cells. It can therefore be interpreted as the exposure of new protein amino groups.

Because there is no reason to suppose that the membranes of transformed cells have three to five times more proteins than those of normal cells (14), two explanations are open. First, transformation of BHK fibroblasts may be accompanied by the appearance of one or more new proteins that are particularly rich in amino groups, or other possible groups reacting with TNBS. The almost identical electrophoretic pattern for the surface proteins from C13 and B4 cells, however, is against this interpretation. Alternatively, transformation is more likely followed by a general rearrangement of the outer surface of the membrane leading to exposure of amino groups that are sterically inaccessible or nonreacting in the same proteins exposed on the surface of the normal cells. A rearrangement of this type can be understood when account is taken of the possible effect on the membrane structure of all the pleiomorphic changes that are a feature of the so-called "transformed phenotype": deletion of protein molecules (15,16), changes in glycoprotein carbohydrate moieties (17,18), altered glycolipid synthesis (19–21), decreased level of sulfated mucopolysaccharides (22,23), and a possible alteration in the fluidity of the lipid bilayer (24,25).

ACKNOWLEDGMENTS

This work was supported by funds from Italian National Research Council (CNR). The authors are indebted to Dr. G. Filogamo for encouragement and support.

REFERENCES

1. Comoglio, P. M., Tarone, G., Prat, M., and Bertini, M. (1975): *Exp. Cell Res.* (in press).
2. Tarone, G., Prat, M., Comoglio, P. M. (1973): *Biochim. Biophys. Acta,* 311:214.
3. Bonsall, R. W., and Hunt, S. (1971): *Biochim. Biophys. Acta,* 249:281.
4. Gordesky, S. E., and Marinetti, G. V. (1973): *Biochem. Biophys. Res. Commun.,* 50:1027.
5. Gordesky, S. E., Marinetti, G. V., and Love, R. (1975): *J. Membr. Biol.,* 20:111.
6. Vidal, R., Tarone, G., Peroni, F., and Comoglio, P. M. (1974): *FEBS Lett.,* 47:107.
7. Weinsten, P. H., Marsh, J., Glick, M. C., and Warren L. (1969): *J. Biol. Chem.,* 244:4103.
8. Prat, M., Tarone, G., and Comoglio, P. M. (1975): *Immunochemistry,* 12:9.
9. Talmage, D., and Claman, H. (1967): In: *Methods in Immunology and Immunochemistry,* Vol. 1, p. 391, edited by C. Williams and M. Chase. Academic Press, New York.
10. Axen, R., Porath, J., and Ernback, S. (1967): *Nature,* 214:1302.
11. Follet, E. A., and Goldman, R. D. (1970): *Exp. Cell Res.,* 59:124.
12. Freedman, R. B., and Radda, G. R. (1968): *Biochem. J.,* 108:383.
13. Eisen, H., Gray, W., Little, R., and Simms, E. (1967): In: *Methods in Immunology and Immunochemistry,* vol. 1, p. 351, edited by C. Williams and M. Chase. Academic Press, New York.
14. Scott, R., and Carter, R. (1971): *Nature* [*New Biol.*], 233:219.
15. Ghamber, C., Kiehn, D., Hakomori, S. (1974): *Nature* [*New Biol.*], 248:413.
16. Hogg, W. (1974): *Proc. Natl. Acad. Sci. USA,* 71:489.
17. Warren, L., Fuhrer, J., and Buck, C. A. (1972): *Proc. Natl. Acad. Sci. USA,* 69:1838.
18. Van Beck, W., Smets, L., and Emmelot, L. (1973): *Cancer Res.,* 33:2913.
19. Siddiqui, B., and Hakomori, S. (1970): *Cancer Res.,* 30:2930.
20. Sakyama, H., Gross, H., and Robbins, P. W. (1972): *Proc. Natl. Acad. Sci. USA,* 69:872.
21. Steiner, S., Breiman, P. J., and Melnick, J. L. (1973): *Nature,* 245:19.
22. Goggins, J. F., Johnson, G. S., and Pastan, I. (1972): *J. Biol. Chem.,* 247:3759.
23. Chiarugi, V., Vannucci, S., and Urbano, P., *Biochem. Biophys. Acta,* 345:283.
24. Nicolson, G. (1971): *Nature* [*New Biol.*], 233:244.
25. Nicolson, G. (1973): *Nature* [*New Biol.*], 243:218.
26. Maizel, J. V. (1971): *Methods Virol.* 5:315.

Membranes and Disease, edited by L. Bolis, J. F. Hoffman, and A. Leaf. Raven Press, New York, © 1976.

The Role of Cell-Surface Sialic Acid in Lectin-Induced Agglutination of Novikoff Hepatoma Cells

G. Neri,* M. C. Giuliano, D. C. Hixson, and E. F. Walborg, Jr.

Regina Elena Institute for Cancer Research Rome, Italy, and The Department of Biochemistry, The University of Texas System Cancer Center, M. D. Anderson Hospital and Tumor Institute, Houston, Texas 77025

Several lines of evidence indicate that the plasma membrane plays a central role in the expression of the neoplastic state (1,2). In particular, sialic acid, which is widely represented at the cell surface as a component of both glycoproteins and, to a lesser extent, glycolipids, seems to be directly implicated in a number of phenomena which distinguish normal from cancer cells. Warren and co-workers (3) have demonstrated the presence in the plasma membrane of virally transformed cells, of a class of fucose-containing sialoglycopeptides, probably related to the increased activity of a specific sialyltransferase (4). These particular glycopeptides are not only present in the plasma membrane, but can also be identified in all intracellular membranes tested (3). The phenomenon seems to be a general one, in the sense that these observations have been extended to *in vivo* tumors of both animals and man (5,6).

Another phenomenon of general observation is the inability of the tumor-bearing animal to reject the tumor, although there is ample evidence for the existence of tumor-associated antigens which elicit an immune response by the host. One mechanism whereby cancer cells may escape immunosurveillance is the masking of tumor-associated antigens by cell-surface glycoproteins (7), such as the mucin-like peripheral sialoglycoprotein, present on TA3 mouse mammary carcinoma cells, which masks the expression of hystocompatibility antigens and thereby permits transplantation in allogeneic hosts (8). On the other hand, incubation of different kinds of tumor cells with neuraminidase, a treatment which cleaves sialic acid from cell-surface glycoproteins, enhances tumor cell antigenicity by unmasking tumor-associated rejection antigens (9,10).

On the basis of these observations, it seemed important to investigate the role of sialic acid in lectin-induced cytoagglutination. Burger (11) and

**Present address:* The Catholic University School of Medicine, Department of Human Genetics, 00168 Rome, Italy.

Burger and Goldberg (12) reported that neuraminidase treatment of L1210 mouse leukemia cells resulted in a marked decrease in their agglutinability by wheat germ agglutinin (WGA). Nicolson et al. (13) observed that neuraminidase treatment of polyoma transformed hamster fibroblasts reduced their agglutinability by WGA and concanavalin A (Con A), whereas agglutinability by *Ricinus communis* agglutinin I (RCA_I) was slightly enhanced.

The present chapter describes the effect of neuraminidase treatment on the agglutination of Novikoff hepatoma cells by Con A, WGA, RCA_I, *Ricinus communis* agglutinin II (RCA_{II}), and soybean agglutinin (SBA).

EXPERIMENTAL METHODS AND RESULTS

Neuraminidase digestion (1 hr at 37°C in phosphate-buffered saline containing 0.22 mM $CaCl_2$, at pH 7.0) resulted in the release of 0.29 ± 0.06 (SD) μmoles sialic acid/ml packed cells. The same amount of sialic acid was released when digestion was performed in 0.05 M acetate buffer containing 0.9% NaCl and 0.1% $CaCl_2$, at pH 5.5, the optimum for neura-

TABLE 1. *Effect of neuraminidase digestion on the agglutination of Novikoff cells by Con A and WGA*

	Lectin concentration for cytoagglutination (μg/ml)[a]			
	Con A[b]		WGA[c]	
Conditions	Threshold[d]	½ Maximal[d]	Threshold[d]	½ Maximal[d]
Intact	1.5	7.0	0.42	3.4
Incubated without neuraminidase	1.5	6.5	0.42	3.4
Incubated with neuraminidase[e]	1.4	5.6	0.44	3.4

[a] Differences between these and previously reported values (16,25) are due to several factors: (i) differences in the cytoagglutination assay, particularly avoidance of drying the lectins in the flocculation plates prior to addition of the cell suspension; (ii) differences in lectin preparations, e.g., the WGA utilized in these studies was more highly purified, and (iii) observer variation in scoring the degree of agglutination.

[b] Con A, a twice-crystallized product dissolved in saturated NaCl, was obtained from Miles-Yeda, Rehovot, Israel.

[c] WGA was prepared from wheat germ lipase (Pentex Biochemicals, Inc., Kankakee, Ill.) by the method of Nagata and Burger (26). The active fraction resolved on carboxymethyl-cellulose was utilized.

[d] The degree of agglutination (ordinate), scored on a scale from 0 to +4, was plotted vs the lectin concentration (log scale on the abscissa). See Neri et al. (16). The intersection of the agglutination line with the abscissa represents the lectin concentration at threshold agglutination. The concentration of lectin to achieve 50% maximal agglutination is defined as the value of the abscissa at +2 agglutination.

[e] Digestion performed at pH 7.0 using 12.5 units of *Vibrio cholerae* neuraminidase (Behrinwerke, A. G. Marburg/Lahn, Germany) per milliliter of a cell suspension containing 15 × 10^6 cells/ml.

TABLE 2. *Effect of neuraminidase digestion on the agglutination of Novikoff cells by RCA_I, RCA_{II}, and SBA*

Conditions	Presence of saccharide	Lectin concentration for cytoagglutination (μg/ml)					
		RCA_I[a]		RCA_{II}[a]		SBA[b]	
		Threshold[c]	½ Maximal[c]	Threshold[c]	½ Maximal[c]	Threshold[c]	½ Maximal[c]
Intact	None	0.45	1.9	5.6	36	>120	
Intact	1 mM Lactose	0.85	20				
Intact	0.5 mM Lactose			14	>140		
Incubated without neuraminidase	None	0.55	2.0	9.0	46	>120	
Incubated with neuraminidase[d]	None	0.29	1.1	0.73	2.9	4.8	19
Incubated with neuraminidase[d]	5 mM GalNac			6.5	21	>120	

[a] RCA_I and RCA_{II} were prepared according to the methods of Tomita et al. (27) and Nicolson and Blaustein (28).
[b] SBA was prepared essentially according to the methods of Liener (29) and Lis et al. (24).
[c] See Table 1, footnote *d*.
[d] See Table 1, footnote *e*.

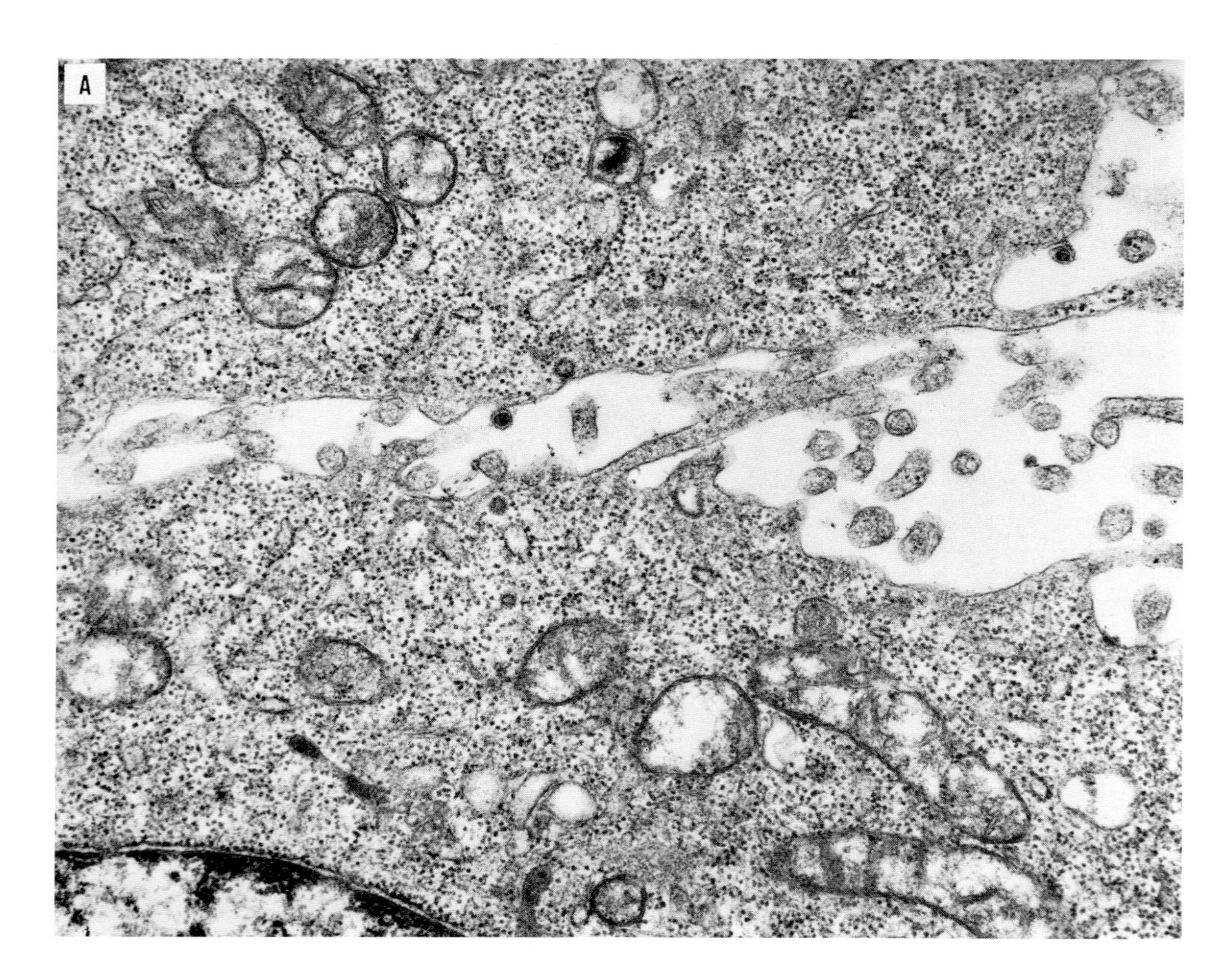
A

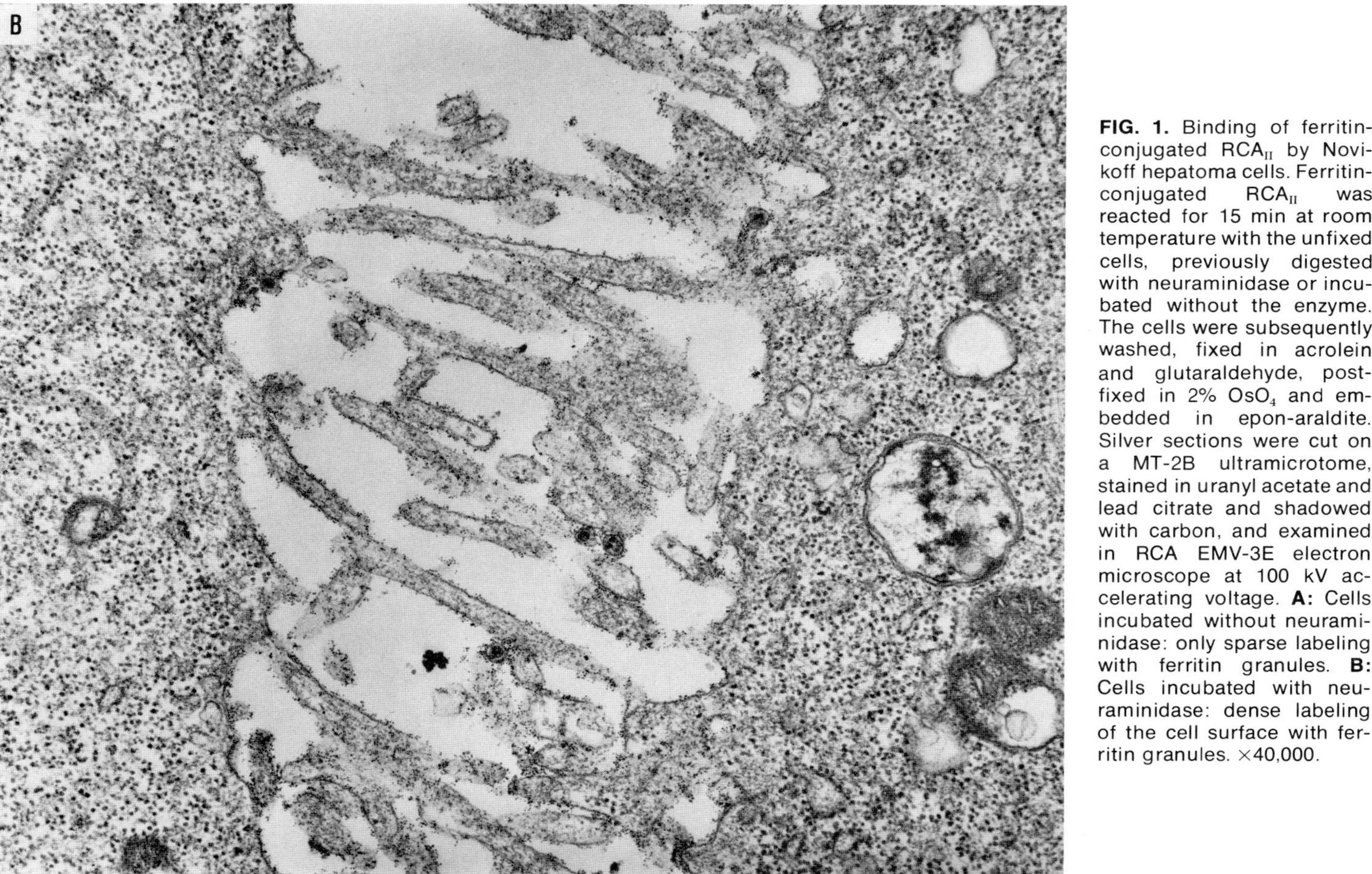

FIG. 1. Binding of ferritin-conjugated RCA_{II} by Novikoff hepatoma cells. Ferritin-conjugated RCA_{II} was reacted for 15 min at room temperature with the unfixed cells, previously digested with neuraminidase or incubated without the enzyme. The cells were subsequently washed, fixed in acrolein and glutaraldehyde, postfixed in 2% OsO_4 and embedded in epon-araldite. Silver sections were cut on a MT-2B ultramicrotome, stained in uranyl acetate and lead citrate and shadowed with carbon, and examined in RCA EMV-3E electron microscope at 100 kV accelerating voltage. **A:** Cells incubated without neuraminidase: only sparse labeling with ferritin granules. **B:** Cells incubated with neuraminidase: dense labeling of the cell surface with ferritin granules. ×40,000.

TABLE 3. *Effect of papain digestion on the agglutination of Novikoff cells by RCA_I, RCA_{II}, and SBA*

Conditions	Lectin concentration for cytoagglutination (μg/ml)					
	RCA_I[a]		RCA_{II}[a]		SBA[b]	
	Threshold[c]	1/2 Maximal[c]	Threshold[c]	1/2 Maximal[c]	Threshold[c]	1/2 Maximal[c]
Intact	0.45	1.9	5.6	36	>120	
Incubated without papain	0.65	2.0	5.5	26	>150	
Incubated with papain[d]	0.14	0.5	3.7	11	>150	

[a] See Table 2, footnote *a*.
[b] See Table 2, footnote *b*.
[c] See Table 1, footnote *d*.
[d] Digestion was performed as described (25), using 3 units of twice-crystallized papain (Worthington Biochemical Co., Freehold, N.J.) per milliliter of a cell suspension containing 15×10^6 cells/ml.

minidase action (14). Sialic acid was determined by the method of Aminoff (15) using *N*-acetylneuraminic acid as standard.

The effect of neuraminidase digestion on lectin-induced agglutination of Novikoff cells was investigated. Removal of the neuraminidase-labile sialic acid had no effect on the agglutinability of Novikoff cells by Con A or WGA, as reported in Table 1. A different situation was observed when other lectins were employed. The effect of neuraminidase digestion on the agglutination of Novikoff cells by RCA_I, RCA_{II}, and SBA is shown in Table 2. The intact cells are agglutinated by RCA_I and RCA_{II}, but are not agglutinated by SBA, at concentrations as high as 120 μg/ml. Agglutination, where found, was saccharide-specific, as shown by inhibition by low concentration of the specific sugar inhibitors. Treatment of the cells with neuraminidase at pH 7.0 resulted in a twofold increase in agglutination by RCA_I, while agglutination by RCA_{II} increased 12-fold. Similarly, the cells became strongly agglutinable by SBA.

This drastic change in RCA_{II}-induced agglutination was correlated with an increase in the binding of ferritin-conjugated RCA_{II} (Fig. 1). Conjugation was accomplished by mixing recrystallized horse spleen ferritin (Pentex Biochemicals, Inc., Kankakee, Ill.) and the lectin in presence of 0.2 M galactose at pH 6.8, and slowly adding 0.25% glutaraldehyde to a final concentration of 0.025%. Conjugation was allowed to proceed for 6 hr at room temperature, and the reaction was terminated by addition of lysine at a final concentration of 2.0 mg/ml. The conjugated lectin was purified from unconjugated material on a 20–35% continuous sucrose gradient. Neuraminidase-treated cells were much more densely labeled by the ferritin RCA_{II} conjugate, as compared to intact cells. Binding was reduced by approximately 50% when labeling was performed in the presence of 0.2 M *N*-acetyl-D-galactosamine, and totally inhibited when performed in presence of 0.2 M lactose.

The effect of neuraminidase treatment on lectin-induced cytoagglutination was compared with the effect caused by papain treatment. Digestion of Novikoff cells with papain released an average of 0.085 μmoles glycopeptide-bound sialic acid/ml packed cells. This value represents about one-third of the neuraminidase-labile sialic acid. It was shown previously that agglutination of Novikoff cells by Con A or WGA is slightly, but consistently, enhanced by papain treatment (16). The effect of papain digestion on RCA_I-, RCA_{II}-, and SBA-induced cytoagglutination is shown in Table 3. There is a fourfold increase in agglutination by RCA_I, a twofold increase in agglutination by RCA_{II}, while agglutination by SBA seems to be unaffected in the sense that the cells remain nonagglutinable, even after treatment with papain.

DISCUSSION AND CONCLUSIONS

Accumulating evidence indicates that cell-surface sialic acid modulates the antigenicity and lectin-induced agglutination of tumor cells.

The presence of sialic acid at the cell periphery seems to be responsible, at least in some cases, for the decreased antigenicity of the tumor cell, thus interfering with the immune reactions of the host. In fact, injection of neuraminidase-treated tumor cells into susceptible acceptors protects the host from the subsequent challenge by the same tumor (9,10). The effect of neuraminidase may be explained in two ways: the exposure of new saccharide determinants at the nonreducing termini of cell-surface oligosaccharide moieties (17), or steric alteration of the plasma membrane-associated macromolecules, induced by removal of a substantial portion of the cell-surface anionic groups (10). In addition, tumor cell-surface sialoglycoproteins bear receptors for Con A, RCA_I, and WGA (18), lectins which have been utilized as molecular markers for neoplastic transformation (19,20).

In this investigation five lectins possessing different saccharide specificities were employed to monitor cell-surface modifications which accompany neuraminidase treatment of rat hepatoma cells. The lectins employed were Con A, specific for α-D-gluco- or α-D-mannopyranosyl residues (21); WGA, specific for 2-acetamido-2-deoxy-D-glucose residues (22); RCA_I, specific for D-galactopyranosyl residues (23); RCA_{II} specific for D-galacto- and 2-acetamido-2-deoxy-D-galactopyranosyl residues (23); and SBA, specific for 2-acetamido-2-deoxy-D-galactopyranosyl residues (24). The most prominent effect of removal of sialic acid from the surface of Novikoff cells appeared to be a marked increase in agglutination by RCA_{II} and SBA. The observation that neuraminidase digestion produced only a twofold increase in agglutination of Novikoff cells by RCA_I, but a 12-fold increase in agglutination by RCA_{II}, suggested that neuraminidase had altered the expression of 2-acetamido-2-deoxy-D-galactose at the cell surface. This conclusion, based on the fact that both RCA_I and RCA_{II} possess specificity for D-galactosyl residues, whereas RCA_{II}, in addition, possesses specificity for 2-acetamido-2-deoxy-D-galactose, was confirmed by the observation that only neuraminidase-treated Novikoff cells were agglutinated by SBA, a lectin which demonstrates high specificity for 2-acetamido-2-deoxy-D-galactose (24).

The goal of these investigations is to contribute to the chemical characterization of the tumor cell periphery and hopefully aid in identifying those alterations related to the altered antigenicity, loss of growth control, and aberrant social behavior exhibited by malignant cells.

ACKNOWLEDGMENTS

This work was supported by research grants from the National Cancer Institute (CA 11710), The Robert A. Welch Foundation (G 354), The Paul and Mary Haas Foundation, The Mary Josephine Hamman Foundation, and General Research Support Grant 5-S01-RR05511 from the National Institutes of Health.

REFERENCES

1. Emmelot, P. (1973): *Eur. J. Cancer,* 9:319.
2. Abercrombie, M., and Ambrose, E. J. (1962): *Cancer Res.,* 22:525.
3. Warren, L., Fuhrer, J. R., and Buck, C. A. (1973): *Fed. Proc.,* 32:80.
4. Warren, L., Fuhrer, J. P., and Buck, C. A. (1972): *Proc. Natl. Acad. Sci. USA,* 63:1418.
5. Van Beek, W. P., Smets, L. A., and Emmelot, P. (1973): *Cancer Res.,* 33:2913.
6. Van Beek, W. P., Smets, L. A., and Emmelot, P. (1975): *Nature,* 253:457.
7. Currie, G. A., and Bagshawe, K. D. (1969): *Br. J. Cancer,* 23:141.
8. Sanford, B. H., Codington, J. F., Jeanloz, R. W., and Palmer, P. D. (1973): *J. Immunol.,* 110:1233.
9. Rios, A., and Simmons, R. L. (1973): *J. Natl. Cancer Inst.,* 51:637.
10. Simmons, R. L., Rios, A., Ray, P. K., and Lundgren, G. (1971): *J. Natl. Cancer Inst.,* 47:1087.
11. Burger, M. M. (1968): In: *Biological Properties of the Mammalian Surface Membrane,* edited by L. A. Manson, p. 78. Wistar Institute Press, Philadelphia, Pa.
12. Burger, M. M., and Goldberg, A. R. (1967): *Proc. Natl. Acad. Sci. USA,* 57:359.
13. Nicolson, G. L., Lacorbiere, M., and Eckhart, W. (1975): *Biochemistry,* 14:142.
14. Ada, G. L., French, E. L., and Lind, P. E. (1961): *J. Gen. Microbiol.,* 24:409.
15. Aminoff, D. (1961): *Biochem. J.,* 81:384.
16. Neri, G., Smith, D. F., Gilliam, E. B., and Walborg E. F., Jr. (1974): *Arch. Biochem. Biophys.,* 165:323.
17. Springer, G. F., Desai, P. R., and Adye, J. C. (1974): *Ann. NY Acad. Sci.,* 134:312.
18. Walborg E. F., Jr., Davis, E. M., Gilliam, E. B., Smith, D. F., and Neri, G. (1976): In: *Cellular Membranes and Tumor Cell Behavior,* edited by Russell W. Cumley, p. 337. Williams & Wilkins, Baltimore.
19. Burger, M. M. (1973): *Fed. Proc.,* 32:91.
20. Nicolson, G. L. (1974): *Int. Rev. Cytol.,* 39:89.
21. Poretz, R. D., and Goldstein, I. J. (1970): *Biochemistry,* 9:2890.
22. Allen, A. K., Neuberger, A., and Sharon, N. (1973): *Biochem. J.,* 131:155.
23. Nicolson, G. L., Blaustein, J., and Etzler, M. E. (1974): *Biochemistry,* 13:196.
24. Lis, H., Sela, B. A., Sachs, L., and Sharon, N. (1970): *Biochim. Biophys. Acta,* 211:582.
25. Wray, V. P., and Walborg, E. F., Jr. (1971): *Cancer Res.,* 31:2072.
26. Nagata, Y., and Burger, M. M. (1972): *J. Biol. Chem.,* 247:2248.
27. Tomita, M., Kurokawa, T. K., Onozaki, K., Osawa, T., and Ukita, T. (1974): *Ann. NY Acad. Sci.,* 134:312.
28. Nicolson, G. L., and Blaustein, J. (1972): *Biochim. Biophys. Acta,* 266:543.
29. Liener, I. E. (1953): *J. Nutr.,* 49:527.

Membranes and Disease, edited by L. Bolis, J. F. Hoffman, and A. Leaf. Raven Press, New York, © 1976.

Chemical Changes in Neoplastic Cell Membranes

L. Warren and C. A. Buck

Wistar Institute of Anatomy and Biology, Philadelphia, Pennsylvania 19104

Studies of normal and malignant cells have revealed numerous differences in their function and behavior. Analysis of many of these differences strongly suggests that cell membranes are involved (1–4). Indeed, it is possible that a critical qualitative change in a membrane protein may have to take place if a cell is to become malignant.

It is widely believed that one or a small number of qualitative changes take place to initiate malignancy (5)—a mutation (spontaneous, or induced by chemicals or radiation), or, what might be its functional equivalent, the usurption of some function of the cell gene by that of an oncogenic virus. The consequence of this discrete, qualitative alteration is a cascade of changes which would appear to be quantitative in nature. The qualitative change in the DNA may result in a corresponding change in RNA and in one or more proteins. From then on, the one change may lead to five, and each of these may lead to many more so that the afflicted cell could differ from the normal in many functions, almost in a random fashion.

The cascade of events could begin with the alteration of an RNA, such as a tRNA, involved in the synthesis of many proteins. Multiple consequences could arise from a mutant protein involved in nuclear function. The key protein might be a critical enzyme whose malfunctioning leads to a succession of changes. In this chapter we will discuss chemical changes in neoplastic cell membranes in the context of the above speculation. Perhaps the critical change takes place in one or more membrane proteins in one or more membrane systems of the cell (1–4). Many, if not most, of the important enzymes are associated with membranes, and their functioning may be sensitive to the state of their lipoprotein environment. If the membrane is altered, the characteristic of many enzymes or the stability of various membrane bound RNAs (4) may be slightly altered with multiple consequences, which, in turn, might have further secondary consequences.

It should be emphasized that if malignancy is caused by an altered gene function, the effects are minimal; otherwise the alteration would be lethal. We are confronted by a situation in which an initial pulse leads to a number of interrelated changes probably of a quantitative nature, which, however, permit the cell to prosper. Of the numerous changes manifested by the malignant cell, many are probably of no consequence, while others are crucial for the expression of malignancy. In other words, there may be a hierarchy of changes. The malignant cell has come to a new steady-state equilibrium consistent with viability, although it may have lost its specialized function. This cell is particularly important to study because it does not seem to interact properly with its neighbors and it persists in dividing. Obviously, these properties are of critical importance to the survival of the *host;* we are compelled to examine this tangled process.

What are some of the changes in function that suggest that membrane structure and composition are altered? There appears to be a decreased intercellular adhesiveness between malignant cells which could facilitate their spread (6). Although the true nature of this adhesiveness is not really known, it must certainly involve the surface membrane of the cell. It has been shown that malignant cells have lost the property of "contact inhibition of motion" (7). Whatever the precise nature of this phenomenon is the surface membrane of the cell must be involved. There is a rapidly growing literature on the appearance of membrane-bound antigens in malignant cells. These antigens represent a structural change in the cell surface (8). Malignant cells manifest altered transport characteristics of essential metabolites, mainly an increase in the number of transport sites in the surface membrane (9,10). Finally, differences between normal and malignant cells in their agglutinability by plant lectins is well known (11). This list of membrane-associated functional differences between normal and malignant cells could be extended.

The chemical differences in the membranes may be quantitative or qualitative in nature. The central, most frustrating problem in our understanding of malignancy is that observed differences between normal and malignant cells as yet cannot be evaluated with precision. Is a particular difference important or is it merely associative? Though quantitative differences (e.g., enzyme levels or relative amounts of a protein in equivalent membrane fractions) may or may not be important, it seems likely that a qualitative difference, such as a change in amino acid sequence of a membrane protein of a malignant cell, could be very significant in the mechanism of malignant cell behavior.

What follows is a brief review of the changes in the lipids, glycolipids, proteins, and glycoproteins of membranes when cells become malignant.

LIPIDS

In general, comparative data on the neutral and phospholipids of normal and malignant cells is not extensive (12). In recent years, however, there has been a quickening of interest because the nature of the lipids of a membrane can determine its fluidity (13), and this membrane characteristic has been invoked to explain the difference in agglutinability of normal and malignant cells by plant lectins. Most of the comparative analytic data suggest that there are few differences and that these are relatively small (14–16), although it is possible that small (average) differences if localized could be important. To date, no specific tumor lipids (or glycolipids) have been found (12). When chick embryo fibroblasts in culture are transformed by Rous sarcoma virus (wild type or temperature-sensitive mutants) there are few changes in the phospholipids, except for some increase in their 18 : 1 fatty acid (oleic) and a decrease in their 20 : 4 fatty acids (arachidonic) (14). The latter may be of special significance because of its involvement in the synthesis of prostaglandins. It is probable that such minimal changes are unrelated to altered membrane lipid fluidity and membrane plasticity. Some studies have shown a difference in these characteristics in normal and transformed cells (17). Other studies on the fluidity and inherent flexibility of membrane lipids of chick embryo fibroblasts before and after transformation show no differences (15,18), which would seem to eliminate lipid membrane fluidity as an explanation for the different agglutinating responses of these cells to lectins. On the other hand, it has been shown that leukemic cells of the human and the mouse contain significantly less unesterified cholesterol than their normal counterparts (19,20). As shown by the fluorescence polarization method, using fluorescent labeled concanavalin A (Con A), the microviscosity of the surface membrane of malignant cells is reduced considerably, i.e., membrane fluidity is increased. In these studies the drop in cholesterol content and the associated increase of membrane fluidity have been considered an important part of the malignant change (20). It may be of significance that leukemic patients are hypocholesterolemic (21,22).

Greater unsaturation of the fatty acids of membrane lipids could increase membrane fluidity, while sphingolipids tend to increase microviscosity. Obviously, much work remains to be done in this field.

GLYCOLIPIDS

Earlier work by Rapport and his co-workers (23) resulted in the isolation of lactosyl ceramide (cytolipin H) from human epidermoid carcinoma. This material functioned as a hapten reacting with antisera against several types of human tumor, as compared to antisera against normal tissue. Another lipid hapten, ceramide tetrahexoside (cytolipin R), was isolated from rat lymphosarcoma.

In recent years there has been a reawakening of interest in glycolipids in cancer (24,25), beginning with the work of Hakomori and Murakami (26), who compared a control cell in tissue culture, BHK_{21}/C_{13}, with a highly malignant cell line transformed by polyoma virus. They showed that the malignant cell contained more glycolipids with shorter carbohydrate chains (reduced sialic acid content) and less with longer chains. This led to the interesting generalization that malignant cells do not appear to complete the synthesis of the carbohydrate component of their glycolipids. This generalization is not without exceptions, but may account for the findings of Rapport et al. In some transformed cells, incomplete ganglioside synthesis has been explained in part by the reduction in activity of the glycosyl transferase responsible for the addition of *N*-acetylgalactosamine to hematoside (27–29). Other transformed cells have only a small capability of transferring sialic acid to lactosyl ceramide (30). It appears that different enzyme blocks are induced by various oncogenic viruses, chemical carcinogens, and X-ray (24,31). On the other hand, spontaneously transformed cells growing in culture have unaltered glycolipid patterns (32). The formation of some glycolipids is growth-dependent, forming solely or in greater amount at high cell density (plateau) (33,34). Transformed cells fail to synthesize these more complex glycolipids even when grown to saturation cell densities (33,34).

Recently, a family of six fucoglycolipids has been detected in rat and baboon cells. When these cells were transformed by an RNA oncogenic virus, the proportion of smaller fucolipids increased at the expense of the larger (35).

The significance of the glycolipid changes in cell transformation is unknown at the present time. Although they are discrete specific information-bearing molecules, it is difficult to relate malignant behavior to them because their normal functions are unknown. They are more exposed on the surface of transformed cells than control (36), but are probably present in all the membrane systems of the cell. Thus, a change in their biosynthesis or degradation could affect the function of many membrane systems.

PROTEINS AND GLYCOPROTEINS

As discussed previously, there is good reason to believe that there are quantitative or qualitative changes, or both, that take place in the membrane glycoproteins when the cell becomes malignant. The appearance of new surface antigens is indicative of change (8,37). More direct evidence has been obtained by a variety of workers who have specifically labeled the surfaces of normal and malignant cells and electrophoresed either isolated membranes or the entire cell on polyacrylamide gel. Membrane protein and glycoprotein bands were detected by autoradiography or staining. From these studies it has become evident that a number of differences exist in the macromolecular populations of control and transformed cells (38–49). Upon transformation, some bands appear or become more intense, whereas others decrease or disappear altogether. Variation from study to study may depend on the type of cell and the methodology employed as well as the state of growth of the cell. Since the identity and function of these proteins is unknown, it is particularly difficult to evaluate the changes upon transformation. However, the partial or total disappearance of a major component from the cell surface, a glycoprotein of MW 250,000 (250 K) has been observed by several workers (38,39,49,50). It is not mucopolysaccharide or collagen and can be readily removed from normal cells with trypsin (50). Although the removal of the 250 K glycoprotein by trypsin has been linked with stimulation of the cell to divide, it has recently been shown that bromelin or chymotrypsin eliminates the 250 K glycoprotein without mitogenic effect (55). On the other hand, the proteolytic enzyme thrombin is mitogenic, but does not cause disappearance of the 250 K glycoprotein (51). The presence of the 250 K glycoprotein is cell cycle-dependent since there is less of it in actively dividing cells or in normal cells blocked in mitosis (50). The suggestion has been made that it is reduced in malignant cells by endogenous proteases, which are known to be increased in these cells (52–54). Controlled proteolysis of the cell surface has been shown to stimulate cell division (55,56) and facilitate agglutination of control cells by lectins (11). These effects and others (increased rate of nutrient transport, reduced levels of cyclic AMP) induced by proteases suggest that increased surface proteolysis may be an important part of the transformation process. The large surface glycoprotein (250 K) is particularly susceptible to proteolysis. It may form a mesh spanning various macromolecules in the fluid surface membrane, keeping them apart. The removal of the mesh by added proteolytic enzymes or upon transformation would permit the membrane proteins and glycoproteins to move about and to form patches when exposed to lectins. The cells would then become agglutinable (11,50). It has also been postulated that this large glycoprotein may be important because it may be functionally linked to proteins within and underlying the plasma membrane (see ref. 19), or may possibly form a link between cells.

Our own particular work in this field has, to date, been primarily methodological. We wish to make as direct a comparison as possible between corresponding proteins of the same membrane systems of control and transformed cells. Thus, we have fractionated control BHK_{21}/C_{13} cells and their malignant counterpart C_{13}/B_4 transformed by Rous sarcoma virus. Surface membranes, endoplasmic reticulum, mitochondria, lysosomes, and nuclear membranes were dissolved with detergent solution, sodium dodecyl sulfate (SDS), containing 2,3-dimercaptoglycerol, and electrophoresed on polyacrylamide gel. Upon staining with Coomassie blue 30 to 40 sharp bands can be seen in each sample, which can be accurately compared when control and malignant membrane proteins are run side by side. We have found, in fact, numerous quantitative changes in the amounts of protein and glycoproteins in various membrane systems when cells become malignant. The significance of these is difficult to assess. In SDS-polyacrylamide gel proteins separate largely on the basis of size as all conformation is lost and charge per unit length (imparted by SDS) is constant. If the proteins differ by one or more amino acids, this would probably not be detected by SDS-polyacrylamide gel electrophoresis. We are now also using Triton X 100: urea:glacial acetic disc gel electrophoresis which separate proteins on the basis of charge. By various preparative disc gel electrophoresis procedures we can now obtain sufficient quantities of homogeneous membrane proteins for fingerprinting. We hope in the near future to have compared certain of the quantitatively major membrane proteins, especially some which appear to be common to various membrane systems of the cell.

GLYCOPEPTIDES FROM MEMBRANE GLYCOPROTEINS

In the past five years we have been comparing the carbohydrate components of the glycoproteins of normal and transformed cells (57–60). Double-label experiments have been carried out in which control cells are grown in the presence of ^{14}C-L-fucose (or other sugar), while transformed cells are grown in the presence of the ^{3}H counterpart. These isotope precursors are incorporated into glycoproteins of the surface and internal membranes. The cells are treated with trypsin, which removes part of the surface glycoproteins (trypsinate). The "trypsinates" from normal and malignant cells are mixed, digested with pronase, which reduces the polypeptide to a constant minimum, leaving the carbohydrate component intact. The digest is chromatographed on a long column of Sephadex G50; each tube of the eluate is counted for ^{14}C and ^{3}H and the data are processed and plotted by computer. We have found that there is a population of fucose-containing glycopeptides of MW 4,300 (peak A) in increased amount in the transformed cell compared to the control; the difference is quantitative. The population of glycopeptides enriched in the membrane glycoproteins of transformed cells is complex, but can be at least partially resolved by high-voltage electrophoresis or by ion-exchange chromatography. It is particularly rich in sialic acid, as compared to that of other glycopeptide populations, which are found in approximately equal quantities in the membrane glycoproteins of normal and malignant cells (peak B) (61). A sialyl transferase has been found that can transfer sialic acid (NAN) from its active form (CMPNAN) to desialylated peak A. This membrane-bound transferase is present in 3 to 11 times greater amount in transformed than in control cells and may therefore be responsible, in part, for the synthesis of peak A glycopeptides (61). A fucosyl transferase capable of transferring L-fucose from GDP-L-fucose to defucosylated peak A is also more than doubled in activity in transformed cells.

If the glycopeptides are treated with neuraminidase of *Vibrio cholerae, Clostridium perfringens,* or influenza virus, the peak A glycopeptides lose approximately 3 sialic acid residues per mole, the peak B glycopeptides lose 1, and they both elute as a single peak from columns of Sephadex G50. The elution patterns of the glycopeptides derived from control and transformed cells become superimposable (61). All of these glycopeptides contain D-mannose, *N*-acetyl-D-glucosamine, D-galactose, L-fucose, and sialic acid and are linked to the polypeptide chain by an alkali-stable β-glycosylamine linkage between asparagine and *N*-acetyl-D-glucosamine (62).

At the present time, we are attempting to determine which membrane glycoproteins contain peak A glycopeptides and which contain the B types. We now know that a larger, more water-soluble, polar glycoprotein has only peak A glycopeptides, while smaller ones contain both A and B types. We are also investigating whether an increase of peak A glycopeptides upon transformation results from greater synthesis of peak A-rich glycoproteins, or whether the number of peak A carbohydrate sidechains per glycoprotein increases in malignant cells.

The difference in the amount of peak A glycopeptides in control and transformed cells in log phase of growth is marked, but is only quantitative. Formation of peak A glycopeptides is growth-dependent. Little of the peak A glycopeptides are seen in the membrane glycoproteins of the cells not in log phase of growth (58). It would seem that a growth-dependent process is highly exaggerated in the transformed cell.

The increase of peak A glycopeptides is found not only in the surface membrane glycoproteins, but in those of the internal membranes as well (nucleus, ER, mitochondria, lysosomes (60). Thus the change that takes place in the carbohydrate moiety of membrane glycoproteins appears to be coordinate.

The shifts in population of glycopeptides described occur in 5 species of cells (human, rat, mouse, chick, and hamster), transformed by a variety of RNA- and DNA-oncogenic viruses, as well as in chemically and spontaneously transformed fibroblasts (63). The membrane glycoproteins of chick embryo fibroblasts transformed by T5, a temperature-sensitive mutant of Rous sarcoma virus, contain far more peak A glycopeptides at permissive temperatures (36°C) where malignancy is manifested than at a nonpermissive temperature (40°C) (64), and the sialyl transferase discussed above is correspondingly increased only at the permissive temperature (65). Enrichment of peak A glycopeptides has been described in various human leukemic cells (66).

Recently, a marked increase in peak A glycopeptides has been found in surface and internal membrane glycoproteins of solid, B16 mouse melanoma tumors found in the lung, liver, peritoneal cavity, and skin (67). The glycopeptides of normal mouse tissues and cells in culture before and after transformation by oncogenic viruses were used as controls. In these experiments the double-label approach was used; some tumor-bearing mice were injected with ^{14}C-L-fucose, while ^{3}H-L-fucose was given to others. The elution patterns of the melanoma glycopeptides from columns of Sephadex G50 were remarkably similar to those of Rous virus-transformed baby hamster kidney cells (C_{13}/B_4). The elution pattern of material derived from B16 melanoma cells grown in culture is remarkable, in that virtually all of the glycopeptides migrate in the peak A area.

Stable lines of B16 melanoma have been developed which, when injected into mice, form a small or large number of tumors in the lung (15-fold difference from equal numbers of injected cells) (68). The glycopeptide populations of these cells grown *in vivo* and *in vitro* show no significant, reproducible differences as seen by Sephadex G50 chromatography (67).

CONCLUSIONS

The differences in the carbohydrate component of the membrane-bound glycoproteins of normal and malignant cells dividing at equal rates are marked, but are essentially quantitative in nature, and the inherent problems of evaluating the relevance to the malignant process remains. Work on the subfractionations purification and analysis of peak A and B glycopeptides is continuing and their biosynthesis is being investigated. On the other hand, working in the dark with only a few leads, comparisons are being made of various membrane proteins with the distant hope of finding a qualitative change of genetic origin.

ACKNOWLEDGMENTS

This work was done with the support of Grants 1 R01 CA 13992–02 and 5 R01 CA 14985–02 from the USPHS and Grants BC 16D and PRP-28 from the American Cancer Society.

REFERENCES

1. Pardee, A. B. (1964): *Natl. Cancer Inst. Monogr.,* 14:7.
2. Kalckar, H. M. (1965): *Science,* 150:305.
3. Wallach, D. F. H. (1968): *Proc. Natl. Acad. Sci. USA,* 61:868.
4. Pitot, H. C. (1974): *J. Natl. Cancer Inst.,* 53:905.
5. Knudson, A. G., Jr. (1973): *Adv. Cancer Res.,* 17:317.
6. Weiss, L. (1967): *The Cell Periphery, Metastasis and other Contact Phenomena.* North-Holland Publishing Co., Amsterdam.
7. Abercrombie, M., and Heaysman, J. E. M. (1954): *Exp. Cell Res.,* 6:293.
8. Baldwin, R. W., Kersey, S. H., and Bagshawe, K. D. (1973): *Adv. Cancer Res.,* 18:1973.
9. Cunningham, D. C., and Pardee, A. B. (1969): *Proc. Natl. Acad. Sci. USA,* 64:1049.
10. Hatanaka, M., Todaro, G. J., and Gilden, R. V. (1970): *Int. J. Cancer,* 5:244.
11. Burger, M. M. (1971): *Curr. Top. Cell Regul.,* 3:135.
12. Bergelson, L. D. (1972): *Prog. Chem. Fats Other Lipids,* 15:1–54.
13. Singer, S. J. (1973): *Science,* 180:983.
14. Yau, T. M., and Weber, M. J. (1972): *Biochem. Biophys. Res. Commun.,* 49:114–120.
15. Gaffney, B. J., Branton, P. E., Wickus, G. G., and Hirschberg, C. G. (1974): In: *Viral Transformation and Endogenous Viruses,* edited by A. S. Kaplan, pp. 97–115. Academic Press, New York.
16. Horwitz, A. F., Hatten, M. E., and Burger, M. M. (1974): *Proc. Natl. Acad. Sci. USA,* 71:3115–3119.
17. Barnett, R. E., Furcht, L. T., and Scott, R. E. (1974): *Proc. Natl. Acad. Sci. USA,* 71: 1992–1994.
18. Gaffney, B. J. (1974): *Proc. Natl. Acad. Sci. USA,* 72:664–668.
19. Shinitzky, M., and Inbar, M. (1974): *J. Mol. Biol.,* 85:603–615.
20. Inbar, M., and Shinitsky, M. (1974): *Proc. Natl. Acad. Sci. USA,* 71:2128–2130; 4229–4231.
21. Muller, G. L. (1930): *Medicine,* 9:119–174.

22. Bases, R. E., and Krakoff, I. H. (1965): *J. Reticuloendothel. Soc.,* 2:8–14.
23. Rapport, M. M. (1969): *Ann. NY Acad. Sci.,* 159:446.
24. Brady, R. O., and Fishman, P. H. (1974): *Biochim. Biophys. Acta,* 355:121.
25. Critchley, D. R. (1973): In: *Membrane Mediated Information,* Vol. 1, *Biochemical Functions,* edited by P. W. Kent, pp. 20–47. M. T. P. Publ., Lancaster, England.
26. Hakomori, S., and Murakami, W. T. (1968): *Proc. Natl. Acad. Sci. USA,* 59:254.
27. Cumar, F. A., Brody, R. A., Kolodny, E. H., McFarland, V. W., and Mora, P. T. (1970): *Proc. Natl. Acad. Sci. USA,* 67:757.
28. Fishman, P. H., McFarland, V. W., Mora, P. T., and Brady, R. O. (1972): *Biochem. Biophys. Res. Commun.,* 48:48.
29. Mora, P. T., Cumar, F. A., and Brady, R. O. (1971): *Virology,* 46:60.
30. Den, H., Schultz, A. M., Basu, M., and Roseman, S. (1971): *J. Biol. Chem.,* 246:2721.
31. Brady, R. O., Borek, C., and Bradley, R. M. (1969): *J. Biol. Chem.,* 244:6552–6554.
32. Mora, P. T., Brady, R. O., Bradley, R. M., and McFarland, V. W. (1969): *Proc. Natl. Acad. Sci. USA,* 63:1290–1296.
33. Hakomori, S. (1970): *Proc. Natl. Acad. Sci. USA,* 67:1741.
34. Robbins, P. W., and Macpherson, I. A. (1971): *Nature,* 229:569.
35. Steiner, S., Brennan, P. J., and Melnick, J. L. (1973): *Nature [New Biol.],* 245:19–21.
36. Gahmberg, C. G., and Hakomori, S. (1974): *Biochem. Biophys. Res. Commun.,* 59:283–291.
37. Alexander, P. (1972): *Nature,* 235:137.
38. Hynes, R. O. (1973): *Proc. Natl. Acad. Sci. USA,* 70:3170.
39. Hogg, N. M. (1974): *Proc. Natl. Acad. Sci. USA,* 71:489.
40. Wickus, C. G., and Robbins, P. W. (1973): *Nature [New Biol.],* 32:80.
41. Stone, K. R., Smith, R. E., and Joklik, W. K. (1974): *Virology,* 58:86.
42. Bussel, P. H., and Robinson, W. S. (1973): *J. Virol.,* 12:321.
43. Sheinin, R., and Onodera, K. (1972): *Biochim. Biophys. Acta,* 274:49.
44. Greenberg, C. S., and Glick, M. C. (1972): *Biochemistry,* 11:3680.
45. Chiarugi, V. P., and Urbano, P. (1972): *J. Gen. Virol.,* 14:133.
46. Gahmberg, C. G., and Hakomori, S. (1973): *Proc. Natl. Acad. Sci. USA,* 70:3329.
47. Gahmberg, C. G., Kiehn, D., and Hakomori, S. (1974): *Nature,* 248:413.
48. Ruoslahti, E., Vaheri, A., Kuusela, P., and Linder, W. (1973): *Biochim. Biophys. Acta,* 322:352.
49. Wickus, G. C., Branton, P. E., and Robbins, P. W. (1974): In: *Control of Proliferation in Animal Cells,* edited by B. Clarkson and R. Baserga, pp. 541–546. Cold Spring Harbor Lab., Cold Spring Harbor, N.Y.
50. Hynes, R. O. (1974): *Cell,* 1:147.
51. Teng, N. N. H., and Chen, L. B. (1975): *Proc. Natl. Acad. Sci. USA,* 72:413.
52. Bosmann, H. B. (1972): *Biochim. Biophys. Acta,* 264:339–343.
53. Schnebli, H. P., and Burger, M. M. (1972): *Proc. Natl. Acad. Sci. USA,* 69:3825–3827.
54. Unkeless, J. C., Tobia, A., Ossowski, L., Quigley, J. P., Rifkin, D. B., and Reich, L. (1973): *J. Exp. Med.,* 137:85–111.
55. Burger, M. M. (1970): *Nature,* 227:170–171.
56. Sefton, B. M., and Rubin, H. (1970): *Nature,* 227:843–845.
57. Buck, C. A., Glick, M. C., and Warren, L. (1970): *Biochemistry,* 9:4567.
58. Buck, C. A., Glick, M. C., and Warren, L. (1971): *Biochemistry,* 10:2176.
59. Buck, C. A., Glick, M. C., and Warren, L. (1971): *Science,* 172:169.
60. Buck, C. A., Fuhrer, J. P., Soslau, G., and Warren, L. (1974): *J. Biol. Chem.,* 249:1541.
61. Warren, L., Fuhrer, J. P., and Buck, C. A. (1972): *Proc. Natl. Acad. Sci. USA,* 69:1838.
62. Warren, L., Fuhrer, J. P., Buck, C. A., and Walborg, E. F., Jr. (1974): In: *Membrane Transformations in Neoplasia,* edited by J. Schultz and R. E. Block, Vol. 8, p. 1. Academic Press, New York.
63. Emmelot, P. (1973): *Eur. J. Cancer,* 9:319.
64. Warren, L., Critchley, D., and Macpherson, I. (1972): *Nature,* 235:275.
65. Warren, L., Fuhrer, J. P., and Buck, C. A. (1973): *Fed. Proc.,* 32:80.
66. Van Beek, W. P., Smets, L. A., and Emmelot, P. (1975): *Nature,* 253:457.
67. Warren, L., Zeidman, I., and Buck, C. A. (1975): *Cancer Res.,* 35:2186.
68. Fidler, J. J. (1973): *Nature [New Biol.],* 242:148.

Membranes and Disease, edited by L. Bolis, J. F. Hoffman, and A. Leaf. Raven Press, New York, © 1976.

Membrane Biochemical Studies in Myotonic Muscular Dystrophy

Stanley H. Appel and Allen D. Roses

Division of Neurology, Duke University Medical Center, Durham, North Carolina 27710

One of the significant problems for molecular neurobiology is to develop a coherent picture of how membranes actually perform their physiological tasks and to apply such knowledge to the devastating diseases of nerve and muscle which afflict man. In particular, what are the molecular species responsible for the selective transport of solutes in graded and regenerative electrical activity, and how are these systems perturbed in human disorders? Our own laboratory has been involved in characterizing the membranes of mammalian skeletal muscle, not only as a model for studying membrane functions in excitable tissue, but also for understanding biochemical features of those human muscular disorders in which the pathophysiology has been studied. As a model of an alteration of excitable properties of muscle, we have concentrated on the human disease, myotonic muscular dystrophy (MyD), inherited as an autosomal, dominant trait (1,2). The metabolic defect in this disorder is widespread and involves many organ systems. Cataracts, frontal balding, bony abnormalities, smooth muscle, skeletal muscle, cardiac muscle, endocrine tissue, red blood cells, central nervous system, and gamma globulin metabolism are clearly involved (1,2). Physiological investigations have suggested a membrane abnormality as the underlying metabolic defect. No abnormalities in muscle contractile tissues or "relaxing factor" have been demonstrated. Repetitive depolarization of myotonic muscle fibers, even after nerve block or neuromuscular blockade, has led many investigators to localize the defect to muscle sarcolemma.

Although muscle is one of the principal target organs, we initially felt that the presence of atrophy, fibrous tissue, and changes of denervation in muscle might produce multiple secondary biochemical changes. Erythrocytes were used as an easily accessible source of membrane preparations that have no known functional abnormality (3). The polypeptide profile of red cell membranes were identical in myotonic and control preparations. In addition, phospholipid composition, gangliosides, cholesterol, and enzyme activities such as Mg-ATPase, Ca-ATPase, and NaKMg-ATPase were all normal (3).

ALTERED PROTEIN PHOSPHORYLATION IN MYOTONIC ERYTHROCYTES

Our initial experiments demonstrated a significant decrease in the phosphorylation of erythrocyte ghost protein, using endogenous protein kinase of frozen erythrocyte membrane as the enzyme source (Table 1) (3). The initial failure to demonstrate a difference in the endogenous protein kinase activity of fresh erythrocyte ghosts prompted an examination of the techniques of ghost membrane preparation and isolation. Such an approach appeared warranted, since variation of membrane protein phosphorylation with buffer, pH, and method of membrane preparation had been observed with human erythrocytes. The use of ghosts prepared with Mg-free solutions resulted in highly permeable membranes. In this preparation endogenous protein kinase activity was more readily demonstrated, and phosphorylation of band III was significantly decreased in freshly prepared ghosts from patients with myotonic muscular dystrophy (Table 1) (4). This region of the gel is known to consist of several different proteins, including the NaK-ATPase, and component A, a minor glycoprotein. The phosphorylation initially appeared to migrate with component A but fractionation by concanavalin A – Sepharose chromatography separated a subfraction which was phosphorylated from one which was not. Such data suggested that component A may be a diverse group of glycoproteins with possible carbohydrate and apoprotein heterogeneity. Our further investigations are aimed at clarifying the particular glycoprotein being phosphorylated and the differences between the normal and myotonic substrates.

TABLE 1. *Protein phosphorylation: ^{32}P incorporated (pmoles/mg protein)*

	Frozen RBC (3)	Fresh RBC band III (4)	Fresh muscle membrane (6)
Normal	17.5 ± 1.2	2.82 ± 0.17	15.90 ± 1.8
Myotonic	8.7 ± 1.2	2.07 ± 0.22	9.95 ± 1.1

Endogenous membrane protein phosphorylation was assayed in each tissue with $AT^{32}P$ and SDS acrylamide gel electrophoresis according to previously described techniques (3,4,6). All studies demonstrated a statistically significant difference between myotonic tissue and matched controls.

DECREASED MEMBRANE PROTEIN PHOSPHORYLATION IN MUSCLE TISSUE

Following the demonstration of protein phosphorylation in purified fractions of rat muscle membrane (5), the technical capacity presented itself for evaluation of human muscle biopsy material in MyD. In membrane isolated from human biopsy material, we were not able to obtain purified membrane fractions because of the limited amount of tissue available (6). Our biopsies were obtained from six patients undergoing operations for femur or hip repair. None of the controls had any known muscle disease. The membranes employed in the assay were known to contain a variable content of sarcolemma, sarcoplasmic reticulum, and other membrane organelles. However, despite the potential heterogeneity of the membrane preparation from different patients, the SDS–polyacrylamide gel electrophoresis pattern of muscle membrane proteins indicated no gross difference between normal and myotonic preparations (6). Furthermore, activity of adenylcyclase and both Mg- and NaKMg-ATPase were equivalent in normal and myotonic preparations. The membrane protein kinase activity was decreased in all myotonic biopsies (Table 1) (6). Phosphorylation of muscle membrane components was noted in band I of ~90,000 MW, band II of ~50,000 MW, and band III of ~30,000 MW. Phosphorylation of bands II and III was specifically decreased in myotonic membranes compared to controls. While there was considerable variation between the experiments, the protein regions in myotonic membranes were phosphorylated ~50% of their control rate. The 90,000 MW was extremely variable and appeared to disappear with extensive washing of the membrane fractions.

Although these data demonstrate a significant alteration in the endogenous membrane-associated protein kinase activity of erythrocyte and muscle membranes from MyD, we could not make any definitive statement as to whether the enzyme itself, the specific substrates, or the state of the lipid–lipid or lipid–protein interactions within the membrane were responsible for the altered enzymic activity. Whether the altered protein phosphorylation of muscle membranes and red blood cells is the primary genetic defect in the disease process is not known at present. Either a primary or secondary alteration may be associated with an abnormal membrane and either interpretation would support the concept of a diffuse membrane abnormality of myotonic dystrophy.

BIOPHYSICAL STUDIES OF MEMBRANE ABNORMALITY IN MYOTONIC DYSTROPHY

Electron spin resonance spectroscopy was utilized to investigate the possible physical alterations of the membrane. The magnetic measurements of a spin labeled probe intercalated into the membrane provided a measure of membrane fluidity and polarity. Stearic acid methyl esters with nitroxide groups located at the 5, 12, and 16 positions on the fatty acid alkyl chain were used (7). The results of these studies demonstrated that the physical state of the erythrocyte membrane from patients with MyD was different from that of normal controls. At all levels of the membrane probed by the spin label, myotonic membranes were more fluid and less polar than were control membranes. The fluidity difference between normal and myotonic membranes was most apparent near the surface of the membrane, while the polarity difference was approximately constant at various depths within the membrane.

Similar results of increased membrane fluidity have been provided by our recent studies employing nitroxide labeled maleimide (2,2,6,6,-tetramethylpiperidine-1-oxyl-4-maleimide) which has a specific affinity for SH groups of membrane protein (8). Thus, the myotonic red blood cell membranes appear to be altered whether analyzed with lipid or with protein probes.

SCANNING ELECTRON MICROSCOPY OF ERYTHROCYTES

Morphological characteristics of red blood cells were studied using scanning electron microscopy of unmanipulated cells from patients with MyD and from controls (9). A large increase in the number of stomatocytes over that seen in normal controls was demonstrated (Fig. 1). However, similar stomatocytes could be produced in normal cells by adverse conditions, such as washing before fixation or extreme pH, although the changes noted in red blood cells from patients with myopathy were more marked. There is no evidence that erythrocytes of myotonic dystrophic patients were misshapen *in vivo*. The stomatocytic shapes were probably the result of intrinsic biochemical membrane differences that respond to fixation in an abnormal manner *in vitro*.

SPECIFICITY OF BIOCHEMICAL AND BIOPHYSICAL TESTS IN MYOTONIC DYSTROPHY

To monitor the specificity of these findings, similar studies were undertaken in red blood cells from patients with congenital myotonia (MC) as a model of myotonia without dystrophy, and from patients with Duchenne muscular dystrophy (DMD) as a model of dystrophy without myotonia. In the assay for protein kinase activity in red blood cell membrane preparations, there was a significant increase in phosphorylation of bands II and III from patients with DMD as well as the DMD carrier state (10). This pattern of phosphorylation was different from controls and was opposite to that noted in myotonic dystrophy. When red blood cell membranes from two patients with MC were examined, endogenous membrane phosphorylation was found to be identical to control membranes.

With electron spin resonance studies employing stearic acid methyl ester spin labels, patients with DMD were normal. However, patients with MC demonstrated the same significant increase in fluidity noted with MyD cells. Morphologic analysis of red blood cells by scanning electron microscopy

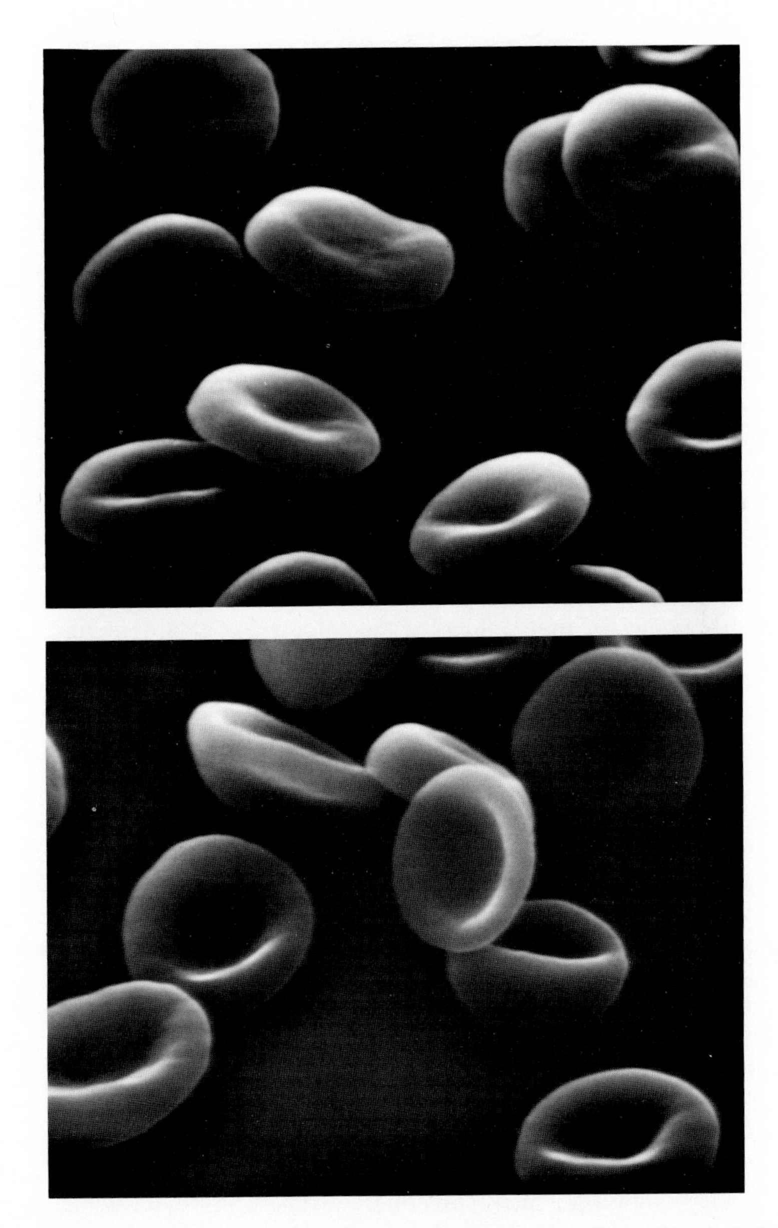

FIG. 1. Scanning electron microscopy of red blood cells. **Top:** Normal control. **Bottom:** Stomatocytes from patient with myotonic muscular dystrophy. [Courtesy of Dr. Sara Miller (9).]

demonstrated a large increase in the number of stomatocytes in all three conditions, MyD, DMD, and MC.

The decrease in protein phosphorylation was noted only in myotonic dystrophy just as the increase in protein phosphorylation was noted only in DMD. The membrane fluidity changes were the same in both MC and MyD, even though the presumed physiological mechanism underlying myotonia is known to be different in the two conditions. Muscle membranes from MC have been reported to demonstrate increased membrane resistance and a decreased conductance to chloride, whereas the membrane resistance and chloride conductance was reported as normal in MyD (11). However, the fluidity changes were noted only in myotonia, since there was no change in membrane fluidity in the nonmyotonic DMD. Finally, abnormal morphology with scanning electron microscopy was found in all three conditions. Thus, the combination of altered protein phosphorylation, electron spin resonance, and scanning electron microscopy permit us to determine which of the three disease states was being examined. However, it is not presently possible to determine whether these results either individually or collectively are specific for these conditions, or whether they might also occur in myopathies or hematological disorders not yet investigated.

MYOTONIA

Myotonia represents an abnormal tendency of the muscle membrane to discharge trains of repetitive action potentials in response to depolarization. It is clinically recognizable as an abnormally persistant contraction of striated muscle either after a willed contraction or a mechanical or electrical stimulation. Persistance of the abnormal electrical activity after nerve block or following curare localizes the defect to the muscle fiber membrane (12,13). However, a molecular understanding of the membrane events responsible for muscle myotonia is not yet available. In human myotonia congenita, goat myotonia, and the drug-induced myotonias, a normal resting membrane potential, an increased membrane resistance, and a decreased chloride conductance have been reported (11). The recent experiments of Adrian and Bryant (14) with goat myotonia and Rudel and Keller (15) with drug-induced myotonia prompted the following explanation of muscle myotonia. An action potential spreading into the transverse tubular system raises the K concentration in the lumen of the tubules. The passage of several action potentials and the increased tubular K results in a cumulative after-depolarization which is normally offset by the entry of chloride into the muscle fiber. If chloride conductance is decreased and cannot counterbalance the depolarizing effect of increased tubular K, repetitive action potentials would result.

The same scheme cannot be employed to explain the myotonia of MyD, since membrane resistance and chloride conductance appear to be normal in MyD (11). Any theory that attempts to explain the myotonia of MyD must take into account these normal membrane properties as well as a decreased resting membrane potential of −62 to −68 mV (16). Two alterations of ion transport in red blood cells from patients with MyD have recently been studied in our laboratory and may be pertinent to the myotonic process. The first represents an alteration in K efflux promoted by Ca; and the second is an alteration in the normal 3:2, Na:K, exchange ratio of the membrane ion pump.

CALCIUM-PROMOTED POTASSIUM EFFLUX

In 1958, Gardos demonstrated that human red cells depleted of energy in the presence of iodoacetate and adenosine are selectively permeable to K upon the addition of Ca (17). Na permeability is unaltered in this situation. Furthermore, in the absence of Ca, no increase in Na or K permeability are noted (17,18). In the studies of Blum and Hoffman (19,20), sensitivity of the Ca-induced system to ouabain paralleled the action of ouabain on the NaK pump, and it was suggested that the molecular mechanism of the NaK pump complex underlies, at least in part, the action of Ca. This interpretation has been questioned by evidence suggesting that ouabain acted to prevent ATP hydrolysis by the pump, thereby leaving traces of ATP in the cells to promote Ca efflux (21,22). In micropuncture studies of *Amphiuma* red cell, Lassen et al. (23) noted a Ca-related increase in K permeability with no metabolic energy depletion. Neither ouabain nor oligomycin altered the hyperpolarizing effect produced in such cells by the Ca-promoted K efflux. Furthermore, the hyperpolarizing effect could be reproduced by direct injections of Ca into the cell. Such studies confirmed the suggestion that Ca at the inside surface of the membrane enhanced K permeability and led to hyperpolarization.

The possible physiological importance of this Ca-promoted K efflux derives from studies of posttetanic hyperpolarization in *Aplysia* neurons (24–26) from studies of the effects of light and metabolic inhibitors on the hyperpolarizing photoreceptor of the snail eye (27), and from the studies of hyperpolarization of the *Amphiuma* red cell noted above (23). In *Aplysia* neurons the membrane K permeability is controlled by the intracellular concentration of free Ca ions. Injected $CaCl_2$ caused an increase in K permeability, which produces a hyperpolarization of the cell (25,26). When EGTA is injected into *Aplysia* neurons, single-action potentials were unaffected. However, with repetitive stimulation the membrane potential rapidly assumed a depolarized steady state, even after the depolarizing stimulus had been terminated and then underwent a period of oscillation and prolonged action potentials (26). The intracellular Ca ion concentration also appears important in the photoreceptor system of the scallop.

Cellular metabolism apparently maintains the Ca concentration at a low level and keeps the membrane depolarized in darkness. Metabolic inhibitors, which are likely to increase intracellular Ca, were noted to have a hyperpolarizing effect due to an increase in K membrane permeability (27).

The potential role of Ca-promoted K efflux as a hyperpolarizing influence, and its possible impairment resulting in depolarization, prompted an investigation of the reaction in MyD. Our initial studies confirmed a rapid K efflux on addition of Ca to cells which are metabolically depleted in the presence of iodoacetic acid and adenosine. Approximately one-third of the intracellular K leaked from the red blood cell after 3 hr of exposure *in vitro* to the inhibitors. Mg did not alter the efflux, and no simultaneous uptake of Na was noted. To confirm the lack of a specific requirement for ATP depletion and the importance of intracellular Ca in the reaction, we employed the Ca ionophore, A23187 (Eli Lilly Co.). When both ionophore and Ca were added to the medium, K leaked rapidly from the red blood cells. In fact, within 15 min, over 70% of the intracellular K had leaked from the cells. Such data would support the hypothesis that Ca promotes K efflux from the internal surface of the membrane; and the role of energy depletion is to sustain a sufficient intracellular Ca concentration, which might otherwise be depleted by active extrusion by a Ca pump.

With this background information, Ca-promoted K efflux was evaluated in red blood cells derived from normal, MyD, and DMD patients. The initial K concentration was identical in the cells derived from all three patient populations (Table 2). Following the introduction of 0.4 mM iodoacetic acid and 1.0 mM adenosine, no K efflux was noted until Ca was introduced into the medium. With the introduction of Ca, K efflux continued for approximately 3 hr. The rate of Ca-promoted K efflux was significantly different in the three populations (Table 3). K efflux was decreased almost 50% in red blood cells from patients with MyD and was increased almost 50% from patients with DuD. The results were similar over a range of Ca concentrations and adenosine concentrations employed. At higher concentrations of iodoacetic acid (1.0 mM), the differences among the three populations of

TABLE 2. *Evaluation of calcium-promoted potassium efflux in RBCs*

	Potassium (mmoles/liter packed cells)		
	N	$\bar{X} \pm$ SEM	
Control	25	90.0 ± 1.3	–
Myotonic dystrophy	20	89.2 ± 1.2	NS
Duchenne dystrophy	4	92.0 ± 2.0	NS

TABLE 3. *Calcium-promoted potassium efflux*

		Net potassium efflux (μmoles/2 hr)	
	N	$\bar{X} \pm$ SEM	*p*
Control	17	35.4 ± 3.0	–
Myotonic dystrophy	14	19.4 ± 2.0	<0.001
Duchenne dystrophy	4	52.0 ± 4.0	<0.001

K efflux was monitored in incubation medium from 60 to 180 min in presence of 0.4 mM iodoacetic acid, 10.0 mM adenosine, and 10.0 mM Ca, and 10% hematocrit. In absence of 10.0 mM Ca, the K efflux from control, myotonic, and Duchenne cells was 4–5 μmoles/2 hr, and the values did not differ statistically from each other.

cells appeared somewhat less, although it was always significantly decreased in myotonic red blood cells and increased in Duchenne red blood cells.

The decreased Ca-promoted K efflux in MyD red cells could be attributed either to a failure of Ca to reach the appropriate internal membrane sites in sufficiently high concentrations, to an accelerated rate of Ca efflux, or to a specific diminution in K efflux. No differences could be demonstrated in the Ca-ATPase or rate of Ca extrusion in normal, myotonic, or Duchenne red blood cells. Thus, an accelerated rate of Ca efflux could not explain the data. The ability of elevated concentrations of iodoacetic acid to lessen the difference between the myotonic and normal populations of cells suggested that entry of Ca, and not efflux of K, might be the critical variable. This conclusion was also supported by studies with the ionophore, A23187. At 1.0 μM to 5μM ionophore and 10.0 mM Ca, K efflux was maximal and no differences could be noted between normal and myotonic cells, either in the initial rate of K efflux, in the maximal extent of K efflux, or in the rate of uptake of ^{45}Ca. All these data suggested that K efflux was normal in the myotonic cells if a sufficient concentration of Ca could be introduced intracellularly. Thus, the primary difference appears to relate to the entry of Ca into the cell or to the internal surface of the membrane.

Unfortunately, as previously noted by Passow (18), at the low Ca permeability of the membrane being studied, it is not possible to demonstrate any entry of ^{45}Ca or significantly raised intracellular Ca concentrations. Thus, it was not possible to demonstrate directly an altered Ca entry in myotonic cells. Furthermore, equilibrium studies of Ca binding to the membrane did not demonstrate any difference in the Ca-chelating capacities of the cells. The relationship of these flux studies to membrane potential have not been delineated. However, by analogy with the studies of Lassen (23), the decreased Ca-promoted K efflux may well diminish any hyperpolarizing effect, and thus provide a depolarizing influence in such cells.

By analogy with the bursting neurons of the *Aplysia,* with repetitive activity the diminution in Ca-promoted K efflux might lead to a diminished hyperpolarizing influence and therefore a more sustained depolarization in myotonic muscle. The potential significance of the reaction in muscle awaits further studies. It is of interest that Samaha and Gergely (28) reported an increased initial rate of active Ca transport and normal Ca-ATPase in fragmented sarcoplasmic reticulum from MyD muscle biopsies. Their data can be readily interpreted as an altered passive Ca leak and might support the concept of an altered transmembrane passage of Ca.

STUDIES OF SODIUM–POTASSIUM PUMP IN MYOTONIC MUSCULAR DYSTROPHY

The ionic pumping mechanism in erythrocytes functions to expel three Na ions for two K ions transported inward from one molecule of ATP hydrolyzed (29,30). Alterations were reported in the NaK-ATPase activity of red blood cells from patients with MyD (31), but other studies, including those from our own laboratory, have failed to confirm these alterations at optimal ionic concentrations (32,4). In studies employing ^{22}Na and ^{42}K, no significant differences between myotonic dystrophic and control red blood cells could be demonstrated in the mean ouabain-insensitive Na efflux, the mean ouabain-sensitive K influx, or the mean ouabain-insensitive K influx (33). In normal cells the ratio of ouabain-sensitive Na efflux to ouabain-sensitive K influx was found to be 1.46 ± 0.08, thereby confirming the 3-Na for 2-K active exchange. When the cells from patients with MyD were examined, the ratio of active Na efflux to active K influx was 1.01 ± 0.06 or a 2-Na for 2-K exchange ratio (Table 4) (33). The difference in stoichiometry in control compared to myotonic cells was primarily related to the impairment in ouabain-sensitive Na efflux (2.33 meq Na/liter red blood cells per hr in controls and 1.64 meq Na/liter red blood cells per hr in myotonic patients). In preliminary studies in DMD and MC, no alteration in the stoichiometry of Na for K exchange was noted.

TABLE 4. *Stoichiometry of the sodium-potassium pump*

	Control $N = 9$	MyD $N = 9$
Mean ouabain-sensitive Na efflux	2.33 ± 0.13	1.64 ± 0.09
Mean ouabain-sensitive K influx	1.62 ± 0.11	1.67 ± 0.12
Na/K ratio	1.46	1.01

Ion fluxes are in meq/liter cells per hr, mean ± SEM.
Data from Hull and Roses (33).

The resting membrane potential is decreased in MyD and the intracellular Na concentration of dystrophic muscle is increased (16). Both of these alterations may predispose to myotonia and may, in fact, be related to an alteration of the NaK pump. At the present time it is unclear whether such an alteration of the pump is present in dystrophic muscle cells. Unfortunately, no data are available on the stoichiometry of the NaK pump in normal muscle. Studies of washout of ^{22}Na in intercostal myotonic muscle suggested a qualitatively normal pump flux of Na, but about twice normal Na concentration (16). The authors suggested that the pump was normal, but that there was an increased permeability with entry of Na into a readily exchangeable as well as into an inexchangeable pool. Ouabain produced the same reduction in active Na injection in both normal and myotonic populations. Despite these findings, it is still possible that an alteration of the pump may exist in MyD, since Hofmann and DeNardo's studies were not capable of defining the many Na pools, the extracellular and intracellular spaces, and the transport and diffusion properties with an accuracy possible in red cell preparations. Clearly, an alteration in the stoichiometry of the pump could be relevant to the decreased resting membrane potential noted in myotonic muscle fibers. However, a similar decrease in resting membrane potential has been noted in DMD; no muscle myotonia is present and no alteration is seen in the stoichiometry of the red blood cell NaK active transport.

RELATIONSHIP OF ION FLUX ALTERATIONS TO MEMBRANE STRUCTURAL CHANGES

At the present time it is unclear whether either the alterations in Ca-promoted K efflux or the alterations of the Na for K exchange ratio in red blood cells are related to the alterations in membrane protein phosphorylation. It is of interest that protein phosphorylation is decreased in red blood cells in MyD and is increased in red blood cells in DMD. Ca-promoted K efflux is similarly decreased in MyD and is increased in DMD. It is thus tempting to suggest that those factors are related either directly or indirectly.

Many critical questions remain to be answered. (i) What is the molecular defect responsible for the dystrophy in MyD and DMD? (ii) What is the inborn error of metabolism? (iii) To what extent does the inborn error either directly or indirectly give rise to the myotonia and the dystrophy? The protein phosphorylation changes, scanning electron microscopy, and electron spin resonance together provide a sensitive and possibly specific assay of the altered membranes. The protein phosphorylation changes are of interest, especially in light of our recent findings in rat muscle where a membrane fraction possibly of transverse tubule origin was the only fraction containing significant endogenous protein phosphorylation (5). A similar localization might be postulated for protein phosphorylation in human muscle tissue. Because of the importance of the transverse tubule in goat myotonia, it is possible that the tubular structures may play a similar role in human MyD. Decreased protein phosphorylation in human muscle may thus be located in the transverse tubule region and may lead to myotonic activity by the effects on altered ion flux. Furthermore, the lowered resting membrane potential may give rise to a situation in which changes of K permeability may lead to repetitive discharges.

At this juncture, our investigations have employed biochemical techniques to explore the pathogenesis of several diffuse membrane disorders in man. It will be of additional value to use the pathological studies to help understand aspects of normal structure and function. Myotonic muscular dystrophy may shed light on the character of repetitive activity and the role of the transverse tubule in electrical events and excitation–contraction coupling in muscle tissue. All of these approaches should provide an opportunity for translating functional alterations of the membrane into biochemical terms and helping to explain normal physiological mechanisms.

ACKNOWLEDGMENTS

We are grateful to Messrs. Winfred Clingenpeel and Michael Herbstreith for technical assistance and to Mrs. Eleanor Chapman for expert secretarial help. We are also grateful to our colleagues, Drs. Allan Butterfield and Sara Miller for collaboration in certain aspects of these studies. This work was supported in part by grants NS07872, NS12213, and NS06233 from the National Institutes of Health and grant 558-D-5 from the National Multiple Sclerosis Society.

REFERENCES

1. Thomasen, E. (1948): *Myotonia, Thomsen's Disease, Paramyotonia and Dystrophia Myotonia,* (Aarhus, Universitetsforlaget, 1948), 251 pp.
2. Caughey, J. E., Myrianthopoulos, N. C. (1963): *Dystrophia Myotonica and Related Disorders,* 282 pp. C C Thomas, Springfield, Ill.
3. Roses, A. D., and Appel, S. H. (1973): *Proc. Natl. Acad. Sci. USA,* 70:1855.
4. Roses, A. D., and Appel, S. H. (1975): *J. Membr. Biol.,* 20:51.
5. Andrew, C. G., Almon, R. R., and Appel, S. H. (1975): *J. Biol. Chem.,* 250:3972.
6. Roses, A. D., and Appel, S. H. (1974): *Nature,* 250:254.
7. Butterfield, D. A., Roses, A. D., Cooper, M. L., Appel, S. H., and Chesnut, D. B. (1974): *Biochemistry,* 13:5078.
8. Butterfield, D. A., Roses, A. D., Chesnut, D. B., and Appel, S. H. (1975): *Trans. Am. Soc. Neurochem.,* 6:163.
9. Miller, S. E., Roses, A. D., and Appel, S. H. (1975): *Arch. Neurol.* (*in press.*)
10. Roses, A. D., Herbstreith, M. H., and Appel, S. H. (1975): *Nature,* 254:351.
11. Lipichy, R. J., and Bryant, S. H. (1973): In: *New Developments in Electromyography and Clinical Neurophysiology,* Vol. 1, edited by J. E. Desmedt, p. 451. Karger, Basel.
12. Bryant, S. H. (1973): In: *New Developments in Electromyography and Clinical Neurophysiology,* Vol. 1, edited by J. E. Desmedt, p. 420. Karger, Basel.

13. Hofmann, W. W., Alston, W., and Rowe, G. (1966): *Electroencephalogr. Clin. Neurophysiol.,* 21:521.
14. Adrian, R. H., and Bryant, S. H. (1974): *J. Physiol. (Lond.),* 240:505.
15. Rudel, R., and Keller, M. (1975): *Proc. 3rd Int. Cong. Muscle Diseases, Newcastle-upon-Tyne (in press.)*
16. Hofmann, W. W., and DeNardo, G. L. (1968): *Am. J. Physiol.,* 214:330.
17. Gardos, G. (1958): *Biochim. Biophys. Acta,* 30:653.
18. Passow, H. (1963): In: *Cell Interface Reactions,* edited by H. D. Brown. Scholar's Library, New York.
19. Blum, R. M., and Hoffman, J. F. (1971): *J. Membr. Biol.,* 6:315.
20. Blum, R. M., and Hoffman, J. F. (1972): *Biochim. Biophys. Res. Commun.,* 46:1146.
21. Lew, V. L. (1971): *Biochim. Biophys. Acta,* 233:827.
22. Lew, V. L. (1971): *Biochim. Biophys. Acta,* 249:236.
23. Lassen, U. V., Pape, L., Vestergaand-Bogind, B., and Bengtsen, O. (1974): *J. Membr. Biol.,* 18:125.
24. Meech, R. W. (1972): *Comp. Biochem. Physiol.,* 42A:493.
25. Meech, R. W. (1974): *Comp. Biochem. Physiol.,* 48A:387.
26. Meech, R. W. (1974): *Comp. Biochem. Physiol.,* 48A:397.
27. Gorman, A. L. F., and McReynolds, J. S. (1974): *Science,* 185:620.
28. Samaha, F. J., and Gergely, J. (1969): *N. Engl. J. Med.,* 280:184.
29. Post, R. L., and Jolly, P. C. (1957): *Biochim. Biophys. Acta,* 25:118.
30. Garrahan, P. J., and Glynn, I. M. (1967): *J. Physiol. (Lond.),* 192:217.
31. Brown, H. D., Chattopadhyay, S. K., and Patel, A. B. (1967): *Science,* 157:1577.
32. Klassen, G. A., and Blostein, R. (1969): *Science,* 163:492.
33. Hull, K., and Roses, A. D. (1976): *J. Physiol. (Lond.)* 254:169.

Membranes and Disease, edited by L. Bolis, J. F. Hoffman, and A. Leaf. Raven Press, New York, © 1976.

Altered Membrane Properties in Myotonia

S. H. Bryant

Department of Pharmacology and Therapeutics, University of Cincinnati, College of Medicine, Cincinnati, Ohio 45267

Myotonia is a clinical syndrome characterized by a failure of relaxation of skeletal muscle following a contraction. Myotonia is most frequently associated with muscular dystrophy in the multisystem disease, myotonia dystrophica, as discussed by Dr. Appel in this volume, but it can occur as the cardinal sign in myotonia congenita of man, and hereditary myotonia of goats (1).

The distinguishing feature of myotonic muscle is the abnormal repetitive firing of action potentials that occurs during and following a tetanic stimulation when forcefully contracting the muscle. The myotonic fibers have an increased excitability to direct stimulation and can fire repetitively in response to constant currents (Erb's reaction) and to mechanical stimulation (percussion response). These latter abnormalities persist after nerve section or curarization, and it has been clear for many years that myotonia is independent of the nerve supply and resides in the properties of the muscle fiber per se.

The hereditary myotonia of goats is transmitted as an autosomal dominant with incomplete penetrance. It is possible then to maintain a colony of goats having severe myotonia involving all of the skeletal muscles with no other discernable disease. For a number of years we have studied preparations of external intercostal muscle fibers *in vitro* using microelectrode techniques. We believe now that the myotonic phenomena in the goat can be almost entirely accounted for by a specific decrease in resting chloride conductance of muscle membrane in an otherwise normal fiber (2).

The electrophysiology of the various human myotonias has not been as well studied as the unique variety in the goat, but indications are that human myotonia congenita may likewise be explained by the low chloride-conductance abnormality. Thus, we have in the myotonia of goats a genetic defect whose main expression is that of a specific alteration of the chloride conductance channel of the skeletal muscle membrane. In the remainder of the chapter I will consider the following: (i) some clinical facts; (ii) the membrane abnormality in myotonia; (iii) newer studies on the properties of the transverse tubular system and excitation–contraction coupling in myotonia; (iv) the control of chloride conductance in normal fibers; and (v) some unexplained membrane phenomena in myotonia.

CLINICAL SYNDROME

Some of the clinical aspects of myotonia in man and goats are worth mentioning to give perspective to the relevant membrane-related phenomena. The classical description of myotonia congenita was by Asmus Thomsen who had the disease and was able to trace the disorder through five generations of his family (3). In collaboration with Dr. R. J. Lipicky, we had the opportunity to study a biopsy from a myotonic descendant of Dr. Thomsen, three generations removed. The electrophysiological behavior resembled what we observed in the myotonic goat.

The Thomsen family disease is transmitted as autosomal dominant, as is the caprine form, and family histories similar to the classical Thomsen's disease are rare. Becker (4) has given evidence that myotonia congenita is more frequently transmitted as a recessive gene, and this form may be more severe, but in many cases there is either no family history of the disease or there may be dystrophic disease in some family members. The genetic situation in the goat is much clearer. All of the myotonic goats known appear to have descended from four affected animals introduced into Tennessee in the southern United States around 1880.

In typical myotonia congenita, the myotonia is detected at birth or shortly afterward and continues in the individual throughout a normal life span. Muscles may hypertrophy, but there is no dystrophy. Patients are usually not incapacitated by the disease, but they can be annoyed by the stiffness that develops when initiating physical activity such as walking, running, or getting up from a seated position. Myotonia can be controlled by drugs such as quinine or procainamide that act by reducing the active sodium currents during an action potential, but the side effects of the drugs are often unpleasant. Myotonia is more severe with the initial attempt at muscular movements of the same muscles. This important phenomenon is referred to as "warm-up."

Myotonia is diagnosed clinically by the electromyogram (EMG), which makes use of the fact that when a myotonic fiber is depolarized by the probing action of the recording electrode when it is inserted into the muscle, it responds with a highly characteristic rhythmic discharge of impulses that increase and decrease in frequency. When this is made audible electronically, the sound is described as that of an attacking "dive bomber." Because the EMG diagnosis can be based on finding a small number of fibers that give the dive-bomber discharge, myotonia has been reported as an incidental finding in numerous diseases and after administration of drugs. The EMG has been the only clinical indication on many occasions (1). The main point is that the EMG detects only the presence of repetitive firing and that this in itself is not sufficient to determine the type of myotonic disease or the nature of the membrane abnormality.

Some other clinical phenomena are: the percusion response, which is an abnormal local contraction of a few seconds duration elicited in the muscle by striking the belly of the muscle with a reflex hammer; the increase in the severity of myotonia by exposure to a cold environment; and the prolonged response of myotonic muscle to depolarization by electric currents, potassium chloride, or a depolarizing neuromuscular blocking agent (e.g., succinylcholine).

In the hereditary myotonia of the goat we see the same clinical picture as in myotonia congenita. If one suddenly disturbs a quietly grazing myotonic goat the initial sudden movement of the animal to get away can lead to a severe myotonic attack in which the goat may fall over or assume a bizarre position for a few seconds with the limbs and body frozen by the strong contractions. The goat recovers, the myotonia diminishes by the warm-up mechanism and a second attack is harder to elicit without a period of rest. The EMG signs, percussion response, the effect of cold, and other signs are also very similar in goat and human myotonia (2).

MYOTONIC MEMBRANE

When I first began my studies on myotonic muscle around 1958, the current thinking then was that myotonic fibers behaved something like nerve fibers in a low-calcium medium, and there was probably some abnormality having to do with the role of calcium in the myotonic membrane. Pharmacological models for myotonia included the effects of drugs, loosely called "veratrinic agents," that could produce abnormal repetitive firing. In our first microelectrode studies of the myotonic goat fibers we were impressed with the relatively normal resting potentials and action potentials to single stimuli, although the falling phase and negative afterpotential were prolonged. The fibers lacked accommodation to sustained stimulating currents, and prolonged repetitive firing could occur accompanied by a slow depolarization of the fiber until the fibers became inexcitable at a membrane potential of about −40 mV. For some years before we learned of suitable methods to quiet the fibers, we were not sure of the origin of the slow depolarization. In a contracting fiber there is depolarization due to injury and tearing of the membrane around the microelectrode, even with flexibly mounted electrodes. This was especially true for the myotonic fiber because of its high membrane resistance, which was the most intriguing of our early measurements. It was in the course of attempting to determine the cause of the high resistance of the myotonic fiber that we began looking at the effects of replacement of chloride ions by less permeant anions. The myotonic fiber had a total membrane conductance about one-third that of normal fibers. If mammalian fibers had similar conductances to frog fibers, in which only one-third of the total conductance was due to potassium (5,6), then we reasoned that chloride conductance must be reduced in the myotonic fiber. Normal goat fibers in the chloride-free medium behaved remarkably similar to myotonic fibers chloride-containing medium. Not only was the membrane resistance equivalent to that of the myotonic fiber, suggesting a lack of chloride conductance, but the fibers had the typical myotonic excitability, including the low threshold, lack of accommodation, repetitive firing, and myotonic after-discharge. At that time I proposed that the basis for myotonia was the congenital absence or reduction of chloride conductance (7). In later studies we confirmed the low chloride permeability of myotonic goat fibers by measurement of chloride fluxes (8) and by measurement of component resting membrane conductances (9). We have also demonstrated the reduced chloride conductance in selected cases of human myotonia congenita (10,11).

In two model myotonias in mammalian fibers, one produced acutely by low concentrations (10^{-5} to 10^{-3} M) of a variety of monocarboxylic aromatic acids, the other produced by chronic administration of certain cholesterol derivatives, the mechanism has also been shown to depend on the block of resting chloride conductance (9,12). Dr. Rüdel discusses the details of the pharmacologically induced myotonias in the following chapter.

Although we proposed that decreased membrane chloride conductance per se could account for the myotonic reaction, we were very cautious at the beginning for several reasons: the mechanism seemed too simple; our impermeant anion substitutes for chloride and chemical inducers could bind calcium or react with the membrane; and in one of the better quantitative models for excitation, the squid axon, chloride conductance was of little importance. Gradually these objections to the theory have been considered and the initial proposal has held up.

A notable contribution was made by Bretag (13), who studied a Hodgkin–Huxley model in which the parameters were adapted to skeletal muscle based on the voltage–clamp studies of Adrian, Chandler, and Hodgkin (14). Under these circumstances, he was able to simulate with the computer repetitive firing and increased excitability by decreasing the chloride conductance (leak conductance term). He could also simulate the action of an antimyotonic drug (e.g., procainamide) in blocking the tendency to repetitive firing, i.e., by decreasing the sodium conductance term. Bretag was also able to induce myotonia in rat muscle using an anion substitute that showed no calcium binding.

A deficiency of the simple Hodgkin–Huxley model for a myotonic surface membrane was that there was no process having a sufficiently slow buildup of depolarization that would permit the myotonic after-discharge following cessation of the stimulus. In the simple model the repetitive firing would occur during a constant current pulse but would stop abruptly when the pulse was turned off, or the repetitive firing could be made to continue indefinitely. There was also no depolarization or after-discharge to a tetanic stimulation.

A solution to this problem was offered by the studies of Adrian and Bryant

(15) in which we were able to demonstrate that the transverse tubular system could act as a suitable sink causing a 1-mV depolarization per impulse that decayed with a time constant of about 0.5 sec. This depolarization disappeared following tubular disruption by glycerol treatment, and the fiber then behaved more or less like the simple computer simulations that include only the surface membrane. The depolarization produced by the tubular membranes is quite probably due to accumulation of potassium ions in the tubules as a result of repetitive action potentials on the tubular membranes. In theory, about 3-mV depolarization should occur across the tubular membrane per impulse. Taking into account the electrical coupling between tubular and surface membranes, one would expect the observed 1 mV/impulse in the myotonic fiber. In the normal fiber this effect is much less caused by the shunting effect of the membrane chloride conductance; also, the accommodation and higher threshold in the normal fiber makes this mechanism unlikely to produce an after-discharge.

In these studies we overcame the usual difficulty experienced when recording membrane potentials with microelectrodes in a contracting fiber by reducing the contraction with dantrolene sodium. Fortunately the site of action of dantrolene is beyond the tubular system (16), and it does not interfere with current flow from the transverse tubules or the mechanism of myotonic discharge. This method of quieting the fiber will be very useful in future studies of myotonia.

GENERALITY OF THE MEMBRANE DEFECT

Except for the data of Winer et al. (17) and Burns et al. (18), which suggest a lipid abnormality in the red cell membrane and a lower serum cholesterol in myotonic goats, there have been no published studies indicating a generalized defect in membrane or of membrane components. Harvey (19) reported that renal cortical arginine-glycine transamidinase activity was about half of normal in myotonic goats and this appears to be the only enzyme abnormality studied. Although the evidence is sparse, there are some data suggesting a generalized abnormality.

No one, to my knowledge has, looked in other tissues of the myotonic goat for evidence of the low chloride conductance abnormality. Nerve and other excitable tissues appear to be less dependent on the chloride conductance for electrical stability than muscle and it is possible that a block of chloride conductance in these tissues would not give rise to abnormal excitability. We have not looked at other tissues, but we have started to look deeper into the muscle fiber.

Since the transverse tubular membranes are in a sense invaginations of the surface membrane, it is natural to ask whether they share the low chloride conductance abnormality of the surface membrane. Together with Mrs. K. Morgan we have tried to make these measurements by examining the conductance differences between normal and myotonic fibers, with intact and glycerol-disrupted T systems, and in normal and chloride-free media. Our preliminary conclusions based on 12 biopsies are that tubular membranes in the normal goat have the proportion of potassium and chloride conductance, and that chloride conductance is lower in myotonic tubules. Unfortunately, the disruption process can introduce variable degrees of leakage conductance, hence the variance in the measurements is high and the results not as reliable as desired. We are beginning to apply AC Impedance methods to the problem in the near future to allow separation of tubular and surface leakage without the necessity of disruption. Using the disruption method on frog fibers, Eisenberg and Gage reported no chloride conductance of the normal tubular membrane (20). We suspect that mammalian fibers may differ in this respect.

EXCITATION–CONTRACTION COUPLING

The model myotonia produced by blocking the chloride channel of goat fibers with low concentrations of aromatic monocarboxylic acids approximated the naturally occurring myotonia well enough that one wonders whether the disease is caused by a carboxylic acid that accumulated in the fiber. I suggest an accumulation in the fiber because we have not detected acute induction of myotonia by serum, and the myotonia does not "wash away" *in vitro* during several hours of observation. Along these lines, we recently reported a myotonia in avian muscle (21) induced by feeding small amounts of iodide to the birds over some days or by exposing the muscle *in vitro* to iodide ions. The iodide was shown to induce myotonia by reduction of the resting anion conductance as in congenital myotonia. If the muscles were exposed for some hours to the iodide then the myotonia washed away slowly. The point is that our hypothetical myotonia inducer might be an impermeant material accumulated within the fiber or bound to the membrane like the iodide.

A search for such a substance has not been made by direct chemical assay but we have approached the problem from another point of view and have come up with some surprising results. If one examines the long list of monocarboxylic acid myotonia inducers (2) it is seen that these agents generally produce twitch potentiation in skeletal muscle. One would expect then that myotonic fibers have a potentiated twitch if a carboxylic acid inducer were present. Because twitch potentiation is difficult to evaluate in absolute terms, we chose to examine the threshold of mechanical activation in normal and myotonic fibers.

We used the technique described by Adrian et al. (22) in which a point on a muscle fiber is controlled by a two-microelectrode voltage-clamp. The surface membrane potential is step-depolarized by different degrees and the duration of the pulse that gives a just detectable contraction near the electrodes is determined at each pulse. From these data a membrane potential versus pulse duration plot is made from which one can see a "rheobase" and a duration-dependent region of the curve. On this basis the carboxylic acids lowered mechanical threshold from −58 mV of normal goat fibers in a concentration-dependent manner until the threshold was nearly equal to the −90 mV holding potential. Dantrolene sodium, an agent that blocks excitation–contraction coupling (16), considerably raised threshold (up to −30 mV) as expected. Myotonic goat fibers unexpectedly had an elevated threshold of −44 mV unlike that produced by the carboxylic acid model of myotonia. We found two myotonia-inducing treatments of normal fibers that more closely resemble hereditary myotonia in elevating the threshold of mechanical activation: substitution of sulfate ion for chloride in the bath, and chronic pretreatment of a normal goat by 20,25-diazacholesterol.

Our tentative conclusions from our study of the excitation–contraction coupling mechanisms suggest that the membrane system involved here is abnormal and that the carboxylic acid model does not duplicate the condition, but the diazacholesterol model (12) does. From these data one might suggest that surface charge of the tubular membranes or the voltage-dependent charge movement mechanism (23) is affected in the disease.

REGULATION OF CHLORIDE CONDUCTANCE

A high chloride conductance is necessary for a stable skeletal muscle membrane and its absence leads to myotonia. One may ask, then, if there are factors or possibly some regulating mechanisms for maintaining the normal chloride conductance of the skeletal muscle membrane. In frog muscle (24) chloride conductance does not depend upon an intact innervation, but we have recently reported (25,26) that in goat gastrocnemius fibers and rat fast fibers that chloride conductance decreases to zero after an initial delay of a few days following denervation. Since denervation blocks the influence of nerve impulses as well as chemical factors transported by the motor nerve, we have begun pharmacological studies using agents such as vinblastine and colchicine that can be applied locally to the motor nerve and can block fast axoplasmic transport without blocking transmission of nerve impulses (27). Dr. Conte-Camerino demonstrated that colchicine and vinblastine applied to the motor nerves of rats can have different effects on the resting conductances. In the concentrations used, colchicine blocked chloride conductance and reduced potassium conductance, whereas vinblastine raised potassium conductance without affecting chloride conductance. These data suggest that there are more than one factor involved in the control of chloride conductance.

Myotonia might be explained as the absence of the factor that maintains the chloride conductance, but it is not known whether there are fewer chloride channels or whether existing chloride channels are blocked. The diazacholesterol studies in the goat (28) suggest a role of cholesterol in maintaining chloride conductance. Interference with cholesterol synthesis could produce myotonia, but it should be noted that there are agents that block cholesterol synthesis without producing myotonia (29).

The general properties of the chloride channel in mammalian muscle are like those in frog skeletal muscle (30). In unpublished studies we have found that the chloride conductance is turned off at pH 4 and is saturated at pH 9 with an inflection (pK_a ?) around 7.1 at 38°C. Chloride conductance is increased by bicarbonate concentration independent of pH, and is decreased by the foreign anions and cations used in the Hutter and Warner studies (31,32). A voltage dependence of the chloride conductance as described in frog muscle (33) has not been investigated in mammalian fibers. An important difference between mammalian and frog skeletal muscle is the lack of effect of the carboxylic acid myotonia inducers on the chloride conductance of the frog membrane.

UNEXPLAINED PHENOMENA

Many of the phenomena associated with myotonia, some of which have been mentioned above, are without a satisfactory explanation. There may already be explanations from membrane studies on other systems. I will simply list some of the more important phenomena for your consideration (2).

1. The increase in severity of myotonia in a cold environment
2. The "warm-up" (34)
3. The abolition of myotonia by water deprivation (35)
4. The percussion response
5. The increased mechanical threshold in the myotonic fiber
6. The regulation of chloride conductance by the motor nerve
7. The action of the myotonia inducers (carboxylic acids and diazacholesterol) in blocking chloride conductance in mammalian fibers, and the lack of effect of the carboxylic acids on frog muscle
8. The biochemical nature of the genetic defect

SUMMARY AND CONCLUSIONS

The myotonias are characterized by alterations of the skeletal muscle membranes that lead to abnormal repetitive firing of action potentials. In the hereditary myotonia of goats, which closely resembles *myotonia congenita* in man, the electrophysiology of the myotonic fiber can be largely explained as a consequence of an inherited low membrane chloride conductance.

Although the electrophysiological aspects of myotonia can be studied in a model myotonia produced by carboxylic acids that block chloride conductance in normal fibers, there are differences. New evidence was presented that, in naturally occurring myotonia, the threshold for mechanical activation of the fibers was higher than normal, whereas in carboxylic acid-induced myotonia it is lower than normal. These data do not favor the hypothesis that a carboxylic acid metabolite is the cause of myotonia. Diazacholesterol-induced myotonia more closely resembled the natural variety in having a higher mechanical threshold.

The nature of the chloride channel in mammalian muscle has not been characterized. Some of the factors shown thus far that are involved in the regulation of chloride conductance of normal mammalian muscle are: bicarbonate concentration, extracellular pH, and factors from the motor nerve some of which are carried by the fast transport system. It has not been shown whether the chloride channels are blocked or absent in hereditary myotonia.

Many membrane phenomena associated with myotonia such as the water-deprivation effect, the percussion response, the cold effect, and warm-up are so far unexplained. Although the myotonia in the goat is inherited as autosomal dominant, the possible generality and nature of a specific biochemical lesion has not been systematically investigated. Thus, myotonia as a model of an inherited membrane disease is a fertile area for considerable biophysical and biochemical research.

REFERENCES

1. Walton, J. N. editor (1974): *Disorders of Voluntary Muscle,* 3rd ed. (Churchill) Livingston, Edinburgh and London.
2. Bryant, S. H. (1973): In: *New Developments in Electromyography and Clinical Neurophysiology,* p. 420, edited by J. E. Desmedt. Karger, Basel.
3. Thomsen, J. (1876): *Arch. Psychol. Genève,* 6:702.
4. Becker, P. E. (1967): In: *Proceedings 3rd International Congress Human Genetics.* Johns Hopkins Univ. Press, Baltimore, Md.
5. Hodgkin, A., and Horowicz, P. (1959): *J. Physiol. (Lond.),* 148:127.
6. Adrian, R. H., and Freygang, W. H. (1962): *J. Physiol. (Lond.),* 163:61.
7. Bryant, S. H. (1962): *Fed. Proc.,* 21:312.
8. Lipicky, R. J., and Bryant, S. H. (1966): *J. Gen. Physiol.,* 50:89.
9. Bryant, S. H., and Morales-Aguilera, A. (1971): *J. Physiol. (Lond.),* 219:367.
10. Lipicky, R. J., Bryant, S. H., and Salmon, J. H. (1971): *J. Clin. Invest.,* 50:2091.
11. Lipicky, R. J., and Bryant, S. H. manuscript in preparation.
12. Rüdel, R., and Senges, J. (1972): *Pflügers Arch. Eur. J. Physiol.,* 331:324.
13. Bretag, A. (1973): In: *New Developments in Electromyography and Clinical Neurophysiology,* edited by J. E. Desmedt. Karger, Basel.
14. Adrian, R. H., Chandler, W. K., and Hodgkin, A. L. (1970): *J. Physiol. (Lond.),* 208:607.
15. Adrian, R. H., and Bryant, S. H. (1974): *J. Physiol. (Lond.),* 240:505.
16. Ellis, K. O., and Bryant, S. H. (1972): *Nauyn-Schmiedebergs Arch. Pharmacol.,* 275:83.
17. Winer, N., Klachko, D. M., Baer, R. D., Dale, H. E., and Burns, T. W. (1965): *Clin. Res.,* 13:326.
18. Burns, T. W., Dale, H. E., and Langley, P. L. (1965): *Clin. Res.,* 13:235.
19. Harvey, J. C. (1969): *Johns Hopkins Med. J.,* 125:270.
20. Eisenberg, R. S., and Gage, P. W. (1969): *J. Gen. Physiol.,* 53:279.
21. Morgan, K., Entrikin, R. K., and Bryant, S. H. (1975): *Am. J. Physiol. Am. J. Physiol.* 229:1155.

22. Adrian, R. H., Chandler, W. K., and Hodgkin, A. L. (1969): *J. Physiol.* (*Lond.*), 204:207.
23. Schneider, M. F., and Chandler, W. K. (1973): *Nature,* 242:244.
24. Hubbard, S. J. (1963): *J. Physiol.* (*Lond.*), 165:443.
25. Bryant, S. H., and Conte-Camerino, D. *J. Neurobiol.* (*in press*).
26. Conte-Camerino, D., and Bryant, S. H. *J. Neurobiol.* (*in press*).
27. Albuquerque, E. X., Warnick, J. E., Tasse, J. R., and Sansone, F. M. (1973): *Exp. Neurol.,* 37:607.
28. Burns, T. W., Dale, H. E., and Langley, P. L. (1969): *Am. J. Physiol.,* 209:1227.
29. Winer, N., Klachko, D. M., Baer, R. D., Langley, P. L., and Burns, T. W. (1966): *Science,* 153:312.
30. Woodbury, J. W., and Miles, P. R. (1973): *J. Gen. Physiol.,* 62:324.
31. Hutter, O. F., and Warner, A. E. (1967): *J. Physiol.* (*Lond.*), 189:403.
32. Hutter, O. F., and Warner, A. E. (1967): *J. Physiol.* (*Lond.*), 189:445.
33. Hutter, O. F., and Warner, A. E. (1972): *J. Physiol.* (*Lond.*), 227:275.
34. Rüdel, R., and Senges, J. (1973): *Pflügers Arch. Eur. J. Physiol.,* 341:121.
35. Hegyeli, A., and Szent-Gyorgi, A. (1961): *Science,* 133:1011.

Membranes and Disease, edited by L. Bolis, J. F. Hoffman, and A. Leaf. Raven Press, New York, © 1976.

The Mechanism of Pharmacologically Induced Myotonia

Reinhardt Rüdel

Physiological Institute of the Technical University, Munich, West Germany

Among a variety of chemical agents that can generate repetitive activity in mammalian skeletal muscle three groups have been found to induce a state which closely resembles clinical myotonia (1): impermeant anion substitutes for chloride in the extracellular medium, the monocarboxylic aromatic acids, and diazocholesterol and its analogs. Substances of either group are applied in experimental work designed to study the membrane defect leading to the symptom of myotonia. Impermeant anions and the monocarboxylic aromatic acids have the advantage of acting within minutes after application and of producing myotonia even *in vitro* (2–4). The cholesterol derivatives, when administered for several weeks, generate a chronic state of myotonia (5).

In this chapter I will first describe to what extent pharmacologically induced myotonia resembles clinical myotonia (6). I will then try to explain many of the peculiarities of myotonia as likely to result from potassium accumulation in the tubular system of muscle cells which have a reduced membrane conductance to chloride. Finally, I would like to report some preliminary results concerning the possible effect of myotonia-inducing drugs on the properties of muscle spindles. For more details and methods the papers listed in the bibliography should be consulted.

CLINICAL MYOTONIA

The clinical condition called myotonia seems to present two closely related phenomena (7): prolonged muscular contraction following neural, electrical or mechanical stimulation ("peripheral" or "muscular" myotonia) and a cramplike muscular stiffness on voluntary movement which involves central mechanisms ("after-spasm"). In a sequence of movements, as when a patient starts climbing stairs, it is usually not the first step which is most impaired, but stiffness increases with the second and third steps. Thereafter the condition improves, a phenomenon called "warm-up."

When in a myotonic patient a motor nerve is stimulated at low frequency the compound muscle action potential, as recorded by electromyography, often shows a decline in amplitude. With a short burst of high frequency stimuli, repetitive activity can be elicited in the afflicted muscle fibers which outlasts stimulation for many seconds. Such a "myotonic run" usually has a characteristic increase and decrease in frequency. It is generally believed that it is this repetitive activity which causes peripheral myotonia.

Lipicky and Bryant (8) investigated the properties of the cell membranes of muscle fibers from myotonic patients. Of all parameters tested electrophysiologically only the specific membrane resistance presented a major abnormality, namely, a twofold increase. This was shown to be due to a marked reduction of the membrane conductance to chloride.

PHARMACOLOGICALLY INDUCED MYOTONIA

Laboratory animals fed on a diet containing 20,25-diazocholesterol (9) or injected with 2,4-dichlorphenoxyacetic acid (10) develop muscular stiffness which closely resembles myotonia in man. Prolonged muscular contraction can be demonstrated on mechanical or electrical stimulation and repetitive activity can be recorded electromyographically.

When excised diaphragm from normal rats is bathed in a Krebs' solution containing 2 to 2.5 mM 2,4-dichlorphenoxyacetate (11) or in a solution in which 70–95% of the chloride content has been replaced by the impermeant anion isethionate (4), repetitive activity following stimulation can be recorded with a concentric needle electrode and this gives rise to long-lasting contractions (Fig. 1). Myotonic runs and prolonged contraction become shorter when stimulation is repeated after a 1-sec interval. This may be regarded as an equivalent of warm-up. As in muscles from myotonic patients the major alteration in the membrane parameters of muscles made myotonic by monocarboxylic aromatic acids (2) and by 20,25-diazocholesterol treatment (4) was shown to be a marked reduction of the membrane conductance to chloride.

INTRACELLULAR RECORDINGS OF MEMBRANE POTENTIAL

A study as to how the reduced chloride conductance impairs electrical stability of the cell membrane requires intracellular recordings of membrane

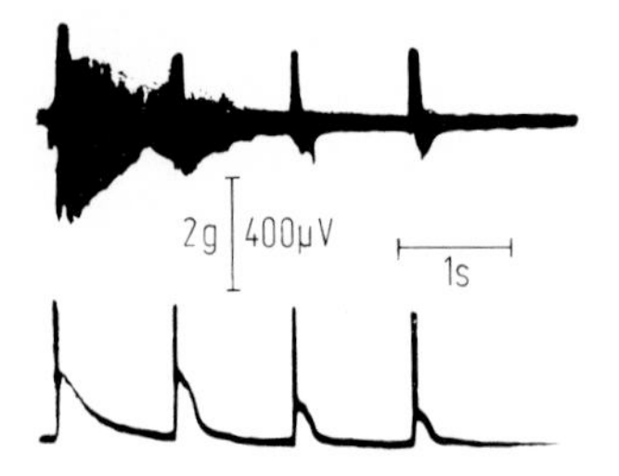

FIG. 1. Simultaneously recorded electromyogram (*upper trace*) and mechanogram (*lower trace*) from a strip of rat diaphragm made myotonic by substitution of isethionate for chloride in the bathing solution. Stimulation rate 1/sec. Myotonic response gets smaller because of "warm-up" (4).

potentials from myotonic muscle fibers. These measurements are not easy to conduct because the strong movement accompanied by the repetitive activity tends to dislodge the recording microelectrode. This difficulty can be overcome by adding to the bathing solution a substance, dantrolene sodium, which uncouples excitation and contraction without affecting electrical events at the surface membrane of the cell (12). Typical action potentials recorded intracellularly from a muscle fiber made myotonic by addition of 2 mM 2,4-dichlorphenoxyacetate to a dantrolene-containing bathing solution are shown in the upper traces of the records in Fig. 2. In Fig. 2A, following a single stimulus, an action potential starts from a resting potential of −75 mV. Its up- and down-stroke speeds are quite normal, as judged from the first time derivatives (lower trace), and so is its overshoot. However, following it, the membrane shows a large after-depolarization which decays more slowly than in a nonmyotonic fiber (compare to controls illustrated in Fig. 3E). When two action potentials were stimulated at 10-msec intervals this after-depolarization was increased (Fig. 2B). After three stimuli the after-depolarization was even larger, so that a threshold was reached at which a myotonic run was triggered (Fig. 2C).

The whole course of myotonic runs is illustrated in Fig. 3A-D together with a control record (Fig. 3E) in which a nonmyotonic fiber was stimulated at a rate corresponding to the mean frequency seen in myotonic runs (100 Hz). In all the myotonic fibers following the last stimulated action potential this pronounced after-depolarization was observed which lead to the threshold for repetitive activity. During the runs the potential between two subsequent spikes fell to less negative values and concomitantly the upstroke velocity and the size of the action potentials decreased. This depolarization either continued through the entire run as in Fig. 3A and D or it gave way

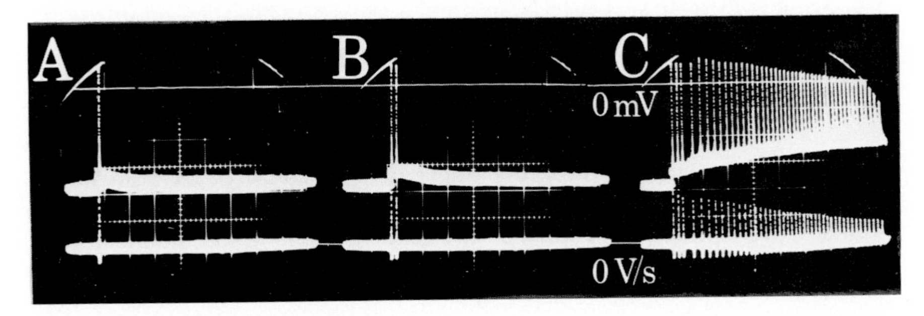

FIG. 2. Intracellularly recorded action potentials (*upper traces,* 20 mV/div) and their first time derivatives (*lower traces,* 200 V/sec-div) from a rat diaphragm fiber made myotonic by the addition of 2 mM 2,4-dichlorphenoxyacetate to the bathing solution. **A:** Single stimulus. **B:** 2 stimuli, 10 msec apart. **C:** 3 stimuli, 10 msec apart. *Time calibration:* 50 msec/div.

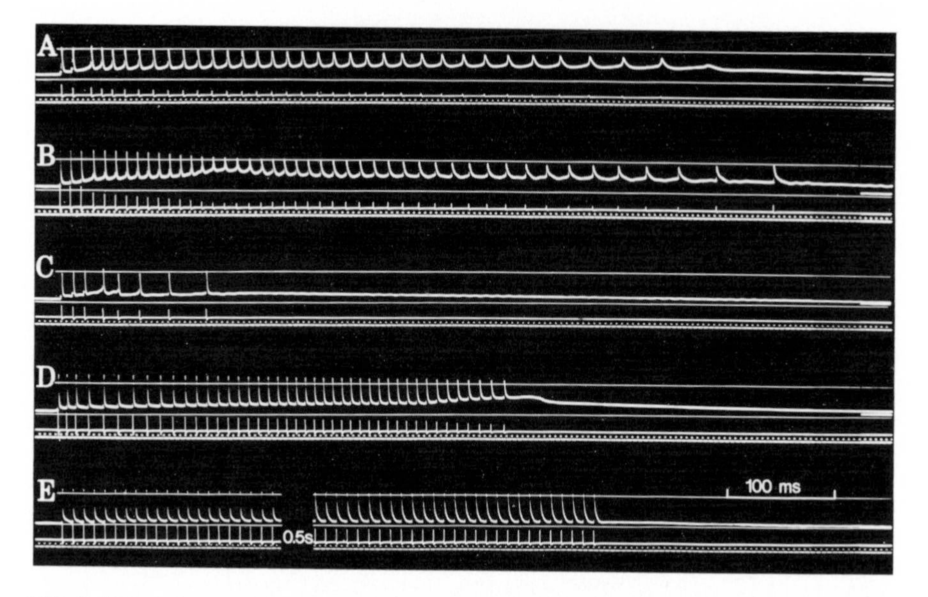

FIG. 3. Myotonic runs from 4 different rat diaphragm fibers made myotonic with 2,4-dichlorphenoxyacetate (A–D) and control from nonmyotonic fiber stimulated for 1 sec at 100 Hz (E). *Upper traces:* Membrane potential *V*, calibration 0 and −80 mV given by thin lines; *middle traces:* time derivative *dV/dt*, calibration 0 and 400 V/sec given by thin lines; *lower traces:* time marks, 5 msec interval. The potential attained by the membrane 1 sec after the last action potential is indicated at the end of each upper trace (14).

to a repolarization as in B or the run illustrated in Fig. 4. Whichever was the case seemed to be dependent on the distance of the recording electrode from the pacemaker site. The latter appeared not to be always constant throughout a run as indicated by the variability of the onsets of the action potentials. In Fig. 3A the pacemaker was close to the electrode all the time because each action potential was preceded by a slow depolarization. In Fig. 3B, however, it seems to have moved away toward the end of the run because the upstrokes of the action potentials became increasingly abrupt. On the contrary, in Fig. 3D the pacemaker seems to have come closer to the electrode during the run. Such wandering of the pacemaker is in fact expected according to computations of instabilities in mathematical models of muscle fibers with lowered conductance (Adrian, *private communication*).

In conclusion, an increased and prolonged after-depolarization following each action potential is a characteristic feature of the myotonic membrane. Depolarization seems to be greatest at the site which becomes pacemaker.

TUBULAR POTASSIUM ACCUMULATION

Investigating myotonic goat muscles Adrian and Bryant (13) came to the conclusion that tubular potassium accumulation resulting from the initially driven activity in combination with the reduced membrane conductance to chloride can account for the increased after-depolarization. The same mechanism is likely to operate in pharmacologically induced myotonia (14).

In a skeletal muscle fiber each action potential alters the potassium concentration by about 0.3 mM (15). In a driven nonmyotonic fiber accumulation of potassium in the tubules only slightly affects the surface membrane potential (Fig. 3E) because normally the surface membrane conductance is much greater than is tubular conductance. The reduction of the surface membrane conductance in myotonic fibers lets tubular depolarization spread to the surface membrane where it can reach the threshold for repetitive activity.

Freygang, Goldstein, and Hellam (16) have devised a method to assess that potassium is accumulated in the tubular system during activity: injection of a square pulse of negative current into a muscle fiber causes hyperpolarization of the membrane with a slow component which is due to the depletion of the tubular system of potassium. When the current pulse is injected immediately following activity, the slow component should be pronounced if activity had increased the tubular potassium content. This experiment was carried out with rat diaphragm bathed in a solution containing 2 mM 2,4-dichlorphenoxyacetate (Fig. 4). A control current pulse was injected into a fiber and then a myotonic run was triggered by five extracellular stimuli. Test current pulses were injected into the fiber every half second since the duration of the run could not be predicted. The first membrane hyperpolarization following the myotonic run in Fig. 4 obviously had much more of a

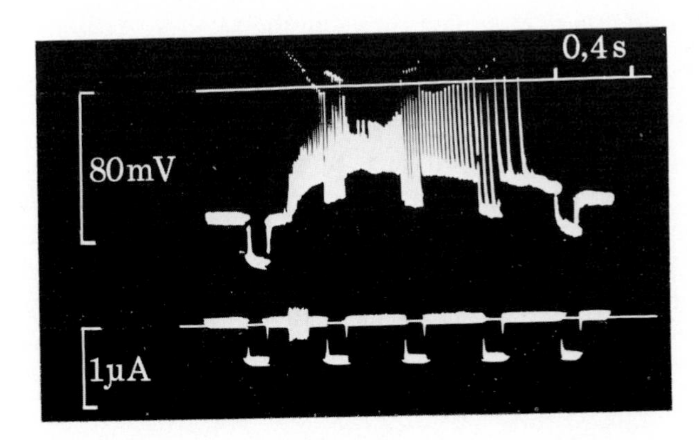

FIG. 4. Injection of hyperpolarizing current pulses of 100 msec duration into a rat diaphragm fiber made myotonic by 2,4-dichlorphenoxyacetate. Separation between current injecting and voltage recording electrodes about 0.1 mm. The myotonic run was elicited by a train of extracellular stimuli applied after the first current pulse. Some of the spikes are hardly visible, because of underexposure (14).

slow component than the one during the test pulse, indicating that indeed the tubular system had been filled with potassium during the run.

The mechanism of muscular myotonia thus seems to be the following: several driven action potentials are required to increase the tubular potassium concentration to a degree that depolarization of the surface membrane reaches threshold for repetitive activity. This might be the explanation for why myotonic patients are less stiff at the beginning of a sequence of movements. Repetitive activity then causes further depolarization of the membrane so that the Na carrying system gets more and more inactivated. This inactivation brings the run to an end.

The hyperpolarizations induced during the run in Fig. 4 caused the frequency of the repetitive activity to decrease and the size of the action potentials to increase. This suggests that the typical changes in these two parameters observed also during electromyographically recorded myotonic runs are mainly a consequence of the increased after-depolarization.

EFFECT OF MONOCARBOXYLIC AROMATIC ACIDS ON MUSCLE SPINDLES

While the *in vitro* experiments give information on the mechanism of peripheral myotonia, an investigation of muscular stiffness on willed effort has to be carried out with whole animals because after-spasm seems to involve a central mechanism. As Denny-Brown and Nevin (6) have pointed out after-spasm is likely to be reflexly determined by some persistent effect in the muscle proprioceptors. To investigate whether or not the properties of intrafusal muscle fibers can be influenced by drugs that induce myotonia in extrafusal fibers single muscle spindles in the gastrocnemius muscle of anesthetized cats were identified by tieing the severed Achilles tendon to a sine-wave generator and by recording the output in the isolated corresponding primary afferent Ia-fiber (Grüsser, Pellnitz, and Rüdel, *unpublished*). In two cases also the efferent γ-fiber could be isolated. When 2,4-dichlorphenoxyacetate or anthracene-9-carboxylic acid was injected into the femoral artery, at 8 mg/kg, a concentration that produces severe myotonia in goats (1) and at 16 mg/kg no change in the spindle response to mechanical pull or to stimulation of the γ-fiber could be observed while at the same time long lasting myotonic runs were recorded from extrafusal gastrocnemius fibers with a concentric needle electrode. Search for an effect of monocarboxylic aromatic acids on intrafusal fibers which might explain after-spasm, has therefore not yet been successful.

REFERENCES

1. Bryant, S. H. (1973): In: *New Developments in Electromyography and Clinical Neurophysiology,* Vol. 1, p. 420, edited by J. E. Desmedt. Karger, Basel.

2. Bryant, S. H., and Morales-Aguilera, A. (1971): *J. Physiol.,* 219:367.
3. Rüdel, R., and Senges, J. (1972): *Pflügers Arch.,* 331:324.
4. Rüdel, R., and Senges, J. (1972): *Naunyn-Schmiedebergs Arch. Pharmacol.,* 274:337.
5. Burns, T. W., Dale, H. E., and Langley, P. L. (1965): *Am. J. Physiol.,* 209:1227.
6. Landau, W. M. (1952): *Neurology,* 2:369.
7. Denny-Brown, D., and Nevin, S. (1939): *Brain,* 62:341.
8. Lipicky, R. J., and Bryant, S. H. (1973): In: *New Developments in Electromyography and Clinical Neurophysiology,* Vol. 1, p. 451, edited by J. E. Desmedt. Karger, Basel.
9. Winer, N., Martt, J. M., Somers, J. E., Wolcott, L., Dale, H. E., and Burns, T. W. (1965): *J. Lab. Clin. Med.,* 66:758.
10. Eyzaguirre, C., Folk, B. P., Zierler, K. L., and Liliental, J. L. (1948): *Am. J. Physiol.,* 155:69.
11. Senges, J., and Rüdel, R. (1972): *Pflügers Arch.,* 331:315.
12. Ellis, K. O., and Bryant, S. H. (1972): *Naunyn-Schmiedebergs Arch. Pharmacol.,* 274: 107.
13. Adrian, R. H., and Bryant, S. H. (1974): *J. Physiol.,* 240:505.
14. Rüdel, R., and Keller, M. (1975): In: *Recent Advances in Myology,* p. 334, edited by W. G. Bradley, D. Gardner-Medwin, and J. N. Walton. Excerpta Medica, Amsterdam.
15. Adrian, R. H., and Peachey, L. D. (1973): *J. Physiol.,* 235:103.
16. Freygang Jr., W. H., Goldstein, D. A., and Hellam, D. C. (1964): *J. Gen. Physiol.,* 47:929.

Membranes and Disease, edited by L. Bolis, J. F. Hoffman, and A. Leaf. Raven Press, New York, © 1976.

Effect of Colchicine and Vinblastine Treatment on the Conductances of Rat Skeletal Muscle Fibers

D. Conte-Camerino and S. H. Bryant

Istituto di Farmacologia, Università di Bari, Bari, Italy, and Department of Pharmacology and Therapeutics, University of Cincinnati, College of Medicine, Cincinnati, Ohio 45267

The decreased chloride conductance of the sarcolemmal membrane of skeletal muscle fibers of goats or humans with hereditary myotonia can account reasonably well for the abnormal excitability and repetitive firing characteristic of this disease. To date the biochemical lesion of the membrane has not been identified, but there is the suggestion that membrane lipid composition may be abnormal (2). One of the approaches that we are now exploring is to gain a better understanding of the physiological factors that control or influence the normal chloride conductance in mammalian fibers. In this way we hope to get further leads as to the etiology of the myotonic membrane, or at least uncover useful phenomena the analysis of which will tell more of the nature of membrane chloride conductance.

We recently found that denervation of goat gastrocnemius fibers leads to a delayed block of sarcolemmal chloride conductance; we then found this to be true for the denervated fibers of the rat extensor digitorum longus (3,4). The dependence of the mammalian muscle membrane upon innervation in order to maintain the high chloride conductance is different from the situation in the frog wherein chloride conductance is relatively independent of innervation. In our denervation studies the accompanying paralysis and absence of action potentials made it impossible to decide whether the block of the chloride conductance was due to the loss of supply of neural substances or lack of membrane stimulation.

In the studies reported in this chapter we have applied two pharmacological agents, colchicine and vinblastine, to the motor nerve supplying the rat extensor muscle and have observed the effects on membrane chloride conductance. These agents in appropriate concentrations produce many of the effects of denervation without block of motor activity or muscle fiber contraction. Because these agents produce their effects by blocking microtubular transport in the axon, we hoped to determine whether chloride conductance was influenced by transported "trophic" factors (6).

METHODS

Two groups of chronically treated rats were prepared. In one group, from 100 to 450 μg of colchicine were injected by microsyringe under the epineurium of the peroneal nerve (5). In the other group, silicon polymer cuffs containing 0.1% vinblastine were placed around the peroneal nerve (7). In either case after the surgical procedures on the nerve the rats had normal use of the extensor muscles during the treatment period.

After 7–10 days of the colchicine treatment or after 9–17 days of the vinblastine treatment the extensors of the treated and control legs of each rat were removed and placed in a double muscle chamber for measurement and comparison of their electrical parameters.

Electrical measurements were made with two microelectrodes, one for delivering current and the other for recording the membrane potentials. We determined resting potentials, action potentials, excitability, and cable parameters. The cable parameters were calculated in the usual way from the electrotonic potentials generated by square-wave hyperpolarizing constant current pulses recorded at two or more distances from the current electrode (1). In addition the cable parameters were measured in both normal physiological solution and in a chloride-free (methylsulfate-substituted) physiological solution. Neglecting the contribution of ionic conductances other than those of potassium and chloride the component conductances of the membrane are obtained in the following way: the reciprocal of the membrane resistance in chloride-free solution is the potassium conductance G_K and the total conductance in normal solution G_m minus G_K is the chloride conductance G_{Cl}.

RESULTS

Except for a slight paralysis that lasted about a day following the colchicine treatment the drugs did not interfere with normal contraction of the muscle, yet abnormal changes in membrane properties characteristic of denervation were observed during the time the measurements were made *in vitro*. The colchicine-treated preparations all had fibers that were capable of action potential production in normal physiological solution and following addition of 10^{-6} M TTX. The vinblastine treated fibers, likewise, were able to produce action potentials in TTX, but the TTX resistance diminished after 13 days.

In the colchicine-treated fibers the membrane resistance increased from a control value of $350 \pm 14\ \Omega \cdot cm^2$ to $950 \pm 46\ \Omega \cdot cm^2$, whereas in chloride-free solution there was little change, the control and treated fibers averaging $11{,}918 \pm 1{,}963\ \Omega \cdot cm^2$ and $14{,}796 \pm 2{,}362\ \Omega \cdot cm^2$, respectively. In contrast, the membrane resistance of the vinblastine-treated fibers, when measured in normal solution, remained the same ($315 \pm 13\ \Omega \cdot cm^2$ and $305 \pm 27\ \Omega \cdot cm^2$,

TABLE 1. *Component resting conductances of rat EDL fibers following colchicine and vinblastine treatment*

Preparation	Days	N	G_m (μmhos/cm²)	N	G_K (μmhos/cm²)	G_{Cl} (μmhos/cm²)
			Colchicine Treated Fibers			
Mean of 9 controls		68	3,206 ± 146	52	213 ± 30	2,993 ± 112
Mean of 5 colchicine-treated	7–10	62	1,182 ± 54[a]	37	116 ± 15[a]	1,066 ± 44[a]
			Vinblastine Treated Fibers			
Mean of 5 controls		24	3,297 ± 133	21	276 ± 56	3,021 ± 104
Mean of 6 vinblastine-treated	9–17	35	3,947 ± 268	27	473 ± 66[b]	3,474 ± 206

The columns from left to right are as follows: Preparation; days following colchicine and vinblastine treatment; N, number of fibers for G_m and G_K, respectively; G_m, G_K, and G_{Cl} are total membrane conductance, potassium conductance, and chloride conductance, respectively.

[a] Significantly different from mean of control group ($p < 0.01$).

[b] Significantly different from mean of control group ($p < 0.05$).

for control and treated, respectively), but in the chloride-free solution the treated fibers had only half the resistance of the control, 3,230 ± 407 Ω·cm² compared with 7,412 ± 1,625 Ω·cm², respectively.

The component conductances calculated from these resistant values for colchicine and vinblastine treatment are given in Table 1. It can be seen that colchicine lowered both G_{Cl} and G_K, whereas vinblastine produced no change in G_{Cl} but increased G_K.

The mean unselected resting potentials for both control groups were the same in chloride-containing or chloride-free media and the combined average was 60 mV. Resting potentials from surface fibers after penetration with two electrodes for the cable measurements were often 5–20 mV smaller that those obtained from subsurface fibers penetrated by a single electrode. Since analysis of the cable measurements did not indicate any significant differences in the means when we selected for high membrane potentials (>70 mV), we chose not to select fibers on the basis of potential so that we could better compare the experimental values with those of the controls. The mean resting potentials from colchicine- and vinblastine-treated preparations were lower ($p < 0.001$) than the mean of control fibers by 8 mV and more.

DISCUSSION

We have given evidence in this study that the resting membrane conductances of a fast mammalian muscle may be under the control of factors transported by the motor nerve. We further suggest that these factors can be influenced in specific ways by pharmacological agents such as colchicine and vinblastine that can alter microtubular transport. The effects of these agents probably depend on the concentration used and the method of application to the nerve (6), but under the particular conditions we chose to use, it appears possible to either increase or decrease G_K, or decrease G_{Cl}, without affecting nerve or muscle impulses. In the denervated mammalian fiber where there is loss of conducted impulses, we have found G_K to increase and G_{Cl} to decrease after 10 days. On the basis of our findings, we could postulate at least two factors carried by microtubular transport, one that maintains the high chloride conductance of the membrane, and another that keeps potassium conductance low.

REFERENCES

1. Bryant, S. H. (1969): *J. Physiol.*, 204:539.
2. Bryant, S. H. (1973): *New Dev. EMG Clin. Neurophysiol.*, 1:420.
3. Bryant, S. H., and Camerino, D. (1975): *J. Neurobiol.* (in press).
4. Camerino, D., and Bryant, S. H. (1975): *J. Neurobiol.* (in press).
5. Kreutzberg, G. W. (1969): *Proc. Natl. Acad. Sci. USA*, 62:722.
6. Ochs, S. (1974): *Ann. NY Acad. Sci.*, 228:202.
7. Robert, E. D., and Oester, Y. T. (1970): *J. Pharmacol. Exp. Ther.*, 174:133.

Membranes and Disease, edited by L. Bolis, J. F. Hoffman, and A. Leaf. Raven Press, New York, © 1976.

Transmembrane Transport of Inorganic Phosphate and Its Implication in Some Diseases

R. W. Straub, J. Ferrero, P. Jirounek, G. J. Jones, and A. Salamin

Département de Pharmacologie, Ecole de Médecine, CH-1211 Geneva 4, Switzerland

A great deal is known about the role of phosphates in metabolism, and in the structure of biological membranes and nucleic acids; on the other hand, the mechanism by which inorganic phosphate is taken up or released by cells has so far received little attention.

The phosphate turnover in healthy individuum is regulated by a complicated interplay between vitamin D and its metabolites (1), parathyroid hormone, calcitonin, and cyclic AMP (2). The equilibrium necessary for normal turnover can thus be subject to a variety of disturbances. Some of these can be corrected empirically by increasing or decreasing the amount of the regulating compounds in the body, but an experimental approach to their mode of action and to a more rational use of them will be possible only when the mechanism of phosphate transport is better understood.

Further, a familial, X-chromosome-linked form of insufficiency of phosphate uptake from the intestine (3) and of the reabsorption of phosphate in the kidney (4) has been described; this insufficiency is the cause of a vitamin D-resistant form of hypophosphatemic rickets. The phosphate transport in other tissues is also affected (4) and the transport system is also resistant to the action of parathyroid hormone. An understanding of this interesting lesion clearly depends on a better knowledge of the phosphate transport mechanism.

One reason why not much work has been done so far on the transport of inorganic phosphate may be that measurements of phosphate fluxes across the surface membranes of living cells seem to be difficult, due to the presence of several ionic forms of inorganic phosphate, the involvement of phosphate in many metabolic reactions, and the relative smallness of the phosphate fluxes (5,6) in comparison to those of other ions more frequently studied.

In order to overcome some of these difficulties we have used a preparation of small nonmyelinated nerve fibers, which by their large surface to volume ratio (7,8) allow the transmembrane fluxes to be measured relatively easily. Studies on the influx of inorganic phosphate into this tissue have shown that the influx per unit surface area of membrane is small; at 1 mM extracellular phosphate it amounts to only a few fmoles/cm^2-sec (9,10). Metabolic incorporation of phosphate, on the other hand, is much faster; the slowness of the influx is thus not caused by a slow rate of metabolism, but by a barrier to the influx of phosphate at the surface membrane (9). Increasing the extracellular phosphate concentration was found to increase the influx, with a tendency to saturation. Further, a large proportion of the influx depends on the presence of extracellular Na, this part saturates with increasing phosphate concentrations; on the other hand, the Na-independent influx is approximately proportional to the external phosphate. The influx of both the divalent and the monovalent phosphate ions appears to be Na-sensitive (11). The influx is slightly decreased in the presence of ouabain (10), or, when in high Na solutions, K is omitted. The influx is inhibited by arsenate.

In summary, then, a large proportion of the uptake of phosphate appears to be mediated by a specific Na-dependent, saturable transport system.

In this chapter, some recent studies on the efflux of phosphate and its relation to the influx will be described.

METHODS

The experimental methods, the apparatus, and the solutions were similar to those described in detail elsewhere (9). In brief, for loading the tissue with radioactive phosphate, a desheathed cervical vagus nerve from a rabbit was mounted in a tube opposite the window of a β-counter which was connected to a rate meter and a pen recorder. At the beginning of an experiment, during an equilibrating period of 1 hr, the tube was perfused with inactive Locke containing a given phosphate concentration. Then ^{32}P-labeled Locke, with the same phosphate concentration, was applied and the overall radioactivity of the preparation and the medium was recorded. The loading period lasted 2–3 hr. Preparations in which the uptake of radiophosphate differed appreciably from that of earlier experiments were discarded.

For measuring the efflux, the preparation was kept in the apparatus and washed with inactive solution. The effluent was collected during 5- or 10-min periods and counted. At the end of the experiment the nerve preparation was removed, weighed, dried, reweighed, and then dissolved in nitric acid and also counted.

RESULTS AND DISCUSSION

Efflux of Phosphate at 0.2 mM External Phosphate

In a first series of experiments nerves were loaded with radiophosphate by exposure to labeled Locke with 0.2 mM phosphate. They were then washed with inactive 0.2 mM phosphate Locke and the effluent collected and counted. The experiments showed that the efflux of radiophosphate decreased continuously, rapidly first, and then more slowly.

A better description of the rate of efflux is obtained if the fraction of total radioactivity lost per minute is calculated. This is done by dividing the amount of radioactivity collected during 1 min by the total radioactivity in the tissue during each collection period. In Fig. 1, this fraction is plotted against time after the beginning of perfusion with inactive Locke. Fig. 1, which is typical of the results obtained in many similar experiments, shows an initial rapid decrease of the fraction lost per min, which later reaches a fairly constant value.

In some experiments, samples of the effluent were passed through a chromatography column (9), that allows the separation of the phosphates into mono-, di-, and triphosphates. Counts of the fractions obtained after elution showed that more than 95% of the total radioactivity was in the monophosphate fraction. Further tests for creatine phosphate, which is not separated from orthophosphate by this method, were negative. Thus, the radioactivity of the effluent is present to a large extent as inorganic phosphate.

The fraction lost per minute, which can be considered to be the rate constant of the efflux of radiophosphate was around 0.001 min^{-1} (Fig. 1). This value can be compared to the rate constant of phosphate turnover estimated from the oxygen consumption of this preparation (11b) and the total amount of phosphate in ATP, ADP, AMP, CrP, and in the inorganic fraction (12). The rate constant of metabolic turnover calculated in this way amounts to

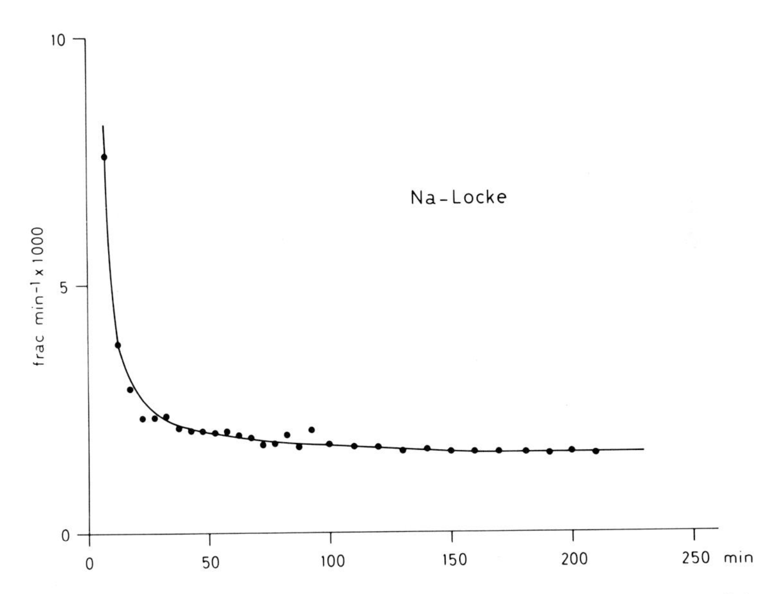

FIG. 1. Efflux of inorganic phosphate from desheathed rabbit vagus nerve, at 0.2 mM extracellular phosphate. The preparation was loaded with radiophosphate by exposure to labeled Locke with 0.2 mM phosphate and then washed with inactive Locke. The fraction of total radiophosphate lost per minute is plotted on the ordinate, the time after the beginning of the washing on the abscissa. Temperature 37°C, pH 7.40.

at most 0.1 min^{-1}. Thus, both influx and efflux of inorganic phosphate appear to be limited by the rate of transmembranal passage.

Efflux at Different Extracellular Phosphate Concentrations

In a second series of experiments, the effect of changing the external phosphate concentrations was studied. In these experiments the nerves were loaded at a given phosphate concentration and the efflux then measured at the same concentration. Figure 2 shows the results obtained. It is evident that, although the curves at different phosphate concentrations show the same general aspect, at low external phosphate concentration the rate constant of the efflux of radiophosphate is much smaller than at high external phosphate concentration. Further, at low phosphate concentrations a fairly constant level of fractional loss of radiophosphate is rapidly attained, while at high concentrations the fractional loss decreases continuously.

The rate of efflux of radiophosphate can be compared to the rate with which isotopic equilibrium is reached in influx experiments. Since, in the influx experiments, isotopic equilibrium is approached along an approximately exponential curve, its time constant was measured, and the corresponding rate constant then calculated. In a previous paper (9), such measurements made at different extracellular phosphate concentrations have been described. Figure 3 shows a comparison of the rate constants at differ-

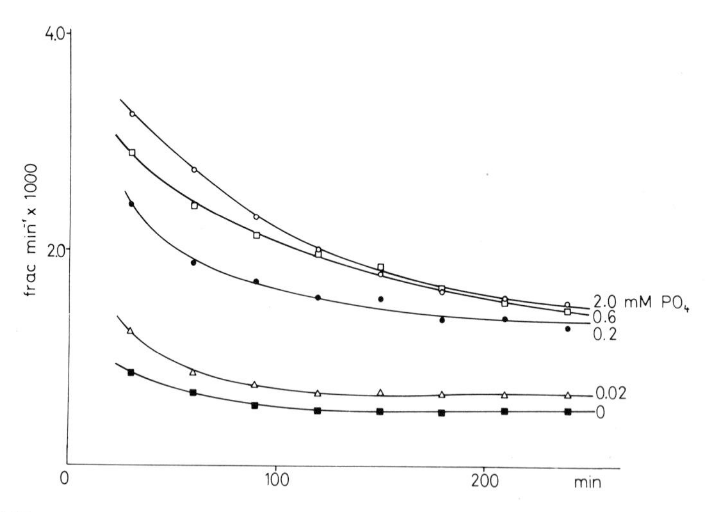

FIG. 2. Efflux of inorganic phosphate from desheathed rabbit vagus nerve at different extracellular phosphate concentrations. The preparations were loaded with radiophosphate by exposure to labeled Locke with different extracellular phosphate concentrations, and then washed with inactive Locke of the same phosphate concentrations. The fraction of total radiophosphate lost per minute is plotted against time after beginning of washing. Each point represents the mean of at least 3 measurements. Temperature 37°C, pH 7.40.

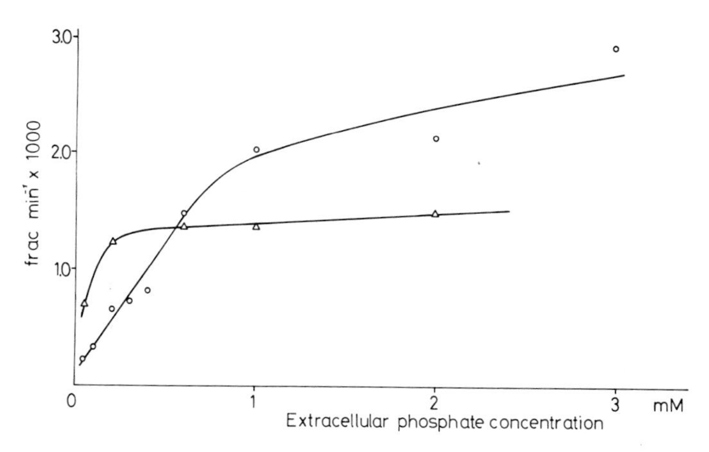

FIG. 3. Rate constants of phosphate efflux taken from experiments of Fig. 2 (△) and rate constants of phosphate efflux determined from the rate of approach to isotopic equilibrium in influx experiments (○) plotted against extracellular phosphate concentration. Note that the same rate constant is found in both sets of experiments when extracellular phosphate is ~0.6 mM. Temperature 37°C, pH 7.40.

ent phosphate concentrations, obtained in both efflux and influx experiments.

If there is no net gain or loss of phosphate, the kinetics of the fluxes of radiophosphate are described by formula 9 given by Hodgkin (13). The formula implies that, in the absence of a net flux, the rate constant with which the isotopic equilibrium is reached in influx experiments should be equal to that found for the rate constant of loss of isotope in efflux experiments. Figure 3 shows that at an extracellular phosphate concentration of approximately 0.6 mM the two rate constants are equal, suggesting that at this extracellular concentration the tissue is in phosphate equilibrium. It is interesting to compare this value with the phosphate concentration of the cerebrospinal fluid: Friedman and Levinson (13b) found in 70 normal subjects a mean value of 1.6 mg P/100 cm^3 (range 1.2–2.0 mg/100 cm^3), which is equivalent to 0.56 mM, and Cohen (14) reports a value of 1.64 mg P/100 cm^3. In the thoracic duct lymph fluid the phosphate concentration is also in the same range, around 1 mM according to Werner (15).

Effect of Na Withdrawal on the Efflux of Phosphate

As mentioned above, it has previously been found that the influx of phosphate is greatly decreased when Na is replaced by choline. It seemed interesting, therefore, to study whether Na withdrawal would also affect the efflux.

Nerves were loaded with radiophosphate during exposure to choline-Locke, they were then perfused with choline-Locke and the efflux was collected. Figure 4 shows an experiment where the efflux was measured in

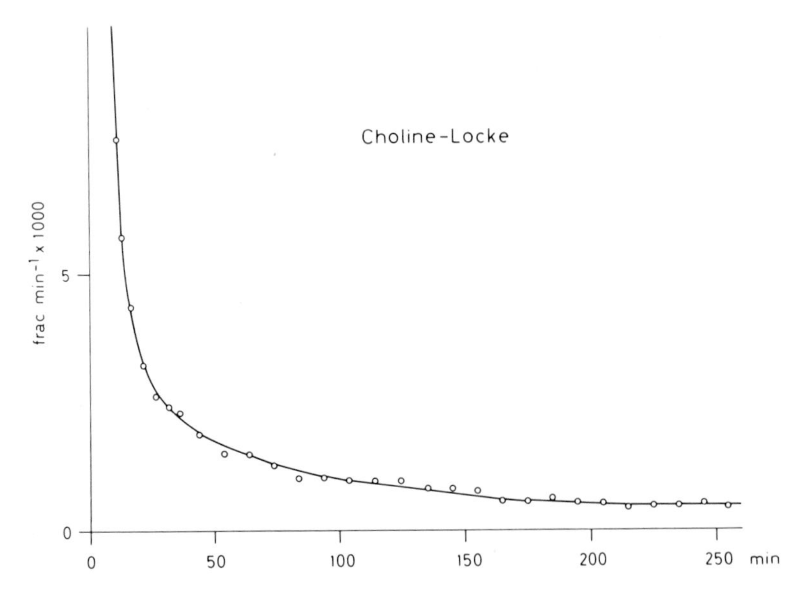

FIG. 4. Efflux of inorganic phosphate in desheathed rabbit vagus nerve in choline-Locke with 0.2 mM phosphate. The preparation was loaded with radiophosphate by exposure to labeled choline-Locke and then washed with inactive choline-Locke. Rate constant of efflux of radiophosphate is plotted against time after beginning of washing. Temperature 37°C, pH 7.40.

0.2 mM external phosphate. The fraction lost per minute became considerably smaller than in the presence of Na (cf. Fig. 1). Mean values, obtained at different times during the efflux period, were, for Na-Locke and choline-Locke, respectively: 0.00158 ± 0.00005 and 0.00143 ± 0.00023 min^{-1}, at 90 min; 0.00143 ± 0.00005 and 0.00093 ± 0.00015 min^{-1}, at 150 min; 0.00140 ± 0.00015 and 0.00065 ± 0.00020 min^{-1}, at 210 min. Exposure to Na-free solution for long periods, therefore considerably lowers the rate of efflux.

A very different effect was seen when, during perfusion with Locke, choline-Locke was applied. In this case the efflux rapidly *increased* after the application of choline-Locke and later slowly declined. Return to Locke produced a fall of the efflux followed by a slow increase (Fig. 5).

A first hypothesis which came to our mind, to explain the contrast between the effect of transient and prolonged Na withdrawal, was that in Locke a large proportion of the radiophosphate coming out of the axons is reabsorbed. The amount of radiophosphate appearing in the perfusion fluid would then be considerably smaller than the true transmembranal efflux. In this case, sudden application of choline-Locke would produce an increase in the amount of radioactivity collected, because of the greatly lowered reabsorption in the absence of extracellular Na. Quite in line with this hypothesis was the observation that lowering the temperature to near zero degrees [which blocks the influx of phosphate; (15a)] abolishes also the

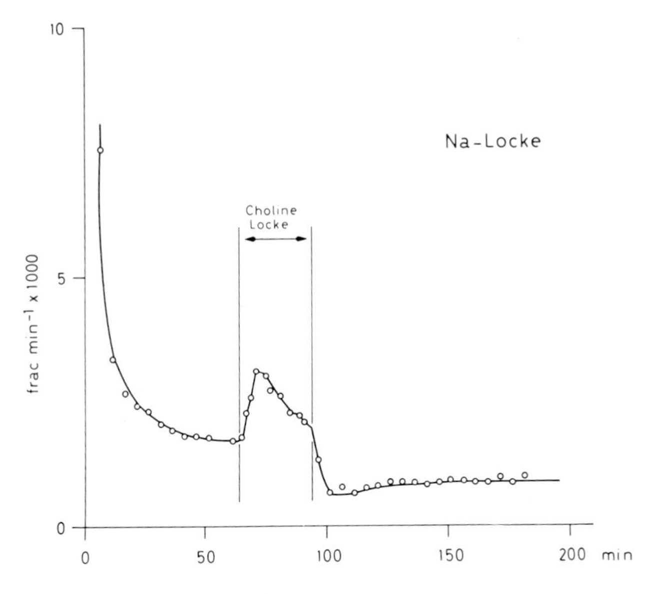

FIG. 5. Effect of Na withdrawal on efflux of phosphate. Plot of fraction of total radioactivity lost per minute shows a rapid transient increase after application of choline-Locke, followed by a somewhat slower fall. Upon return to Locke, a rapid decrease is found which is succeeded by a slow recovery of the rate of efflux. All solutions contained 0.2 mM phosphate, temperature 37°C, pH 7.40.

sudden increase in the efflux upon application of choline-Locke (Fig. 6). Quantitatively however, this hypothesis could not be maintained.

A more probable hypothesis, which also accounts for the increase in efflux seen when the extracellular potassium is removed and for effects of sudden changes in external phosphate concentration (16) would postulate a saturable phosphate transport system which mediates both the influx and the efflux of phosphate. In Locke, with phosphate, Na, and K on both sides of the membrane, the transport system would be distributed between the outside and the inside, thus mediating fluxes, whose actual values depend on the concentration of phosphate, Na, and K on each side of the membrane. Any change in phosphate concentration, or in Na, or in K, would lead to a redistribution of the fraction of the system available for transport in a given direction, and thus account for the seemingly paradoxical effects described in this chapter and by Ferrero et al. (16). The results obtained so far do not allow a complete quantitative description of the kinetics involved. In other Na-dependent transport systems, where quite similar effects are found, models have been worked out which account for the fluxes observed when the concentrations of substrate, Na, or K are changed (17). Although it appears unlikely that the transport of inorganic phosphate is mediated by systems that transport amino acids or sugars, the kinetics of these systems may well be comparable to those of the phosphate transport system.

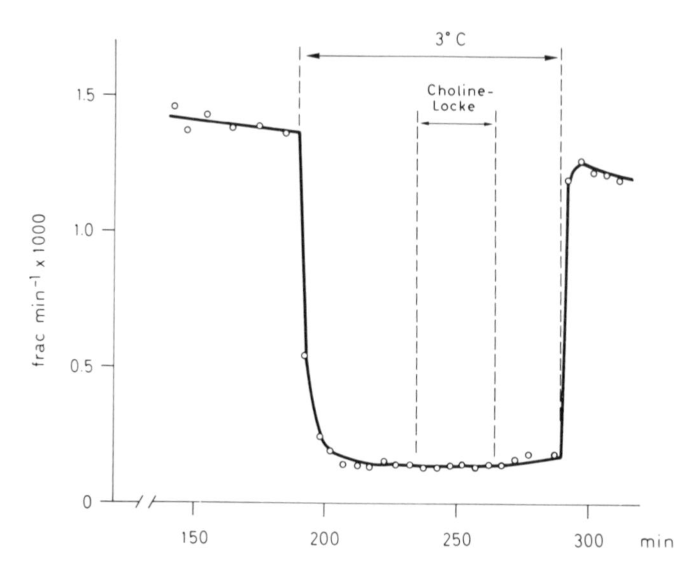

FIG. 6. Effect of lowering the temperature and of replacing Locke with choline-Locke on efflux of phosphate. The fraction of total radioactivity lost per minute is rapidly decreased when the temperature is lowered from 37 to 3°C. At low temperature, the efflux is not affected when Locke is replaced by choline-Locke and when Locke is again applied. On raising the temperature to 37°C, the efflux rate returns to normal. All solutions contained 0.2 mM phosphate, pH 7.40.

An increasing amount of evidence shows that Na-dependent phosphate transport is not limited to rabbit vagus nerve. A number of years ago a Na dependence was found for the influx of phosphate in fertilized sea urchin eggs (18), and Harrison and Harrison (19) observed a Na-dependent phosphate uptake in rat intestine. Our observation in mammalian nonmyelinated fibers, in pike olfactory nerve (20), and those of P. C. Caldwell (*personal communication*) in squid giant nerve fibers, show that, widespread in the nervous system, the phosphate uptake and efflux are Na-dependent. More recently, Na-dependent phosphate transport has been observed for the reabsorption of inorganic phosphate in rat kidney (21), and in cultured tumor cells a Na-dependent influx and efflux has been described by Brown and Lamb (22). A Na-dependent phosphate transport system appears therefore to be a general feature of many types of animal cells.

ACKNOWLEDGMENT

This work was supported by SNSF grant 3.0890.73.

REFERENCES

1. de Luca, H. F. (1974): Vitamin D: The vitamin and the hormone. *Fed. Proc.*, 33:2211–2219.

2. Kurokawa, K., Nagata, N., Sasaki, M., and Nakane, K. (1974). Effects of calcitonin on the concentration of cyclic adenosine 3′,5′-monophosphate in rat kidney *in vivo* and *in vitro*. *Endocrinology,* 94:1514–1518.
3. Short, E. M., Binder, H. J., and Rosenberg, L. E. (1973): Familial hypophosphatemic rickets: Defective transport of inorganic phosphate by intestinal mucosa. *Science,* 179:700–702.
4. Glorieux, F., and Scriver, C. R. (1972): Loss of a parathyroid hormone-sensitive component of phosphate transport in X-linked hypophosphatemia. *Science,* 175:997–1000.
5. Tasaki, I., Teorell, T., and Spyropoulos, C. S. (1961): Movement of radioactive tracers across squid axon membrane. *Am. J. Physiol.,* 200:11–22.
6. Caldwell, P. C., and Lowe, A. G. (1970): The influx of orthophosphate into squid giant axons. *J. Physiol. (Lond.),* 206:271–280.
7. Ritchie, J. M., and Straub, R. W. (1957): The hyperpolarisation which follows activity in mammalian non-medullated nerve fibres. *J. Physiol. (Lond.),* 136:80–97.
8. Keynes, R. D., and Ritchie, J. M. (1965). The movement of labelled ions in mammalian non-myelinated nerve fibres. *J. Physiol. (Lond.),* 179:333–367.
9. Anner, B., Ferrero, J., Jirounek, P., and Straub, R. W. (1975): Uptake of orthophosphate by rabbit vagus nerve fibers. *J. Physiol. (Lond.),* 247:759–771.
10. Straub, R. W., Ferrero, J., Jirounek, P., Jones, G. J., and Salamin, A. (1975): Na-dependent transport of orthophosphate in vertebrate non-myelinated nerves at different pH. *Abstr. 6th Int. Congr. Pharmacol., Helsinki,* p. 367.

11a. Ferrero, J., Jirounek, P., Jones, G. J., Salamin, A., and Straub, R. W. (1975): Monovalent and divalent orthophosphate uptake in desheathed rabbit vagus nerve. *Experientia,* 31:709.

11b. Ritchie, J. M. (1967). The oxygen consumption of mammalian nonmyelinated nerve fibres at rest and during activity. *J. Physiol. (Lond.),* 188:309–329.

12. Chmouliovsky, M., Schorderet, M., and Straub, R. W. (1969): Effect of electrical activity on the concentration of phosphorylated metabolites and inorganic phosphate in mammalian non-myelinated nerve fibres. *J. Physiol. (Lond.),* 202:90–92P.

13a. Hodgkin, A. L. (1951): The ionic basis of electrical activity in nerve and muscle. *Biol. Rev.,* 26:339–409.

13b. Friedmann, A., and Levinson, A. (1955): Cerebrospinal fluid inorganic phosphorus in normal and pathologic conditions. *Arch. Neurol. Psychiat.,* 74:424–440.

14. Cohen, H. (1924): The inorganic phosphorus content of cerebrospinal fluid. *Quart. J. Med.,* 17:289–301.
15. Anner, B., Ferrero, J., Jirounek, P., and Straub, R. W. (1973): Inhibition of intracellular orthophosphate uptake in rabbit vagus nerve by Na withdrawal and low temperature. *J. Physiol. (Lond.),* 232:47–48P.

15a. Werner, B. (1966): The biochemical composition of the human thoracic duct lymph. *Acta Chir. Scand.,* 132:63–76.

16. Ferrero, J., Jirounek, P., Jones, G. J., Salamin, A., and Straub, R. W. (1975): Effects of ions and temperature on the efflux of orthophosphate from rabbit vagus nerve fibres. *J. Physiol. (Lond.),* 254:64–65 P.
17. Schultz, S. G., and Curran, P. F. (1970): Coupled transport of sodium and organic solutes. *Physiol. Rev.,* 50:637–718.
18. Chambers, E. L. (1963): Role of cations in phosphate transport by fertilised sea urchin eggs. *Fed. Proc.,* 22:331.
19. Harrison, H. E., and Harrison, H. C. (1963). Sodium, potassium and intestinal transport of glucose, L-tyrosine, phosphate and calcium. *Am. J. Physiol.,* 205:107–111.
20. Straub, R. W., Anner, B., Ferrero, J., and Jirounek, P. (1975): Transport of inorganic phosphates across nerve membranes. In: *Comparative Physiology,* pp. 249–257, edited by L. Bolis, H. P. Maddrell, and K. Schmidt-Nielsen. North-Holland, Amsterdam.
21. Baumann, K., de Rouffignac, C., Roinel, N., Rumrich, G., and Ullrich, K. J. (1975): Renal phosphate transport: Inhomogeneity of local proximal transport rates and sodium dependence. *Pfluegers Arch.,* 356:287–297.
22. Brown, K. D., and Lamb, J. F. (1975): Na-dependent phosphate transport in cultured cells. *J. Physiol. (Lond.),* 251:58–59P.

Membranes and Disease, edited by L. Bolis, J. F. Hoffman, and A. Leaf. Raven Press, New York, © 1976.

The Molecular Basis of Brush-Border Membrane Disease

*R. K. Crane, *D. Menard, *H. Preiser, and **J. Cerda

**College of Medicine and Dentistry of New Jersey, Rutgers Medical School, Department of Physiology, Piscataway, New Jersey 08854 and **University of Florida Medical School, Division of Gastroenterology, Gainesville, Florida 32610*

Brush-border membrane disease was defined 10 years ago (1) in order to clarify relationships among a growing list of congenital and acquired gastrointestinal disturbances characterized by intolerance to specific dietary carbohydrates. Some of these disturbances, like lactose intolerance, are malfunctions of digestion, while others, like intolerance to glucose and galactose, are malfunctions of absorption; both, however, are caused by malfunctions at the brush-border membrane. Accordingly, we defined brush-border membrane disease as an absence or a greatly diminished activity of a specific functional protein element of the brush-border membrane. This concept originated in a substantial body of our earlier work.

In 1960 we described the brush border of the intestinal epithelial cell as a functionally unique subcellular organelle, as well as digestive–absorptive surface (2), and Fig. 1 was used to illustrate the key findings on which this description was based. At that time, the brush border had recently been shown to be the location of the active absorptive process for glucose and galactose (4). Brush borders had also been successfully isolated and found to possess intrinsic hydrolytic digestive capacity at their outer surface (5). Substrate-specific carriers were assumed to provide the means for sugars to cross the brush-border membrane, and Na^+ had been found necessary for the operation of the carriers for glucose and galactose, though not of those for fructose (6). It was recognized that this specific action of Na^+ might well represent the long-sought means by which the intracellular accumulation of the actively absorbed sugars, glucose and galactose, could be energized, and the Na^+ gradient hypothesis based upon the concept of cotransport of substrate and Na^+ was accordingly formulated.

At the first, as indicated in Fig. 1, a brush-border ATPase with Na^+-pump properties was assumed as the energy source for accumulation. This pump is now believed to be a component only of the basolateral membrane and not of the brush-border membrane (7). However, the change of location of the energy source does not change the mechanism of energy coupling.

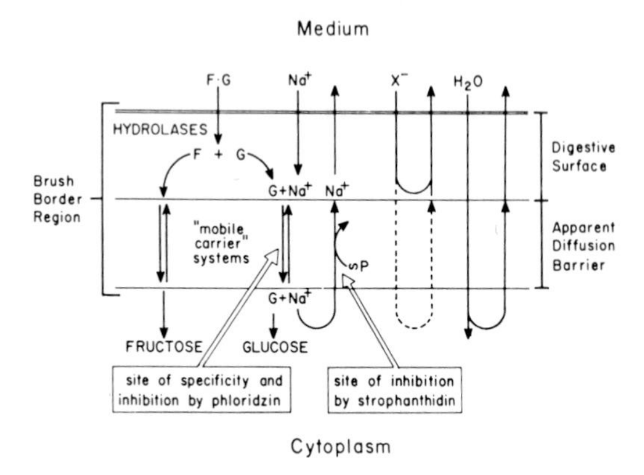

FIG. 1. Model of the intestinal brush-border membrane from Crane (3) as modified from Crane et al. (2).

In the intervening years, the hypothetical statements of Fig. 1 have been tested vigorously and extensively confirmed (8,9).

The brush-border membrane as a site of extensive digestive hydrolysis by intrinsic enzymes has become a well-established concept owing to a substantial number of researchers with nearly as many techniques. A current list of brush-border enzymes is given in Table 1. The list is arbitrary to the extent that entering an activity involves a decision that the evidence for location is sufficient. For each activity listed in Table 1, there are data in the

TABLE 1. *Brush-border enzyme activities*

Oligopeptidase
γ-Glutamyl transferase
Enteropeptidase
Glucoamylase
Maltase
Sucrase
Isomaltase (α-dextrinase)
Lactase
Trehalase
Phlorizin hydrolase (glycosylceramidase)
Alkaline phosphatase

As of 1975, based upon work reported by the following authors who are referenced fully in Crane (9): Alpers, Arias, Blumenfeld, Cohen, Crane, Dahlqvist, Deane, Dempsey, Diedrich, Doell, Eichholz, Frezal, Gartner, Gray, Hadorn, Holmes, Holt, Jos, Kelly, Kretchmer, Lamy, Leese, Lobley, Lojda, Maestracci, Malathi, Miller, Nordstrom, Peters, Preiser, Rey, Rhodes, Rosen, Schmitz, Semenza, and Troesch.

published literature that show the activity to be present in the most highly purified brush-border membrane preparations at a specific activity increase over the starting tissue homogenate about the same as for sucrase. Over the years, other enzymes have been proposed as brush-border enzymes (e.g., 10), but each has failed to meet the established criterion. Some activities listed, particularly oligopeptidase and alkaline phosphatase, are known to represent more than one enzyme species. Others apparently represent a single species. Sucrase and isomaltase occur together in a complex which is isolable and can be separated only with difficulty (11). Lactase and phlorizin hydrolase also occur together in a complex (12). Maltase is contributed by glucoamylase, sucrase, and isomaltase. Altogether the listed enzyme activities provide for the digestive cleavage of oligosaccharides, maltose, and peptides, which are the products of pancreatic amylase and proteases, as well as of the dietary disaccharides, sucrose, lactose, and trehalose. The glycosyl ceramides found in milk are also hydrolyzed as indicated (13).

Cotransport of Na^+ as a mechanism to energize intestinal absorption has turned out to be more extensively utilized than initially foreseen; at least that is the present indication. A current list of active transport systems for organic substrates is given in Table 2. This list contains a wide variety of compounds and represents an extensive search of the literature. It seems reasonable to take the position that active, i.e., energized, absorption of

TABLE 2. *The pathways of energized absorption for organic substrates*

Na^+-dependent pathways
Amino acids, acidic (Schultz, Yu-Tu, Alvarez, and Curran)
Amino acids, basic (Reiser and Christiansen, partial dependency)
Amino acids, neutral (Schultz and Curran)
Ascorbic acid (Stevenson and Brush)
Bile salts (Holt)
Bilirubin (Serranni, Corchs, and Garay)
Biotin (Berger, Long, and Semenza)
Choline (Herzberg and Lerner)
Dipeptides (Addison, Burston, and Matthews)
Glucose and galactose and their analogues (Crane)
Myoinositol (Caspary and Crane)
Riboflavin (Rivier)
Thiamin (Ferrari, Ventura, and Rendi)
Uracil (Csaky)
Na^+ dependency not tested
L-Dopa (Wade, Mearrick, and Morris)
Vitamin K_1 (Hollander)

As of 1975, based upon work reported or discussed by the authors indicated in parentheses, who are referenced fully in Crane (9).

organic substrates by the intestine is, with the few exceptions noted below, Na^+-dependent. Whether this means that absorption is in all cases coupled to the Na^+ gradient by cotransport as is the absorption of glucose, galactose, and amino acids, has not been determined.

As indicated, two cases of energized absorption have been reported with no test of Na^+ dependency and no indication of an alternate mechanism. L-DOPA is one and vitamin K_1 is the other. In the case of L-DOPA it would be reasonable to anticipate Na^+ dependency to be found when tested for; L-DOPA is an analog of neutral amino acids. The case of vitamin K_1 is, however, another matter, especially as the absorption of the related compound, vitamin K_3, appears to be by passive diffusion only (14). There is no present basis for a prediction of mechanism in this case, and it must stand as an unknown for the time being.

The reported active absorption of two other water-soluble vitamins is not listed in Table 2 because their basic mechanisms of absorption may be quite different than above; at least it is the case from appearances. These two vitamins are folate and vitamin B_{12} and, with respect to the compounds listed in Table 2, the obvious difference in their absorption mechanisms is the involvement of binding proteins clearly separate from or perhaps in addition to carriers. The binding protein for folate appears to be an intrinsic part of the brush-border membrane (15) less easily removed than, say, Ca^{2+} binding protein (16). The binding protein for vitamin B_{12}, intrinsic factor, is elaborated by the oxyntic cells of the stomach. In each case, however, complex formation with the binding protein is the presumed first step in absorption. In the case of vitamin B_{12}, there are, additionally, specific ileal sites which recognize and bind the complex (17). For want of more specific knowledge, the absorption of folate, at least at low concentrations, and that of vitamin B_{12} will be considered as binding protein-dependent, rather than as transport protein-dependent.

In theory, the list of brush-border membrane diseases could be at least as long as the unknown number of individual protein species making up the brush-border membrane. More practically speaking, we could expect to find a brush-border membrane disease for each enzyme represented by the activities in Table 1, for each Na^+-dependent or binding protein-dependent transport and for each carrier-dependent, facilitated diffusional transport, such as that of fructose. However, they have not all been seen. Those that have been are listed in Table 3.

There are 5 well-characterized diseases related to enzyme deficit and 4 to a deficiency in membrane transport. One disease folate malabsorption may be presumed to be a deficiency in binding protein. However, clinical differences between cases suggest that there is more than one kind of transport defect for folate (36). Malabsorptive states for vitamin B_{12} and for conjugated bile salts are well known, but these relate to the surgical removal of

TABLE 3. *Brush-border membrane diseases*

Enzyme deficit
Sucrose–isomaltose maldigestion (18)
Congenital lactose maldigestion (19,20)
Adult lactose maldigestion (21–23)
Trehalose maldigestion (24,25)
Enterokinase deficiency (26)
Transport or binding protein deficit
Glucose–galactose malabsorption (27,28)
Neutral amino acid malabsorption (29–31)
Basic amino acid–cystine malabsorption (32–34)
Amino acid-glycine malabsorption (35)
Folate malabsorption (36)

Phlorizin hydrolase (glycosylceramidase) activity goes together with lactase in the same way as sucrase and isomaltase go together and is reduced in adult lactose maldigestion (12).

the ileum with the consequent loss of the regionally located binding sites and transport proteins, respectively (37,38).

It will be recognized that several of the amino acid transport diseases, notably basic amino acid–cystine malabsorption, are seen clinically primarily as defects in reabsorption by the kidney. Cystinuria is an important problem disease in children. However, it may also be recognized that these defects in reabsorption are nonetheless brush-border membrane diseases, even though of the epithelial cells of the proximal convoluted tubule.

During the late 1960s it became ever more clear that further advance in understanding of the brush-border membrane diseases would require having pure human membrane available for study. Consequently, in 1969 we committed the resources of this laboratory to the isolation, purification, and study of human brush-border membranes from normal individuals and from individuals having one of the several identifiable enzyme or carrier deletion diseases. What follows represents the current produce of this long-term commitment.

In line with our previous work on laboratory animals, we took as our initial purpose the isolation of intact human brush borders. However, two considerations quickly showed this not to be the most productive line to follow. We were obviously committed to the study of the brush-border membrane in very small samples such as may be obtained by peroral biopsy from individuals with a brush-border membrane disease, and we could not expect to find the desired variety of such individuals locally. In consequence, we needed to focus on scaling down our methods and on using tissue kept in the frozen state as would be required for long-distance transportation. Hence, we bypassed the isolation of brush borders per se and went directly to the membrane (39). The method finally developed was based on the sequential application of Ca^{2+}, Tris disruption, and gradient centrifugation to homogenates of human mucosal samples (40). Our routine preparations contain no DNA, and traces only of succinate dehydrogenase and NADPH–cytochrome c reductase. (Na^{+}–K^{+})ATPase and β-glucuronidase specific activities are reduced 91% and 86%, respectively, compared to the homogenate, whereas brush-border membrane markers are increased 2,300% and more. A resume of our early biochemical data is given in Table 4. These data indicate that the membranes prepared by this method are respectably pure. Studies with the electron microscope are in agreement. Also, it turns out, the method is equally as applicable to fresh as to frozen tissue and to surgical specimens as to peroral biopsy samples.

With the membrane available, we then proceeded to view its protein components. In order to do this, membranes were dissolved in sodium dodecyl sulfate and subjected to electrophoresis on acrylamide gels (41). Over 20 bands were found in the membrane protein patterns, corresponding to a heterogeneous group of polypeptides of molecular weight ranging from 25,000 to over 400,000. The bands corresponding to the enzymes glucoamylase, lactase–phlorizin hydrolase, sucrase–isomaltase, enteropeptidase, alkaline phosphatase, trehalase, and γ-glutamyl transferase were identified (42). We were ready for our studies.

In March 1974, two patients, siblings aged 11 and 13, were diagnosed from enzyme assays of homogenates of peroral jejunal biopsies as being deficient in sucrase–isomaltase as compared to controls (see Table 5), and it was decided to proceed with a study of their membranes. It is important to make note of the fact that the trace of isomaltase activity in patient J. B. may be attributed to glucoamylase (44), and that true sucrase–isomaltase

TABLE 4. *Resume of the purification of brush-border membranes*[a]

Fraction	Sucrase	Trehalase	Leucyl naphthylamidase	Alkaline phosphatase
		Surgical specimens		
Homogenate	92	32	54	255
Membrane	2,410	721	1,185	5,438
	Average increase of specific activity = 24-fold			
		Peroral biopsies		
Homogenate	90	23	43	224
Membrane	2,735	567	1,158	6,964
	Average increase of specific activity = 29-fold			

[a] Enzyme activities are given as International Units per gram of protein. Data from Schmitz et al. (40).

TABLE 5. *Enzyme-specific activities in homogenates of peroral jejunal biopsies*[a]

	Patients		Controls	
Enzymes	J. B.	M. B.	O. P.	W. Y.
Sucrase	0	0	36.1	31.7
Isomaltase	0.4	0	20.7	17.5
Maltase	30.7	25.5	119.5	133.4
Glucoamylase	8.9	7.5	7.8	8.5
Lactase	48.7	42.5	22.4	16.8
α,α-Trehalase	24.8	30.4	12.6	14.6
Alkaline Phosphatase	170.8	201.4	161.0	69.0
Leucylnaphthylamide hydrolyzing activity	43.0	34.3	56.7	38.3
γ-Glutamyltransferase	22.7	18.9	25.3	26.9

[a] Enzyme activities are given as International Units per gram of protein.
Data from Preiser et al. (43).

activity in these children is probably as frankly nil as we have indicated.

Purified brush-border membranes were prepared from the homogenates and subjected to solubilization with SDS and gel electrophoresis with the results shown in Fig. 2. In this photograph, it is obvious that the sucrase–isomaltase protein band, which is in its normal position in control gel W. Y., is gone in gel M. B. Not so obvious, perhaps, is the fact that the band is not merely gone from that position, the proteins are gone from the gel. A careful search of the original gels showed no new bands. Hence, we concluded that sucrase–isomaltase proteins are not present in the brush-border membranes of these children who have a complete deletion of sucrase and isomaltase activities (43). By way of contrast, we have also studied a case diagnosed as sucrase–isomaltase deficiency in which some residual enzyme activity was present (2.5 IU/g protein). The membranes from this case showed a band on gel electrophoresis in the correct position for sucrase–isomaltase complex at an intensity of staining reduced from the normal intensity about as one might expect.

These are novel findings* which raise questions of considerable interest.

* A presumptively confirmatory short communication has appeared (51). However, we are not as certain as the authors of what has been confirmed. By inspection only of the published photograph of an electrophoretically developed and stained gel we are unable to decide whether Fig. 2 of ref. 51 resembles more our Fig. 2, or the gel pattern we have seen in low sucrase–isomaltase as mentioned above. Unfortunately, enzyme levels on the biopsy actually used to obtain the membranes for the gel patterns are not reported in ref. 51, though it is reported that they were done, nor is the site of the biopsy, duodenum, or jejunum, indicated. When the patient studied was 10 months old she was reported to have no sucrase or isomaltase activity in a duodenal sample (52). However, the patient is now 10 years old and tolerant of low levels of sucrose and starch in the diet (51). It should be ascertained whether this tolerance may not be due to measureable sucrase activity particularly if the present sample is jejunum where activity is higher than in the duodenum (53). It may also be pointed out that the methods we used for Fig. 2 (43) are a good deal more sensitive in revealing protein bands than those reported previously from this laboratory (41) and used in ref. 51.

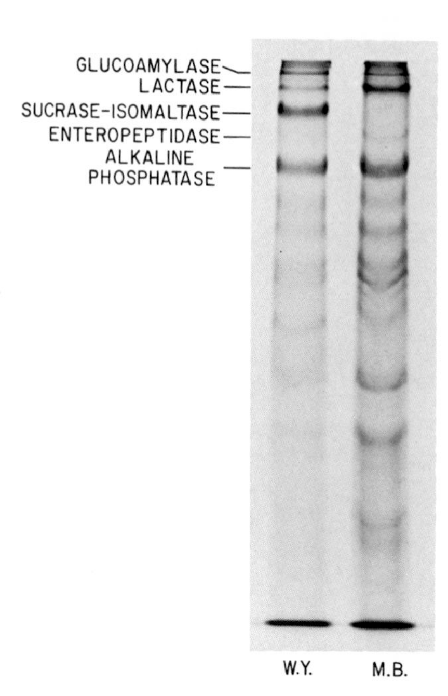

FIG. 2. Comparison of the acrylamide gel electrophoresis band patterns obtained from the brush-border membranes of M. B. and W. Y. (see Table 5).

First, these findings are out of keeping with any suggestion to the effect that sucrase–isomaltase proteins are present in the membrane of the deficient state in the usual quantities, but in altered and less active forms (e.g., ref. 45). Second, if the proteins are not present in altered form, then one may ask what kind of mutation is responsible for the deficient state. Structural gene mutations in man are well known (46), and if sucrase–isomaltase deficiency is a structural gene mutation, the aberrant proteins may be changed in such a way that they either fail to be inserted into the membrane or, if inserted, fail to stay there at least in sufficient amounts to be detected. Also, they may be degraded as rapidly as they are synthesized (47). However, the possibility exists that sucrase–isomaltase deficiency is a control gene mutation and that the proteins are absent from the membrane in individuals such as our patients J.B. and M.B. because they are not made.

Since the publication of these early results on the sucrase–isomaltase deficient siblings, we have had an opportunity to study the other members of the same immediate family. Of three other children, one was found to be deficient in sucrase–isomaltase and the gel pattern from the membranes of this child was, in all observable respects, the same as the others who are deficient. Biopsies from the parents, however, gave different results. In Table 6, the enzyme activities of the parents are compared with those of 13 controls for whose biopsies we had a reasonably complete series of assays.

As will be seen, these data permit the identification of the mother and father in this family as heterozygotes for sucrase–isomaltase deficiency.

In previous studies of families having sucrase–isomaltase-deficient members, the heterozygous state was shown to be identifiable (48,49). It was found that the average specific activity of sucrase and isomaltase is reduced about 50% in parents of deficient children, although there is considerable overlap with control subjects, and this measure alone is of dubious value. Ratios between lactase and sucrase were found to be different from the controls but current appreciation of the wide individual variation in lactase activity and the high incidence of lactase deficiency would not suggest use of this ratio as a standard for comparison. Trehalase/sucrase ratios as used by Kerry and Townley (48) would seem basically more reliable. However, in our studies we have routinely measured six hydrolytic capacities in addition to sucrase as indicated in Table 6. As it turns out, maltase is 39 ± 5% of the sum of these six activities (columns 4–9 in Table 6). Making this same calculation for the mother and father, the percent maltase in both is found to be about half the normal value and well outside the range of chance. By this measure, the mother and father resemble one another very closely and they are both very different from the controls. It appears that we can positively identify the *individual* heterozygote.

The key to use of this method is the larger number of hydrolytic activities measured. Although the ratios lactase/sucrase or maltase/trehalase, for example, may identify the heterozygous state, the use of the larger number of hydrolytic activities reduces the impact of individual variation in any one, and provides a greater depth of security in the conclusions drawn as to a presumed heterozygous individual.

Also, during the past year we have studied brush-border membranes from a number of individuals with low lactase activity and find the situation in this brush-border membrane disease to be possibly more complex than in sucrase–isomaltase deficiency. In the latter, as mentioned before, the major band representing sucrase–isomaltase complex is completely missing in our patients J. B., M. B., and the other sibling mentioned above. In lactase deficiency, the band representing the lactase–phlorizin hydrolase complex is reduced in intensity about as much as lactase is reduced in activity, similar to the case of low sucrase–isomaltase activity mentioned above. This reduced band was found in all individuals studied. However, as indicated by diagrams in Fig. 3, minor bands are sometimes observed above or below the diminished lactase band. Also, as shown by the individual M. P., who was diagnosed as having bacterial overgrowth of the intestine, there can be low lactase activity with a major protein band in the lactase position. Furthermore, an unidentified major protein has been found in the lactase–phlorizin hydrolase position when these activities have been removed from the membrane with a proteolytic enzyme as shown in Fig. 4. When enzyme activities are removed from the membrane, in this case with elastase, the sucrase–

TABLE 6. *Enzyme activities of parents compared with controls*

	Sucrase	Isomaltase	Glucoamylase	Maltase	Lactase	Trehalase	Alkaline phosphatase	Leucylnaphthylamidase	γ-Glutamyltransferase	Sum of measured hydrolytic capacity 4–9	% Maltase
Mother	32.1	16.8	13.9	119.1	60.5	33.6	385.3	39.2	31.2	668.9	18
Father	8.7	7.1	3.7	44.8	8.6	6.5	102.2	23.9	14.8	200.8	22
Controls	36.4 ± 11	–	–	148.3 ± 48	19.3 ± 8	17.4 ± 9	136.4 ± 53	32.1 ± 9.2	26.3 ± 12	379.8	39 ± 5

Enzyme activities are given as International Units per gram of protein.

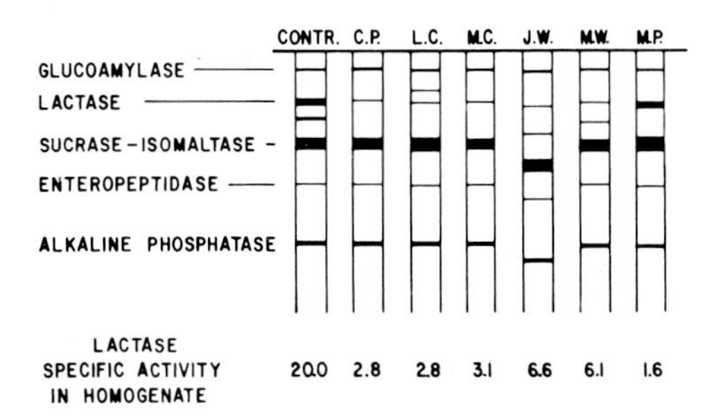

FIG. 3. Diagram comparing the acrylamide gel electrophoresis band patterns obtained from the brush-border membranes of individuals with low-lactase activity compared to a control.

isomaltase band disappears progressively with the enzyme activity, indicating the presence of only one protein complex in this position. However, a protein clearly remains in the lactase–phlorizin hydrolase position even when 95% of lactase and phlorizin hydrolase have been removed. This protein has not been identified with respect to possible enzymic activity. However, the existence of this protein may complicate experiments in which the decision as to meaning must be based upon the presence or absence of a band in the lactase–phlorizin hydrolase position. Nonetheless, it is our impression from the study of a number of gels from low-lactase membranes that

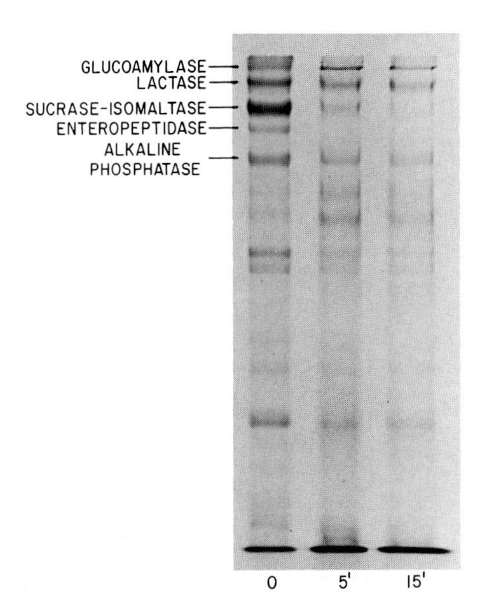

FIG. 4. Comparison of the acrylamide gel electrophoresis band pattern of brush-border membranes from a normal individual before and after treatment with elastase.

this nonlactase protein is reduced in amount along with the lactase–phlorizin hydrolase complex. Consequently, we would predict that the residual protein bands indicated by Fig. 3 will not be found in membranes from individuals with congenital lactose maldigestion. Brush-border lactase is completely deficient in these cases (50).

Studies of additional brush-border membrane diseases are currently projected as appropriate biopsy samples may become available. Work is progressing on developing methods more sensitive and discriminating than those already in use.

ACKNOWLEDGMENTS

Work of this laboratory has been supported by grants from the National Science Foundation and the National Institute of Arthritis, Metabolism and Digestive Diseases. D. M. is the recipient of a Fellowship of the Medical Research Council of Canada.

We thank Jack Welsh and Robert Ringrose for the tissue samples from the individual with low sucrase–isomaltase activity.

REFERENCES

1. Crane, R. K. (1966): *Gastroenterology,* 50:254–262.
2. Crane, R. K., Miller, D., and Bihler, I. (1961): In: *Membrane Transport and Metabolism,* edited by A. Kleinzeller and A. Kotyk, pp. 439–450. Academic Press, New York.
3. Crane, R. K. (1962): *Fed. Proc.,* 21:891–895.
4. McDougal, D. G., Jr., Little, K. D., and Crane, R. K. (1960): *Biochim. Biophys. Acta,* 45:483–489.
5. Miller, D., and Crane, R. K. (1961): *Biochim. Biophys. Acta,* 52:293–298.
6. Bihler, I., Hawkins, K. A., and Crane, R. K. (1962): *Biochim. Biophys. Acta,* 59:94–102.
7. Fujita, M., Ohta, H., Kawai, K., Matsui, H., and Nakao, M. (1972): *Biochim. Biophys. Acta,* 274:336–347.
8. Schultz, S. G., and Curran, P. F. (1970): *Physiol. Rev.,* 50:637–718.
9. Crane, R. K. (1976): In *Physiology,* Ser. 2, Vol. 4: *Gastrointestinal Physiology,* edited by R. K. Crane. MTP International Review of Science, Butterworths, London; University Park Press, Baltimore (*in preparation*).
10. Forstner, G. G., Riley, E. M., Daniels, S. J., and Isselbacher, K. J. (1965): *Biochem. Biophys. Res. Commun.,* 21:83–88.
11. Braun, H., Cogoli, A., and Semenza, G. (1975): *Eur. J. Biochem.,* 52:475–480.
12. Colombo, V., Lorenz-Meyer, H., and Semenza, G. (1973): *Biochim. Biophys. Acta,* 327: 412–424.
13. Leese, H. J., and Semenza, G. (1973): *J. Biol. Chem.,* 248:8170–8173.
14. Hollander, D., and Truscott, T. C. (1974): *Am. J. Physiol.,* 226:1516–1522.
15. Leslie, G. I., and Rowe, P. B. (1972): *Biochemistry,* 11:1696–1703.
16. Bredderman, P. J., and Wasserman, R. H. (1974): *Biochemistry,* 13:1687–1694.
17. Mathan, V. I., Babior, B. M., and Donaldson, Jr., R. M. (1974): *J. Clin. Invest.,* 54:598–608.
18. Auricchio, S., Prader, A., Mürset, G., and Witt, G. (1961): *Helv. Paediatr. Acta,* 16:483
19. Durand, P. (1958): *Minerva Pediatr.,* 10:706–711.
20. Holzel, A., Schwartz, V., and Sutcliffe, K. W. (1959): *Lancet,* 1:1126–1128.
21. Auricchio, S., Rubino, A., Landolt, M., Semenza, G., and Prader, A. (1963): *Lancet,* 2: 324–326.

22. Haemmerli, V. P., Kistler, H. J., Ammann, R., Auricchio, S., and Prader, A. *Helv. Med. Acta,* 30:693–705.
23. Dahlqvist, A., Hammond, J. B., Crane, R. K., Dunphy, J., and Littman, A. (1963): *Gastroenterology,* 45:488–491.
24. Bergoz, R. (1971): *Gastroenterology,* 60:909–912.
25. Madzarovova-Nohejlova, J. (1973): *Gastroenterology,* 65:130–133.
26. Hadorn, G., Tarlow, M. J., Lloyd, J. K., and Wolff, O. H. (1969): *Lancet,* 1:812–813.
27. Meeuwisse, G. W., and Dahlqvist, A. (1968): *Acta Paediatr. Scand.,* 57:273–280.
28. Stirling, C. E., Schneider, A. J., Wong, M. D., and Kinter, W. B. (1972): *J. Clin. Invest.,* 51:438–451.
29. Scriver, C. R. (1965): *N. Engl. J. Med.,* 273:530–532.
30. Hooft, C., Carton, D., and Snoeck, J. (1968): *Helv. Paediatr. Acta,* 23:334–349.
31. Shih, V. E., Bixby, E. M., Alpers, D. H., Bartsocas, C. S., and Thier, S. O. (1971): *Gastroenterology,* 61:445–453.
32. Thier, S. O., Segal, S., Fox, M., Blair, A., and Rosenberg, L. E. (1965): *J. Clin. Invest.,* 44:442–448.
33. Rosenberg, L. E., Downing, S., Durant, J. L., and Segal, S. (1966): *J. Clin. Invest.,* 45: 365–371.
34. Morin, C. L., Thompson, M. W., Jackson, S. H., and Sass-Kortsak, A. (1971): *Biochem. J.,* 134:11–26.
35. Goodman, S. I., McIntyre, C. A., and O'Brien, D. (1967): *J. Pediatr.,* 71:246–249.
36. Lanzkowsky, P. (1970): *Am. J. Med.,* 48:580–583.
37. Booth, C. C., and Molin, D. L. (1959): *Lancet,* 1:18–21.
38. Lack, L., and Weiner, I. M. (1961): *Am. J. Physiol.,* 200:313–317.
39. Welsh, J. D., Preiser, H., Woodley, J. F., and Crane, R. K. (1972): *Gastroenterology,* 62: 572–582.
40. Schmitz, J., Preiser, H., Maestracci, D., Ghosh, B. K., Cerda, J., and Crane, R. K. (1973): *Biochim. Biophys. Acta,* 323:98–112.
41. Maestracci, D., Schmitz, J., Preiser, H., and Crane, R. K. (1973): *Biochim. Biophys. Acta,* 323:113–124.
42. Maestracci, D., Preiser, H., Hedges, T., Schmitz, J., and Crane, R. K. (1975): *Biochim. Biophys. Acta,* 382:147–156.
43. Preiser, H., Menard, D., Crane, R. K., and Cerda, J. J. (1974): *Biochim. Biophys. Acta,* 363:279–282.
44. Kely, J. J., and Alpers, D. H. (1973): *Biochim. Biophys. Acta,* 315:113–122.
45. Dubs, R., Steinmann, B., and Gitzelmann, R. (1973): *Helv. Paediatr. Acta,* 28:187–198.
46. Stanbury, J. B., Wyngaarden, J. B., and Frederickson, D. S. (1972): *The Metabolic Basis of Inherited Disease,* 3rd ed., Chap. 1. McGraw-Hill, New York.
47. Goldberg, A. L., and Dice, J. F. (1974): *Annu. Rev. Biochem.,* 43:835–869.
48. Kerry, K. R., and Townley, R. R. W. (1965): *Aust. Pediatr. J.,* 1:223–235.
49. Rey, J., and Frezal, J. (1967): *Arch. Fr. Pediatr.,* 24:65–101.
50. Asp, N.-G., Dahlqvist, A., Kuitunen, P., Launiala, K., and Visakorpi, J. K. (1973): *Lancet,* 11:329–330.
51. Schmitz, J., Commegrain, C., Maestracci, D., and Rey, J. (1974): *Biomedicine,* 21:440–443.
52. Auricchio, S., Ciccimarra, F., Moauro, L., Rey, F., Jos, J., and Rey, J. (1972): *Pediatr. Res.,* 6:832–839.
53. Newcomer, A. D., and McGill, D. B. (1966): *Gastroenterology,* 51:481–488.

Note added in proof: The absence of sucrase–isomaltase protein from the membrane in sucrase–isomaltase deletion has been confirmed and extended to an absence of sucrase antigen from the entire cell by means of radioimmunoassay (Gray, G. M., Conklin, K. A., and Townley, R. R. W. (1976): *N. Engl. J. Med.* (*in press*).

Membranes and Disease, edited by L. Bolis, J. F. Hoffman, and A. Leaf. Raven Press, New York, © 1976.

Glycosidases of Small Intestinal Brush Borders

G. Semenza

Laboratorium für Biochemie, Eidgenössische Technische Hochschule, Universitätstrasse 16, CH-8006 Zurich, Switzerland

Following the first reports on disaccharide malabsorption in the early 1960s a great number of papers have appeared on the biochemistry, physiology and physiopathology of small intestinal disaccharidases. The goal of this chapter is to present a compact review of the state of our knowledge on the biochemical properties of these enzymes, particularly on what is known on their structure, catalytic mechanism, and role in sugar intestinal transport (for a more extensive review, see ref. 1) with the purpose of providing the theoretical background for a discussion of their physiopathology (for reviews, see refs. 2,3).

Digestive carbohydrases are associated with or are building blocks of the membrane of small intestinal brush borders. Table 1 lists those carbohydrases which have been localized with these structures with a reasonable degree of security and by various approaches (2,4,5). Some of these activities, sucrase in particular, is actually used routinely as a marker for the isolation and purification of brush border membrane from the small intestine.

Small intestinal cells possess other carbohydrase activities also, not associated with the brush border membranes and probably playing no role in the digestion of nutrients (e.g., two more β-galactosidases, a lysosomal α-glucosidase and a β-glucosidase, mannosidase, etc.). In the following we will mainly concentrate on the properties of the sucrase–isomaltase complex and of the β-glycosidase complex.

TABLE 1. *Carbohydrases bound to the membrane of the small intestinal brush border of the "average mammal"*

Enzyme	Complex
1 or 2 maltases–glucoamylases	
1 maltase–sucrase	the sucrase–isomaltase complex
1 maltase–isomaltase	the sucrase–isomaltase complex
1 trehalase	
1 lactase	the β-glycosidase complex
1 glycosyl-ceramidase (= the major phlorizin-hydrolase)	the β-glycosidase complex
1 minor phlorizin-hydrolase	

THE SUCRASE–ISOMALTASE COMPLEX

After solubilization with either detergents (6,7) or by proteolytic treatment (7–9) this complex can be obtained in a homogeneous form from the small intestine of a number of mammals. It can be split into the constituent subunits either by urea or guanidine hydrochloride (10) treatment (both of which, however, produce irreversible denaturation), by selective inactivation and destruction of one of the two subunits (11,12), or by citraconylation (13). The latter is the only procedure which allows the separation of the two subunits in an active form, albeit on an analytical scale only. Neither solubilization of the complex from its original membrane nor its separation into its subunits noticeably change their catalytic properties, as judged by the K_m values and the extent of activation by sodium.

Table 2 presents a comparison of structural and catalytical properties of sucrase and isomaltase. Both enzymes have been suggested to catalyze the

TABLE 2. *A comparison between the sucrase and isomaltase moieties of the sucrase–isomaltase complex from rabbit small intestine*

		Ref.
Molecular weight	Approximately 120,000 each	10
SH groups	None in either (native)	8
Number of polypeptide chains	No evidence for more than one, for either sucrase or isomaltase	11
Number of active sites	One in sucrase subunit, one in isomaltase subunit	11,23
Anomeric form of glucose liberated during hydrolysis	α-Glucopyranose, by both subunits	14,15
Bond split	That between C_1 of glucose and glycosyl-oxygen	16
Kinetic mechanism	Ping-pong BiBi, rapid equilibrium	17,18
Transglucosidase activity	Present in both sucrase and isomaltase	17[a]
Na activation	Increase of k_{cat} (i.e., k_3) by approximately 30%, with slight decrease in K_m.	17,19[a]
Substrate specificity	Both subunits split maltose, maltotriose, maltitol, and a number of aryl-α-glucopyranosides; in addition, sucrase splits sucrose; isomaltase splits isomaltose, isomaltulose	20,21 22[a]
Competitive inhibitors	Tris, glucono-1,5-lactone, α-phenylthioglucopyranoside	18,21[a]
	Dextran is either a pseudosubstrate or a competitive inhibitor, mainly for the isomaltase moiety	22
	Sephadex G-200 retards isomaltase considerably more than sucrase (affinity chromatography)	11,22
Irreversible competitive inhibitor (affinity label)	Conduritol B–epoxide binds irreversibly at the active site of sucrase and of isomaltase forming an ester bond with the β-carboxylate of an aspartate residue; the isomaltase site reacts approximately 8 times faster than the sucrase site	23

TABLE 2. (*Continued*)

		Ref.
Partial sequence around the COO^- at the active sites (the residue bound to the label is *italicized*)	Sucrase: -Ile-*Asp*-Met-Asn-Gln-Pro-Asn-Ser-Ser- Isomaltase: Asn-Gly-Gly-Gln-Ile-*Asp*-Met-	24
A deprotonated group participates in the catalysis	Its pK_a' (from double logarithmic Dixon plots of both sucrase and isomaltase) corresponds to that of the carboxylate reacting with conduritol B-epoxide in each of the two active sites	20,23
Additional groups identified at the active sites	One more COO^- (or COOH), as identified by reaction with R—N=C=N—R′ No positive evidence for other residues	25
Secondary deuterium effect at C_1 of glucopyranosyl residue	The k_D/k_H for *p*-Cl-phenyl-α-glucopyranoside ranges between 1.14 and 1.20 for both activities	21
Hammett-Hansch equations	Sucrase: $\log k_{cat} = 0.253\pi + 0.093\sigma + 2.22$ Isomaltase: $\log k_{cat} = -0.009\pi - 0.394\sigma + 2.52$	21
A proposed minimum mechanism	Protonation of the glycosyl oxygen by an as yet unidentified acid (perhaps a COOH); splitting of the bond between C_1 of glucose and glycosyl oxygen, with formation of an oxocarbonium ion (which is temporarily stabilized by the βCOO^- of the *Asp*) and liberation of the aglycone; final stabilization of the oxocarbonium by a nucleophile from the solvent (OH^- in the case of hydrolysis), with reformation of the original α-configuration at C_1 of glucose	21

[a] In addition, unpublished data from our laboratory.

hydrolysis of α-glucopyranosides via protonation of the glycosyl-oxygen, followed by the temporary formation of an oxocarbonium ion which is stabilized by a carboxylate (21). In both sucrase and isomaltase this carboxylate was identified as the β-carboxylate group of an aspartic acid residue (24).

Small intestinal sucrase and isomaltase activities are subjected to the same or to related biological control mechanism, as shown by their simultaneous appearance during development (26,27), by their close correlation in a random sample of human peroral biopsies (28), and by their simultaneous absence in sucrose–isomaltose malabsorption (29,30), a monofactorial genetic disease (31). Prima facie the sucrase–isomaltase complex is reminiscent, therefore, of hemoglobin, which is composed of two kinds of subunits of similar, but not identical, structure and function and which have arisen in all likelihood by gene duplication. Naturally, our knowledge of the sucrase–isomaltase system is much less advanced than that of the hemoglobin system.

The genetic defect in sucrose–isomaltose malabsorption is discussed extensively in this volume by Crane (32).

THE LACTASE–GLYCOSYL CERAMIDASE (β-GLYCOSIDASE) COMPLEX

Mammalian small intestinal brush border contain at least two major β-glycosidases, i.e., lactase and a hetero-β-glucosidase. The best substrate of lactase is lactose, although several aryl-β-galactosides as well are split with high velocity and rather small K_m values. The natural substrates of the other major β-glycosidase were identified with glycosyl-ceramides (33). Lactase and glycosyl–ceramidase [previously determined as phlorizin-hydrolase (34)] are bound together in a complex reminiscent of the sucrase–isomaltase complex. However, up to now, the two β-glycosidases have been much less extensively characterized, because the β-glycosidase complex is available in much smaller amounts than the sucrase–isomaltase complex (35–37). However, through selective heat inactivation and through mutual inhibition studies (33,38,39) some information is available as to the properties of the individual active sites in the β-glycosidase complex. The aglycone subsite of lactase is rather hydrophilic (contrary to what one may expect, it splits cellbiose), whereas that of glycosyl-ceramidase is rather hydrophobic. The glycone subsites of the two glycosidases have rather broad specificity, lactase splitting some glucosides (and being inhibited competitively by phlorizin) and glycosyl-ceramidase splitting glucosides and galactosides often equally well.

Lactase and glycosyl-ceramidase activities are subjected to the same or to related biological control mechanisms, as shown by the constancy of the ratio between lactase and phlorizin-hydrolase activities in a random sample of peroral human biopsies (39), their simultaneous absence in adult lactose malabsorption (39), a genetic condition quite common in man (40), and similar, although not identical, rates of disappearance of the time of weaning (38,41,42). (The small difference may be due to the presence of the additional minor phlorizin-hydrolase which Kraml et al. (43) reported.) The biochemical correlation between lactase and glycosyl-ceramidase has a physiological significance: lactose and glucosyl- and lactosyl-ceramides (44) all occur in milk, and the milk is the only food before weaning, i.e., at the time when both lactase and phlorizin-hydrolase (i.e., glycosyl-ceramidase) activities are at their highest.

Brush border lactase occurs in mammalian small intestine only, whereas phlorizin-hydrolase is known to occur in the small intestine of other vertebrates as well (41,45). One could entertain the speculation that phlorizin-hydrolase (i.e., glycosyl-ceramidase) may be the philogenetic precursor of mammalian lactase, the gene for lactase having possibly arisen from that of glycosyl-ceramidase by gene duplication.

INTESTINAL ABSORPTION OF FREE MONOSACCHARIDES

Fructose is absorbed by a carrier-mediated system, which is not sodium-dependent and does not lead to accumulation against a concentration gradient (refs. 46 and 47 and references therein). The absorption of glucose, galactose, and similar sugars is sodium-dependent and can lead to accumulation against a concentration gradient. It is now universally accepted that the energy for this accumulation derives from the gradient in electrochemical gradient (48) of sodium which is a cosubstrate (49) for the transport of these monosaccharides. For reviews on the subject the reader is referred to refs. 50–52. Glucose–galactose malabsorption originally described by Lindquist and Meeuwisse (53) and reviewed in the present volume by Rosenberg (54), is a monofactorial recessive disease in which the children cannot absorb free glucose and galactose from the small intestine. A puzzling feature of this condition is that these subjects appear to acquire the capacity of absorbing glucose from the small intestine as they grow. The mechanism for this apparent recovery is unknown. One possibility, however, is suggested by some work from our laboratory on the absorption of glucose and galactose in the hamster: in this species free glucose and galactose are absorbed by two carrier systems which are approximately equally active in the adult animal, but of which one is less active in the first weeks of life (46). If the same holds true in man the genetic absence or inactivity of the system, which would be normally the sole one responsible for glucose absorption at birth, would cause glucose galactose malabsorption. The subsequent normal appearance of the second system for glucose and galactose transport would produce a clinical "recovery" of the malabsorption syndrome. The hypothesis of a possible involvement of two systems in intestinal and renal transport of glucose had actually been put forward before us by Meeuwisse (55).

DISACCHARIDASE-DEPENDENT SUGAR TRANSPORT SYSTEMS

Most of the monosaccharides liberated by the action of membrane-bound disaccharidases are efficiently picked up by the transport systems for monosaccharides (56). However, it was established by Crane's group (57–59) that some of the sugars that are provided as disaccharides can enter by other route(s). These transport system(s) are (i) not (as) accessible to free monosaccharides, (ii) slightly or not sodium-dependent, (iii) less sensitive to phlorizin, and (iv) slightly or not inhibited by Tris, in spite of the corresponding hydrolytic activities being strongly inhibited by it (59).

Although these transport systems are probably of little physiological significance [not more than 5–10% of the monosaccharides seem to utilize this route (59)], they are of considerable theoretical interest, because they have provided the first example of a natural transport system to be reconstituted in artificial membranes [in BLM (black lipid membranes) (60) and in liposomes (61)] from a homogeneous membrane protein: the sucrase–isomaltase complex was solubilized by papain digestion, obtained in homogeneous form and incorporated into lipids. The BLM obtained therefrom had a permeability coefficient for ^{14}C-labeled sucrose (or, rather, for

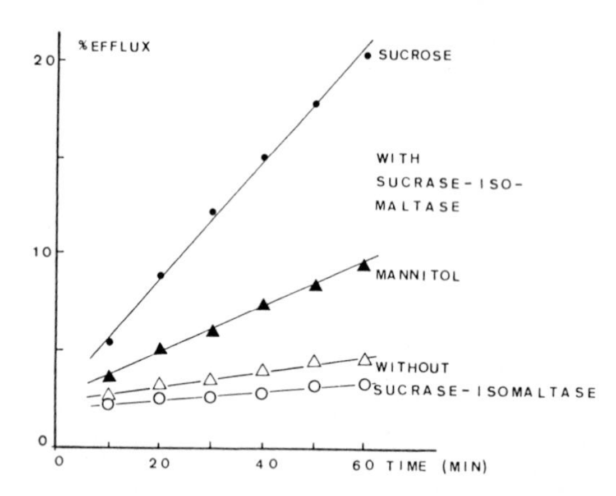

FIG. 1. Efflux of ^{14}C-sucrose (○,●) and of ^{3}H-mannitol (△,▲) from multilamellar liposomes made of lipids alone (open symbols) or of lipids and of sucrase–isomaltase complex (filled symbols). The radioactivity compounds arising from ^{14}C-sucrose and appearing in the outer fluid were identified as fructose, glucose, and sucrose. (From ref. 61.)

the monosaccharides arising from it) which was larger than that for protein-free BLM by some 3 orders of magnitude, at least. The permeability to mannitol, D-glucose or D-fructose was little affected by the sucrase–isomaltase complex, if at all. Similar results were obtained on reconstituting the sucrase-dependent system in multilamellar liposomes (Fig. 1) (61); incorporation of the sucrase–isomaltase complex into the phospholipids produced liposomes with high specific permeability for "sucrose" (or, rather, for the monosaccharides arising from it), which was unaffected by the nature of the cation present and was not inhibited by Tris (Fig. 2).

Conduritol B-epoxide-inactivated sucrase–isomaltase complex failed to produce liposomes with increased permeability for sucrose (Fig. 3). This

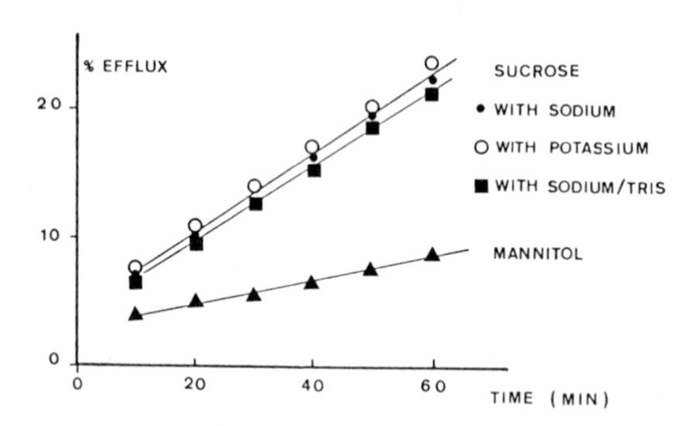

FIG. 2. Efflux of ^{14}C-sucrose and of ^{3}H-mannitol (▲) from multilamellar liposomes in sodium (●) or potassium (○) phosphate buffers (100 mM, pH 6.8), or in sodium buffer in the presence of 37.5 mM Tris (■). The efflux of mannitol was independent of the cations present in the medium. (From ref. 61.)

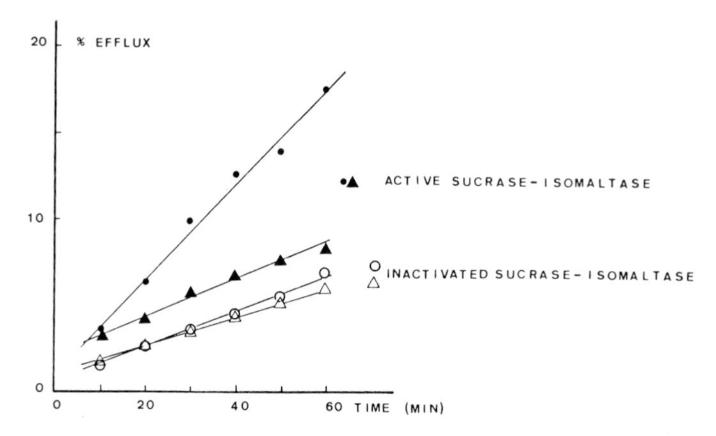

FIG. 3. Efflux of ^{14}C-sucrose (○,●) and of ^{3}H-mannitol (△,▲) from multilamellar liposomes made of lipids and native sucrase—isomaltase complex (filled symbols) or of lipids and conduritol B—epoxide inactivated sucrase—isomaltase complex (open symbols). (From ref. 61.)

observation, along with the fact that the sucrase preparations were homogeneous according to the standard criteria of protein purity, shows that the sucrase-dependent sugar transport system is totally confined within the molecule of the sucrase–isomaltase complex (±, perhaps phospholipids).

A series of control experiments (discussed in detail in ref. 61) ruled out that the increased "sucrose" permeability due to sucrase was due to hydrolysis of the disaccharide in the water phase, followed by permeation of free glucose and fructose across the lipid bilayer(s).

As discussed elsewhere (61), the two most likely groups of mechanisms for the function of sucrase as sugar translocator are as the following.

1. The active site of sucrase may be totally and constantly confined to one side of the membrane. If hydrolysis is vectorial (as it has been suggested for other carbohydrases, e.g., α-amylase), and if the products are liberated in a microenvironment at the surface of (or perhaps even within) the lipid bilayer, from which diffusion into water bulk phase should be slow, a local hyperconcentration of glucose and fructose may ensure, providing the concentration "head" for an apparently increased passive diffusion (Fig. 4A). The free energy of sucrose hydrolysis ($\Delta G° \simeq 7$ kcal/mol) may be sufficient to sustain this local hyperconcentration.

2. The active site of the sucrase involved in transport may have access, either all the time or alternatively, to both sides of the lipid membrane providing either a kind of static, specific pore endowed with enzymic activity or a dynamic translocator (e.g., with alternative gate opening or with actual vectorial movement) (Fig. 4B). Transport would have to take place before the hydrolytic liberation of the first product (fructose), because the

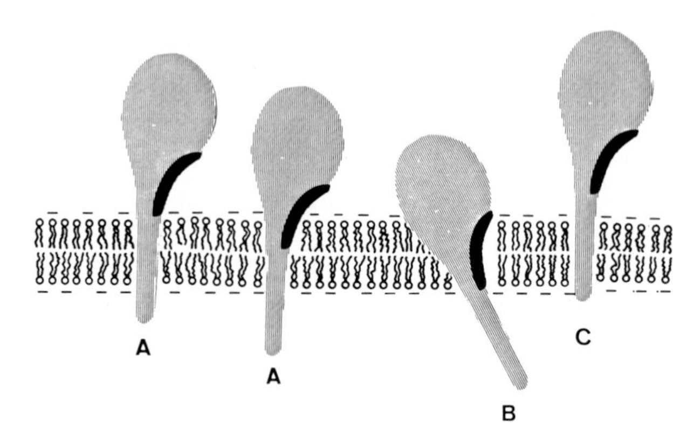

FIG. 4. Possible positions of the sucrase – isomaltase complex (*gray*) in natural or artificial membranes. *Black:* one of the active sites of the complex. For details, see text.

two monosaccharides appear in the trans compartment in equimolar amounts.

No choice can be made at present between mechanisms 1 and 2 (Fig. 4A,B).

Finally, one more point should be discussed. Why do only 5–10% of the sugars provided as disaccharides utilize the disaccharidase transport system (in the presence of Na^+) (59)? Why does Tris inhibit the disaccharidase-dependent transport systems much less than it does their hydrolytic activity? Why does Tris inhibit transport in the reconstituted system even less than in intact cells (Fig. 2)? A possible, although still hypothetical, unifying answer to these questions may perhaps start from the observation (17) that Tris inhibits sucrase at neutral or slightly alkaline pH values, whereas it has little or no effect at acidic pHs. It is also known that the lipid of brush border membranes has an excess of negative charges (62,63). The lipids used to make multilamellar or monolamellar liposomes also had an (even larger) excess of negative charges.

We suggest (Fig. 4) that some 90–95% of the sucrase–isomaltase molecules are positioned in the natural membrane as depicted in Fig. 4C, that is, their active sites would be located outside the domain of the membrane surface charges. These sucrase–isomaltases would not act as sugar translocators and would have the same hydrolytic activity as papain- or triton-solubilized sucrases (identical K_m-values, pH activity curves, energy of activation, Na activation, and Tris inhibition). The monosaccharides arising by their action on glucosides would cross the brush border membrane by way of the carriers for *free* monosaccharides.

On the other hand, some 5–10% of the sucrase-isomaltase molecules would be positioned in the natural membrane as depicted in Fig. 4A,B, whereas in the reconstituted system most of the sucrase–isomaltase molecules would be positioned in these ways; i.e., their active sites would be located either at or within the very surface of the lipid bilayer, or would span across it. These sucrase-isomaltase molecules only would act as sugar translocators. Because of the excess negative charges of the lipid bilayer, the active sites of these sucrase–isomaltase molecules would be in an acidic microenvironment; as a consequence, their activities as translocators (and as hydrolases) would be slightly sensitive, if at all, to the inhibition by Tris and would show different dependence on the pH, as compared with either solubilized sucrase or with the sucrase molecules positioned as in Fig. 4C.

I want to emphasize that this explanation is still hypothetical, that it aims primarily at providing a unifying rationale to a group of puzzling observations, and that the distribution of sucrase-isomaltase molecules between positions A and B and C (Fig. 4) could be either static or dynamic, that is, the same sucrase–isomaltase molecules could also oscillate among positions A + B and C, their *statistical* distribution remaining the same.

REFERENCES

1. Semenza, G. (1976): In *Membrane-Bound Enzymes,* edited by A. Martonosi. (*in press*).
2. Semenza, G. (1969): In *Handbook of Physiology,* Vol. VI, *Alimentary Canal.* p. 2543–2566, edited by C. Code. Amer. Physiol. Soc., Washington, D.C.
3. Dahlqvist, A. (1964): *Il Pensiero Scientifico,* p. 5, edited by P. Durand. Rome, Italy.
4. Louvard, D., Maroux, S., Baratti, J., Desnuelle, P., and Mutafschiev, S. (1973): *Biochim. Biophys. Acta,* 291:747–763.
5. Louvard, D., Maroux, S., Vannier, Ch., and Desnuelle, P. (1975): *Biochim. Biophys. Acta,* 375:236–248.
6. Carnie, J. A., and Porteous, J. W. (1962), *Biochem. J.,* 85:620–629.
7. Sigrist, H., Ronner, P., and Semenza, G. (1975): *Biochim. Biophys. Acta,* 406:433–446.
8. Cogoli, A., Mosimann, H., Vock, C., v. Balthazar, A. K., and Semenza, G. (1972): *Eur. J. Biochem.,* 30:7–14.
9. Kolínská, J., and Kraml, J. (1972): *Biochim. Biophys. Acta,* 284:235–247.
10. Mosimann, H., Semenza, G., and Sund, H. (1973): *Eur. J. Biochem.,* 36:489–494.
11. Cogoli, A., Eberle, A., Sigrist, H., Joss, Ch., Robinson, E., Mosimann, H., and Semenza, G. (1973): *Eur. J. Biochem.,* 33:40–48.
12. Quaroni, A., Gershon-Quaroni, E., and Semenza, G. (1975), *Eur. J. Biochem.,* 52:481–486.
13. Braun, H., Cogoli, A., and Semenza, G. (1975) *Eur. J. Biochem.,* 52:475–480.
14. Semenza, G., Curtius, C.-H., Kolínská, J., and Müller, M. (1967): *Biochim. Biophys. Acta,* 146:196–204.
15. Semenza, G. Curtius, H. Ch., Raunhardt, O., Hore, P., and Müller, M. (1969): *Carbohydrate Res.,* 10:417–428.
16. Stefani, A., Janett, M., and Semenza, G. (1975): *J. Biol. Chem.,* 250:7810–7813.
17. Semenza, G., and Balthazar, A. K. (1974): *Eur. J. Biochem.,* 41:149–162.
18. Janett, M. (1974): Diplomarbeit an der ETH, Zürich.
19. Semenza, G. (1969): *Eur. J. Biochem.,* 8:518–529.
20. Flückiger, R. (1973): Diplomarbeit an der ETH, Zürich.
21. Cogoli, A., and Semenza, G. (1975): *J. Biol. Chem.,* 250:7802–7809.
22. Kolínská, J., and Semenza, G. (1967): *Biochim. Biophys. Acta,* 146:181–195.
23. Quaroni, A., Gershon, E., and Semenza, G. (1974): *J. Biol. Chem.* 249:6424–6433.
24. Quaroni, A., and Semenza, G. (1976): *J. Biol. Chem.* (in press).
25. Braun, H., Cogoli, A., and Semenza, G. (1976): (*to be submitted*).

26. Rubino, A., Zimbalatti, F., and Auricchio, S. (1964): *Biochim. Biophys. Acta*, 92:305–311.
27. Dahlqvist, A., and Lindberg, T. (1966): *Clin. Sci.*, 30:517–528.
28. Auricchio, S., Rubino, A., Tosi, R., Semenza, G., Landolt, M., Kistler, H., and Prader, A. (1963): *Enzym. Biol. Clin.*, 3:193.
29. Semenza, G., Auricchio, S., Rubino, A., Prader, A., and Welsh, J. D. (1965): *Biochim. Biophys. Acta*, 105, 386–389.
30. Auricchio, S., Rubino, A., Prader, A., Rey, J., Jos, J., Frézal, J., and Davidson, M. (1965): *J. Pediatr.*, 66:555–564.
31. Kerry, K. R., and Townley, R. R. W. (1965): *Austr. Paediatr. J.*, 1:223–235.
32. Crane, R. K. (1976): *This volume.*
33. Leese, H., and Semenza, G. (1973): *J. Biol. Chem.*, 248:8170.
34. Malathi, P., and Crane, R. K. (1969): *Biochim. Biophys. Acta*, 173:245–256.
35. Alpers, D. H. (1969): *J. Biol. Chem.*, 244:1238–1246.
36. Schlegel-Haueter, S., Hore, P., Kerry, K. R., and Semenza, G. (1972): *Biochim. Biophys. Acta*, 258:506–519.
37. Wallenfels, K., and Fischer, J. (1960), *Z. Physiol. Chem.*, 321:223–245.
38. Colombo, V., Lorenz-Meyer, H., and Semenza, G. (1973): *Biochim. Biophys. Acta*, 327:412–424.
39. Lorenz-Meyer, H., Blum, A. L., Hämmerli, H. P., and Semenza, G. (1972): *Eur. J. Clin. Invest.*, 2:326–331.
40. Kretchmer, N. (1971): *Gastroenterology*, 61:805–813.
41. Ramaswamy, S., and Radhakrishnan, A. N. (1972): *Comp. Biochem. Physiol.*, 43A:173–179.
42. Birkenmeier, E., and Alpers, D. H. (1974): *Biochim. Biophys. Acta*, 350:100–112.
43. Kraml, J., Kolínská, J., Ellederová, D., and Hiršová, D. (1972): *Biochim. Biophys. Acta*, 258:520–530.
44. Kayser, S. G., and Patton, S. (1965): *Biochem. Biophys. Res. Commun.*, 41:1572–1578.
45. Gimpert-Würth, E. (1972): Diplomarbeit an der ETH, Zürich.
46. Honegger, P., and Semenza, G. (1972): *Biochim. Biophys. Acta*, 318:390–410.
47. Sigrist-Nelson, K., and Hopfer, U. (1974): *Biochim. Biophys. Acta*, 367:247–254.
48. Murer, H., and Hopfer, U. (1974): *Proc. Natl. Acad. Sci. USA*, 71:484–488.
49. Crane, R. K. (1962): *Fed. Proc.*, 21:891–895.
50. Schultz, S. G., and Curran, F. C. (1970): *Physiol. Rev.*, 50:637–718.
51. Hopfer, U., Sigrist-Nelson, and Murer, H. (1976): *NY Acad. Sci.* (in press).
52. Kimmich, G. A. (1973): *Biochim. Biophys. Acta*, 300:31–78.
53. Lindquist, B., and Meeuwisse, G. W. (1962): *Acta Paediatr. Scand.*, 51:674–685.
54. Rosenberg, L. (1976): *This volume.*
55. Meeuwisse, G. W. (1970): Dissertation, University of Lund.
56. Miller, D., and Crane, R. K. (1961): *Biochim. Biophys. Acta*, 52:281–293.
57. Crane, R. K., Malathi, P., Caspary, W. F., and Ramaswamy, K. (1970): *Fed. Proc.*, 29:595.
58. Malathi, P., Ramaswamy, K., Caspary, W. F., and Crane, R. K. (1973): *Biochim. Biophys. Acta*, 307:613–626.
59. Ramaswamy, K., Malathi, P., Caspary, W. F., and Crane, R. K. (1974): *Biochim. Biophys. Acta*, 345:39–48.
60. Storelli, C., Vögeli, H., and Semenza, G. (1972): *FEBS Lett.*, 24:287–292.
61. Vögeli, H., Brunner, J., and Semenza, G. (1976): *in preparation.*
62. Forstner, G. G., Tanaka, K., and Isselbacher, K. J. (1968): *Biochem. J.*, 109:51–59.
63. Millington, P. F., and Critchley, D. R. (1968): *Life Sci.*, 7:839–845.

Membranes and Disease, edited by L. Bolis, J. F. Hoffman, and A. Leaf. Raven Press, New York, © 1976.

Intestinal Hexose Transport in Familial Glucose–Galactose Malabsorption

Leon E. Rosenberg

Department of Human Genetics, Yale University School of Medicine, New Haven, Connecticut 06520

The transcellular, uphill transport of D-glucose across the mucosal epithelium of the mammalian small intestine and proximal renal tubule is mediated by a sodium-dependent, phlorizin-sensitive "carrier" system (1). The postulated carrier, presumably a protein, is thought to be located in the apical brush-border membrane, and to function by combining with sodium ions and with molecules of D-glucose or other structurally similar aldohexoses such as D-galactose at specific receptor sites (2). There is increasing evidence that the uphill transfer of glucose via this ternary complex is produced by the simultaneous facilitated movement of sodium down its electrochemical gradient (3,4). This downhill gradient for sodium across the mucosal brush border is maintained by the cation "pump" system found in the basal and lateral membranes of brush-border epithelial cells which exchanges intracellular sodium for extracellular potassium and which is intimately linked to ouabain-sensitive Na^+, K^+-ATPase. Similar sodium-dependent cotransport systems for many amino acids are also present in gut and kidney.

If this formulation is correct, it follows that the protein components of such carrier systems will be under genetic control and, therefore, subject to mutation. If such mutations alter the brush-border carrier protein specific for glucose and galactose, one might expect to see a specific defect of hexose transport. Similarly, a defect in the carrier mediating neutral amino acid transport would cause selective impairment of transport of this group of substrates. If, on the other hand, the mutation produced a defective Na^+, K^+-ATPase, then the active transport of these cations, as well as the cotransport of hexoses and amino acids, might be impaired. The disorder discussed in this chapter, glucose–galactose malabsorption (GGM), almost surely reflects a mutation of the brush-border carrier protein which mediates the transport of glucose and galactose.

THE CLINICAL ENTITY

In 1962 Lindquist (5), Laplane (6), and their colleagues independently described neonates with very similar findings, namely, profuse watery diarrhea after ingestion of milk or feeds containing lactose, sucrose, glucose, or galactose. Because the diarrhea abated promptly if carbohydrate-free feeds were offered, or if fructose was substituted for glucose and galactose in the diet, these workers proposed that the affected infants suffered from a specific abnormality of glucose and galactose absorption. This formulation was supported by oral loading tests and by intestinal intubation studies (7). Oral glucose tolerance tests resulted in little or no increase in blood glucose and in the prompt excretion of large quantities of free glucose in the diarrheal feces (Fig. 1A). Similar results were obtained after loading with galactose or 3-*O*-methylglucose. In contrast, ingestion of fructose loads was followed by a prompt, normal rise in blood fructose and blood glucose without the appearance of diarrhea or fructose in the feces. Following intestinal intubation and perfusion, affected patients absorbed much less glucose than did controls, and again absorbed fructose normally. Significantly, net absorption of ^{22}Na by patients with GGM was unimpaired.

Subsequently, more than 20 patients with this disorder were reported from Europe, Australia, and the United States (8). Many have died in the first weeks or months of life as a result of dehydration secondary to the profuse diarrhea. Those who have been diagnosed and treated with glucose- and galactose-free diets have thrived and, interestingly, their ability to tolerate glucose and galactose containing foods has improved with age (7,9). Lindquist and Meeuwisse discovered a 35-year-old woman with this condition who had controlled her symptoms from childhood on with a diet self-restricted in the offending hexoses (5,7).

AUTOSOMAL RECESSIVE INHERITANCE

Melin and Meeuwisse (10) provided the first compelling evidence that GGM was inherited. They found four affected females and two affected males in multiple sibships of a highly inbred Swedish pedigree. Based on this informative pedigree, they postulated that the condition was inherited as an autosomal recessive trait, a thesis supported by subsequent reports describing families showing involvement of sibs of both sexes (8), and by the ability to differentiate heterozygous carriers from normals using *in vitro* techniques (11).

IN VITRO TRANSPORT STUDIES

A tetrad of reports in 1966 added important information about the specific nature of the gut transport defect in patients with GGM. Each of these studies employed small pieces of jejunal mucosa, obtained by peroral biopsy, which were incubated in a medium containing ^{14}C-labeled galactose or glucose. Using autoradiographic techniques, Schneider et al. (12) showed that mucosa from a control subject accumulated ^{14}C-galactose to a concen-

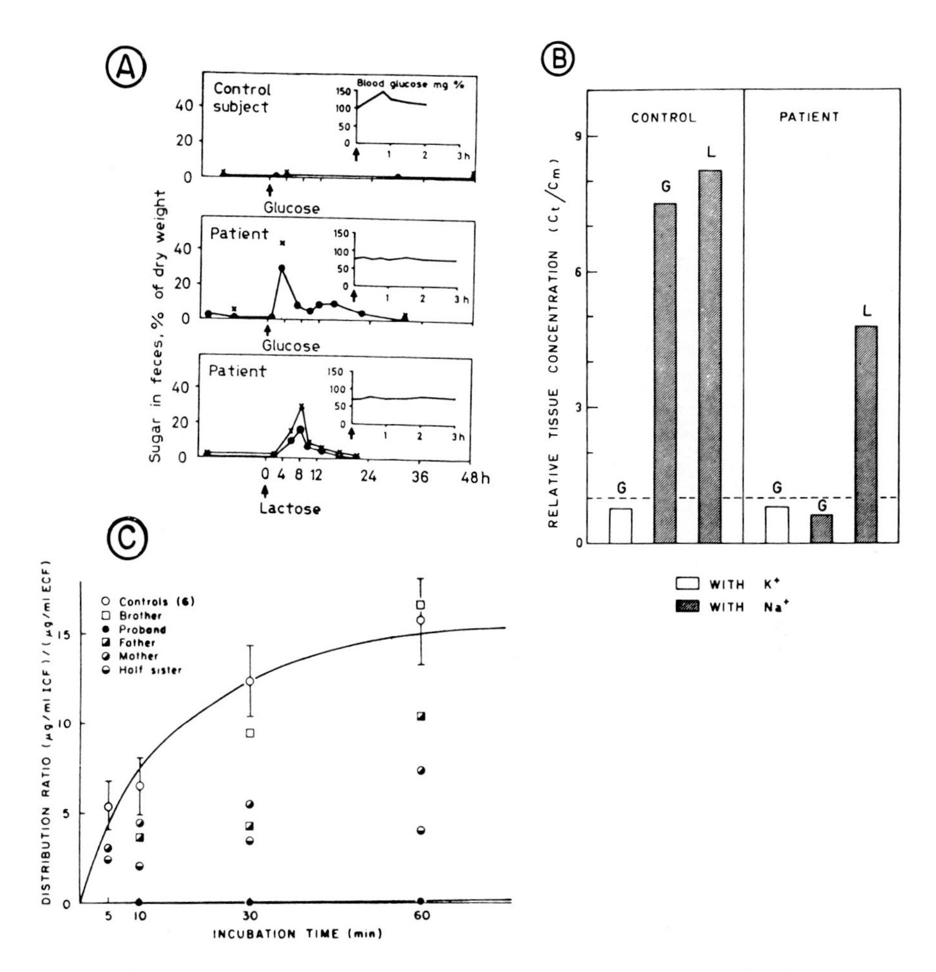

FIG. 1. A: Blood glucose determinations and fecal sugar concentrations after oral loads of glucose in a control subject (*upper panel*) and after glucose (middle panel) and lactose (*lower panel*) in a child with glucose–galactose malabsorption. x = total sugar; ● = glucose. [Reproduced with permission from Lindquist and Meeuwisse (7).] **B:** Accumulation of ^{14}C-D-glucose or ^{14}C-L-leucine by intestinal mucosa from controls (left panel) and a patient with glucose–galactose malabsorption (right panel). Mucosal pieces were incubated for 15 min at 37°C in Krebs-Henseleit buffer, pH 7.4 (*hatched bars*) or in an identical medium in which all sodium was replaced by potassium (*open bars*). Substrate concentrations were 0.1 mM for D-glucose (G) and L-leucine (L). The relative tissue concentration shown on the ordinate is the ratio of the concentration of radioactive material in the tissue (C_t) to that in the incubation medium (C_m). All values are means from two control babies and from two biopsies of the affected child. [Reproduced with permission from Eggermont and Loeb (13).] **C:** Uptake of D-glucose (2 mM) by jejunal mucosa from controls and from several members of a family with glucose–galactose malabsorption. Biopsy specimens were incubated in Krebs-Ringer bicarbonate buffer, pH 7.4, at 37°C for times indicated on the abscissa. Distribution ratios (*ordinate*) represent the ratio of the true glucose concentration in the intracellular fluid (µg/ml ICF) to that in the incubation medium (µg/ml ECF). Values for the individual members of the pedigree represent the mean of at least two observations at each incubation interval. [Reproduced with permission from Elsas et al. (11).]

tration far in excess of that in the incubation medium, and that phlorizin distinctly reduced this accumulation. In contrast, mucosa from a child with GGM failed to demonstrate concentrative uptake of ^{14}C-galactose and showed no inhibitory response to phlorizin, despite appearing completely normal histologically.

Eggermont, Loeb, and their colleagues (13,14) made several interesting observations (Fig. 1B). First, ^{14}C-D-glucose was concentrated eightfold by normal jejunal mucosa incubated in the presence of 140 m*M* NaCl, but no accumulation beyond that of the glucose concentration in the incubating medium was found when KCl was substituted equimolarly for NaCl. Second, tissue from a patient with GGM failed to concentrate ^{14}C-glucose in the presence of either NaCl or KCl. Third, ^{14}C-L-leucine was concentrated to approximately the same extent by mucosa from a control and that from the patient with GGM. Fourth, they found no difference in activity of the brush-border enzymes alkaline phosphatase, maltase, isomaltase, saccharase, lactase, and trehalase obtained from control tissue and that from the patient with GGM.

Similar findings were reported by Meeuwisse and Dahlquist (15). They, too, showed that ^{14}C-glucose uptake by mucosa from two patients with GGM was much impaired and was insensitive to phlorizin inhibition. In addition, they reported that mucosa from GGM patients accumulated L-alanine normally, and had normal total ATPase activity, normal ouabain-inhibited Na^+, K^+-ATPase activity, and normal activity of invertase and disaccharidase.

Two subsequent investigations also deserve mention. Elsas et al. (11) showed: that ^{14}C-glucose uptake by normal jejunum mirrored that of unlabeled glucose; that glucose uptake by normal jejunum was sodium-dependent and markedly inhibited by ouabain, sodium cyanide, and 2,4-dinitrophenol; that glucose uptake by mucosa from a child with GGM was markedly impaired, whereas uptake by mucosa from both of her parents fell between that of the affected child and controls (Fig. 1C); that uptake of ^{14}C-fructose was unimpaired in GGM; and that, using Michaelis–Menten kinetic analysis, the apparent K_m for glucose in tissue from both parents of the affected child was normal, whereas the apparent V_{max} was distinctly reduced (Fig. 2A). Stirling et al. (16) extended their previous autoradiographic results by showing: that ^{3}H-phlorizin uptake by mucosa from a child with GGM was as impaired as was uptake of ^{3}H-galactose (Fig. 2B); that the binding constant (K) for phlorizin was unchanged in mucosa from their GGM patient; and that the estimated number of normal hexose carriers per cell ($\sim 2.8 \times 10^6$) was reduced by 90% in their patient with GGM. Their high resolution autoradiographs convincingly demonstrated the defective uptake of ^{3}H-D-galactose by mucosa from an affected child (Fig. 3).

Finally, it should be pointed out that, although ability to tolerate glucose and galactose improves clinically in patients with GGM, this does not ap-

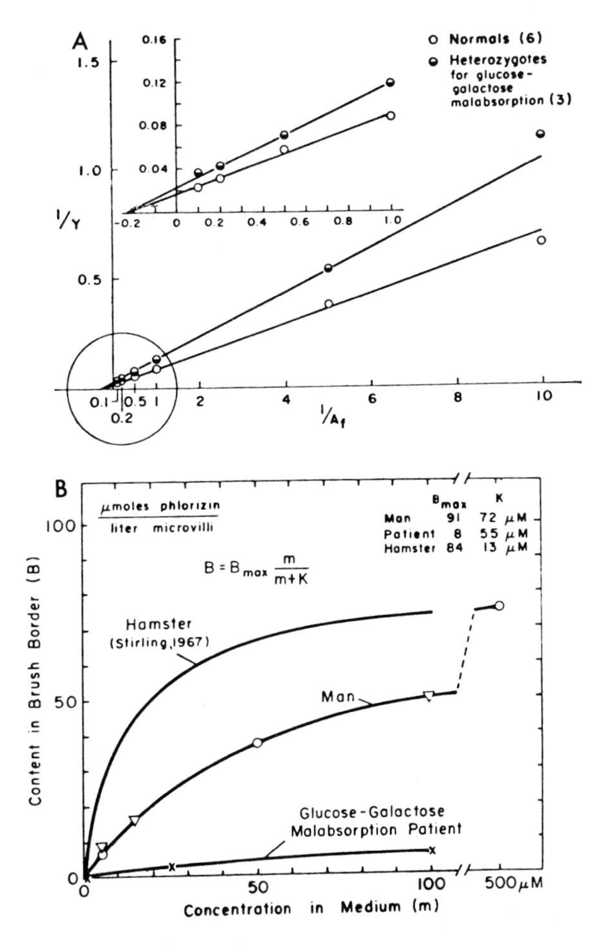

FIG. 2. A: Effect of increasing substrate concentration (A_f) on the mediated uptake (Y) of D-glucose by jejunal mucosa from controls and heterozygous carriers of glucose–galactose malabsorption. Biopsies were incubated for 10 min in ^{14}C-D-glucose concentrations from 0.1–40.0 mM. The data are plotted using the double reciprocal method. Each point represents the mean of at least four observations. The circled portion indicates that part of the plot shown with the expanded scale above. [Reproduced with permission from Elsas et al. (11).] **B:** ^{3}H-Phlorizin binding to jejunal mucosa from controls (O) and a patient with glucose–galactose malabsorption (X). Brush border-bound phlorizin (B), after incubation for 15 min, was calculated from the relative grain density in the microvilli to that in the medium (m). Datum points represent average grain density ratios from two or more sections of individual specimens. Curves were calculated from the adsorption equation shown after the maximum binding, B_{max}, and half-saturation, K, constants were obtained from intercepts of straight lines fitted by eye to reciprocal plots of the data. The hamster curve is from a previous study. [Reproduced with permission from Stirling et al. (16).]

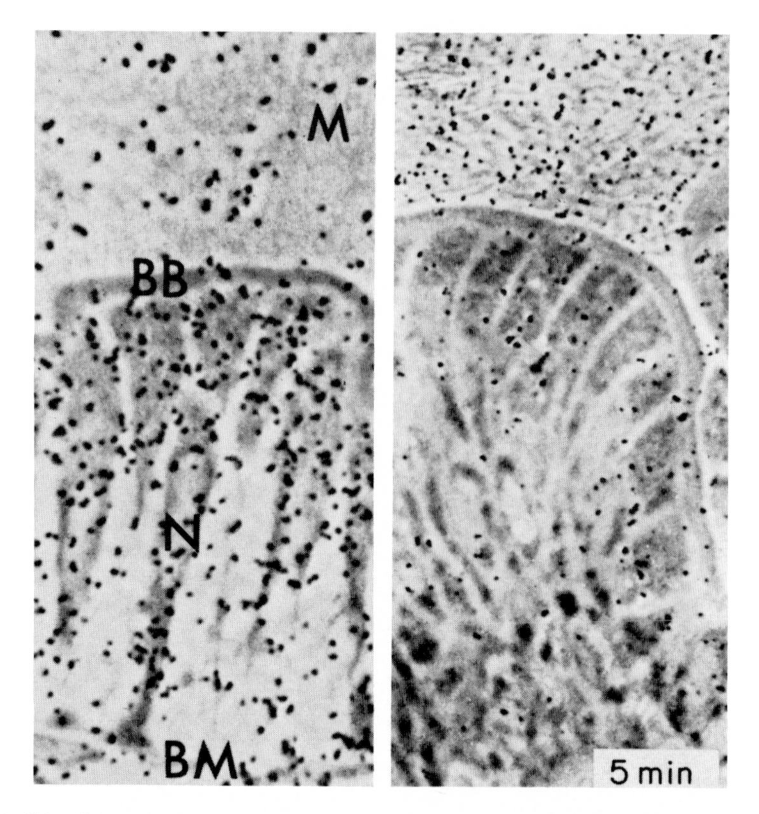

FIG. 3. ^{3}H-D-Galactose radioautographs of columnar, jejunal epithelium from a normal mucosal biopsy (*left panel*) and that from a patient with glucose–galactose malabsorption (*right panel*). Tissues were incubated for 5 min at 37°C in Krebs-Henseleit buffer, pH 7.4, containing 1.0 mM D-galactose. Nuclei (N) divide the cytoplasm into a thick apical portion contiguous with the brush border (BB) and a tenuous basal portion terminating on the indistinct basement membrane (BM). The block dots reflect grain density in tissue or medium (M). *Magnifications:* left panel ×1445; right panel ×1105. Note the dramatic reduction in grains in the tissue from the affected patient compared to that in the control. [From Stirling et al. (16).]

pear to be due to disappearance or amelioration of the mucosal transport defect. Several groups have obtained gut biopsies from affected patients who had "outgrown" their severe diarrheal symptoms (7,9,16). Nevertheless, uptake of labeled glucose or galactose remained as impaired as when these or other patients had been studied at an earlier age.

RENAL GLYCOSURIA IN GLUCOSE–GALACTOSE MALABSORPTION

A number of patients with GGM have been reported to demonstrate glycosuria at normal blood glucose concentrations suggesting that renal tubular transport of glucose may also be impaired in this disorder. Several groups have attempted to quantitate renal tubular reabsorption of glucose using classical glucose titration techniques (11,16–19). Results have varied due to the difficulty in performing this test in children, but two conclusions can be drawn from the data presented: first, most patients with GGM have a reduced renal threshold for glucose excretion; second, the maximal absorptive capacity for glucose reabsorption (T_mG) is unimpaired. These findings suggest that, whereas glucose transport in the renal tubule is abnormal in GGM, the magnitude of the impairment is much less than that demonstrable in the gut.

GLUCOSE TRANSPORT IN OTHER TISSUES IN GLUCOSE–GALACTOSE MALABSORPTION

Several lines of evidence indicate that the hexose transport defect found in gut and kidney in GGM is not found in other tissues. First, intravenous glucose tolerance tests in these patients have yielded normal results (7). Second, uptake of glucose by erythrocytes from patients with GGM is not different than that of controls (20). Third, uptake of glucose and production of $^{14}CO_2$ by cultured skin fibroblasts from a patient with GGM yield normal results (9).

FORMULATION, CONCLUSIONS, AND UNANSWERED QUESTIONS

I will conclude by posing and attempting to answer three questions. First, what is the basic defect in GGM? I believe that all of the *in vivo* and *in vitro* observations are consistent with the hypothesis that the underlying abnormality in this disorder involves a mutation of the postulated "carrier" protein in the brush border of the gut and kidney which mediates the uphill transport of glucose and galactose. Only a mutation of this protein could produce the tissue and substrate specificity which defines this condition. Such a mutation could alter either the quality or quantity of that protein, but none of the data available allow one to choose between these two general possibilities. This question can be answered only by isolation of the carrier protein from normal gut followed by the same sort of detailed biochemical and immunochemical analysis being employed to understand a growing number of other inherited metabolic disorders. Since the characteristics of glucose transport in human jejunum seem to be very similar to those in other well-studied mammalian species, such as hamster, rabbit, and rat (2,11,16), and since reports are beginning to appear which suggest the feasibility of isolation of such carrier proteins from gut or kidney tissue or both (21–23), I believe that we can look forward to a molecular understanding of this disease in the near future.

Second, why do symptoms in GGM decrease with age, whereas the underlying mucosal defect is permanent? Two factors may be at play here.

One involves the expected change in diet with age such that milk (and therefore milk sugar) and other glucose- or galactose-containing foods become much less significant contributors to total nutrition than they are in the early months of life. Another adaptation, and perhaps a more interesting one, may involve the role that the extracellular "shunt" pathway plays in intestinal glucose absorption (4). It is certainly possible that, as the intestine grows and matures, this shunt pathway and its tight junctional anatomic counterpart may be increasingly more important in the net absorption of water and solute across the gut epithelium. A developmental study of this pathway in experimental animals of different ages would be of interest.

Third, why is the defect in GGM more prominent in gut than in kidney and, apparently, not found at all in other tissues? I believe the answer to this question must be related to the differential expression of a common genome in different tissues. There are numerous examples of enzymes and other proteins which are found in one, or a few, tissues but not in others. It seems quite likely to me that glucose transport in gut and kidney may be mediated by a completely different carrier protein than that found in the plasma membranes of other cells, and that, in addition to this carrier shared by kidney and gut, the kidney has an additional carrier(s) not active in the gut (24–26). Such a "mosaic" model of carriers in different tissues can explain the observed tissue differences in GGM and has the attractive feature of being testable.

REFERENCES

1. Crane, R. K. (1968): Absorption of sugars. In: *Handbook of Physiology,* Vol. III, Chap. 69, p. 1323. American Physiological Society, Washington D.C.
2. Lefevre, P. G. (1972): Transport of carbohydrates by animal cells. In: *Metabolic Pathways,* 3rd ed., Vol. VI, edited by L. E. Hokin, p. 385. Academic Press, New York.
3. Schultz, S. G., and Curran, P. F. (1970): Coupled transport of sodium and organic solutes. *Physiol. Rev.,* 50:637.
4. Schultz, S. G., Frizzell, R. A., and Nellans, H. N. (1974): Ion transport by mammalian small intestine. *Annu. Rev. Physiol.,* 36:51.
5. Lindquist, B., Meeuwisse, G. W., and Melin, K. (1962): Glucose–galactose malabsorption. *Lancet,* 2:666.
6. Laplane, R., Polonovski, C., Etienne, M., Debray, P., Lods, J.-C., and Pissaro, B. (1962): L'intolerance aux sucres a transport intestinal actif. *Arch. Fr. Pediatr.,* 19:895.
7. Lindquist, B., and Meeuwisse, G. W. (1968): Glucose–galactose malabsorption. In: *Intestinal Absorption and Malabsorption,* edited by D. H. Shmerling, H. Berger, and A. Prader, p. 92. Karger, Basel.
8. Krane, S. M. (1972): Renal glycosuria. In: *The Metabolic Basis of Inherited Disease,* 3rd ed., edited by J. B. Stanbury, J. B. Wyngaarden, and D. S. Fredrickson, p. 1536. McGraw-Hill, New York.
9. Elsas, L. J., and Lambe, D. W. (1973): Familial glucose–galactose malabsorption: Remission of glucose intolerance. *J. Pediatr.,* 83:226.
10. Melin, K., and Meeuwisse, G. W. (1969): Glucose–galactose malabsorption: A genetic study. *Acta Paediatr. Scand.* [*Suppl.*], 188:20.
11. Elsas, L. J., Hillman, R. E., Patterson, J. H., and Rosenberg, L. E. (1970): Renal and intestinal hexose transport in familial glucose–galactose malabsorption. *J. Clin. Invest.* 49:576.

12. Schneider, A. J., Kinter, W. B., and Stirling, C. E. (1966): Glucose–galactose malabsorption: Report of a case with autoradiographic studies of a mucosal biopsy. *N. Engl. J. Med.*, 274:305.
13. Eggermont, E., and Loeb, H. (1966): Glucose–galactose intolerance. *Lancet,* 2:343.
14. Dubois, R., Loeb, H., Eggermont, E., and Mainquet, P. (1966): Etude clinique et biochimique d'un cas de malabsorption congenitale du glucose et du galactose. *Helv. Paediatr. Acta,* 21:577.
15. Meeuwisse, G. W., and Dahlqvist, A. (1966): Glucose–galactose malabsorption. *Lancet,* 2:858.
16. Stirling, C. E., Schneider, A. J., Wong, M.-D., and Kinter, W. B. (1972): Quantitative radioautography of sugar transport in intestinal biopsies from normal humans and a patient with glucose–galactose malabsorption. *J. Clin. Invest.*, 51:438.
17. Lin, Y.-Y., Anderson, G. J., Tsao, M. U., Moore, B. F., and Giday, Z. (1967): T_m glucose in a case of congenital intestinal and renal malabsorption of monosaccharides. *Pediatr. Res.*, 1:386.
18. Meeuwisse, G. W. (1970): Glucose-galactose malabsorption: Studies on renal glycosuria. *Helv. Paediatr. Acta,* 25:13.
19. Beauvais, P., Vandourt, G., Desjeux, J.-F., LeBalle, J.-C., Girot, J.-Y., and Brissard, H.-E. (1971): La malabsorption congenitale du glucose–galactose. *Arch. Fr. Pediatr.*, 28:573.
20. Meeuwisse, G. W. (1970): Glucose–galactose malabsorption: A study on the transfer of glucose across the red cell membrane. *Scand. J. Clin. Lab. Invest.*, 25:145.
21. Glossman, H., and Neville, D. M. (1972): Phlorizin receptors in isolated kidney brush border membranes. *J. Biol. Chem.*, 247:7779.
22. Thomas, L. (1973): Isolation of *N*-ethylmaleimide-labelled phlorizin-sensitive glucose binding protein of brush border membrane from rat kidney cortex. *Biochim. Biophys. Acta,* 291:454.
23. Murer, H., Hopfer, U., Kinne-Saffran, E., and Kinne, R. (1974): Glucose transport in isolated brush border and lateral-basal plasma membrane vesicles from intestinal epithelial cells. *Biochim. Biophys. Acta,* 345:170.
24. Elsas, L. J., and Rosenberg, L. E. (1969): Familial renal glycosuria: A genetic reappraisal of hexose transport by gut and kidney. *J. Clin. Invest.*, 48:1845.
25. Silverman, M., Aganon, M. A., and Chinard, F. P. (1970): D-Glucose interactions with renal tubule cell surfaces. *Am. J. Physiol.*, 218:735.
26. Busse, D., Elsas, L. J., and Rosenberg, L. E. (1972): Uptake of D-glucose by renal tubule membranes: Evidence for two transport systems. *J. Biol. Chem.*, 247:1188.

Membranes and Disease, edited by L. Bolis, J. F. Hoffman, and A. Leaf. Raven Press, New York, © 1976.

Morphological and Biochemical Characterization of Two Water-Soluble Membrane Proteins Isolated from the Suckling Rat Ileum

E. R. Jakoi, G. Zampighi, and J. D. Robertson

Department of Anatomy, Duke University, Durham, North Carolina 27710

In newborn mammals during the period of suckling, the plasma membrane of the ileal epithelial cells forms a highly invaginated complex consisting of tubular membranes located beneath the microvilli (1–3) (Fig. 1). The entire system is interconnected and open to the lumen, but does not connect directly with the large central vacuole (1,2). These membranes are covered by an ordered array of particles ~7.5 nm in diameter, joined together with a center-to-center separation of ~14.5 nm to form long strips. These strips aggregate laterally in either a square or an oblique two-dimensional lattice, which has been described by negative stain (2,4) and thin section electron microscopy (1,5) (Fig. 1b).

During the suckling stage of development (0–21 days of age), several authors (6–10) have reported that different disaccharidases are highly active in homogenates of epithelial scrapings from the ileum. This elevation of disaccharidase activity falls dramatically at weaning (7–10) or upon the administration of corticosteroids (11–14). Prior to 15–21 days of age, there is little or no digestion of food in the stomach and the duodenum because neither the stomach (15) nor the pancreas (16) is secreting hydrolases. Thus, the milk ingested during this period is most likely digested in part by the lysosomal enzymes found in the ileal homogenates. It occurred to us that the array of particles on the surface of the differentiated plasma membranes might in fact be a lattice of enzymes involved in this digestive process. In particular, it seemed likely that disaccharidases might be located in the lattice. In this case, the surface of the membranes would contain sites for the extracellular digestion of at least the carbohydrate present in the maternal milk immediately before absorption through the membranes into the cytoplasm. In order to determine whether or not any specific enzymic activity was associated with these membranes, and in particular with the particulate coat covering their surfaces, the membranes were isolated and analyzed biochemically and structurally.

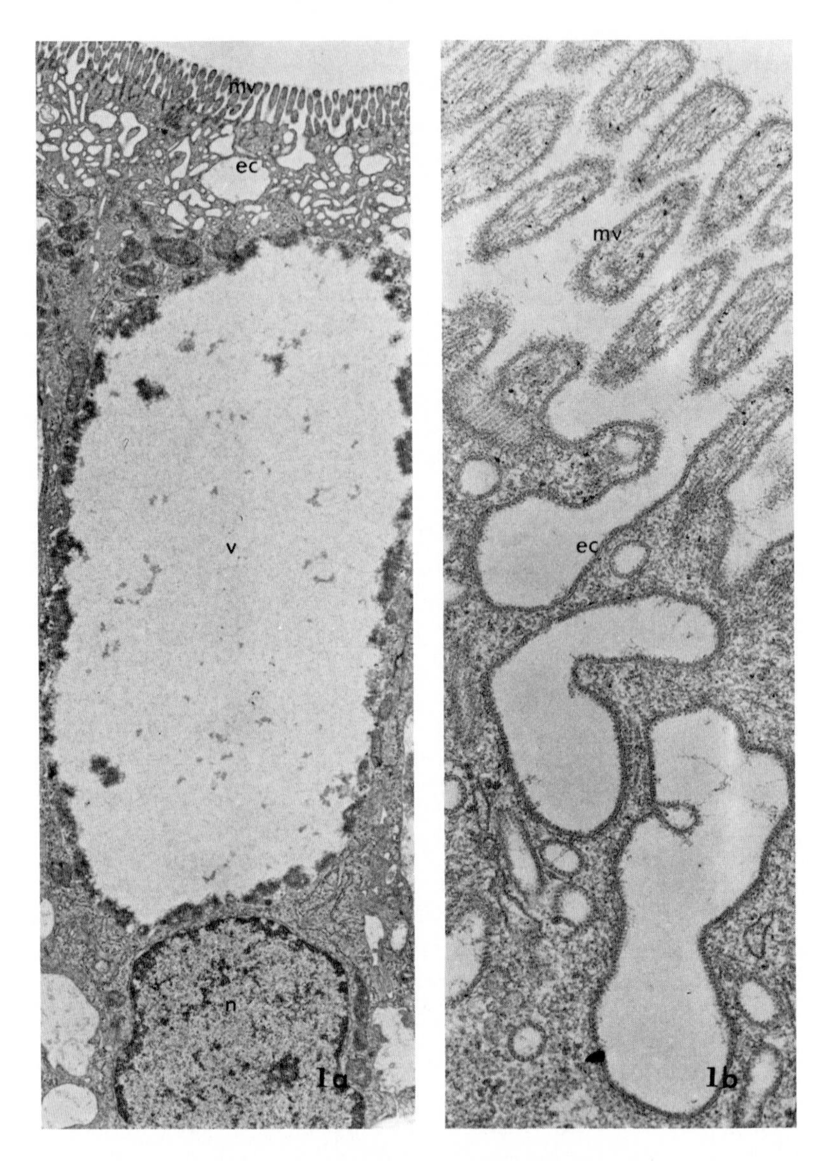

FIG. 1. Electron micrographs of thin sections of ileal epithelial cells from a 10–14-day-old suckling rat. **a:** The relative location of the invaginated plasma membrane complex (ec) is shown with respect to the microvilli (mv), the large central vacuole (v), and the nucleus (n). ×8,550. **b:** A higher-power magnification of the apical region of such a cell in which the particulate covering on the invaginated plasma membranes is seen in transverse and tangential (*en face*) views. ×45,000. [From Knutton et al. (2).]

ENZYMATIC AND STRUCTURAL CHARACTERIZATION OF ISOLATED MEMBRANES

The differentiated plasma membranes from suckling rat ileums were isolated free of other cell membranes by differential flotation through step sucrose gradients (17). The membrane fraction was identified morphologically by negative stain (Fig. 2), and thin section electron microscopy (Fig. 3) to be essentially free of other cell membranes and nonmembranous material. Enzymatic determinations suggested the presence of a microvillar contamination, since alkaline phosphatase activity was present in these fractions and this enzyme has been shown to be located in the ileal microvilli of suckling rats (10,18–20). No significant amounts of α-galactosidase activity, β-galactosidase (pH 5.5) activity, or β-glucuronidase activity were detected in the isolated membrane fractions. Similarly, no acid phosphatase activity was found, i.e., there was no general lysosomal contamination (21,22). However, β-galactosidase (pH 3.5) and *n*-acetyl-β-glucosaminidase were active in these fractions. Thus apparently three enzymatic activities

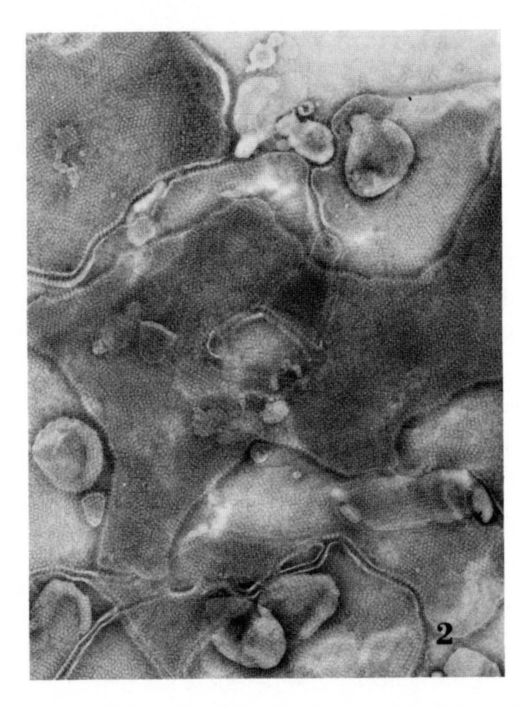

FIG. 2. Negative stain preparation (1% PTA with bacitracin) of the isolated plasma membranes. The membrane surface is covered by a particulate array which retains the same lattice spacings seen *in vivo.* Some microvillar membrane contamination is also seen in these fractions, which at times is continuous with the particulate-coated membranes. ×37,500.

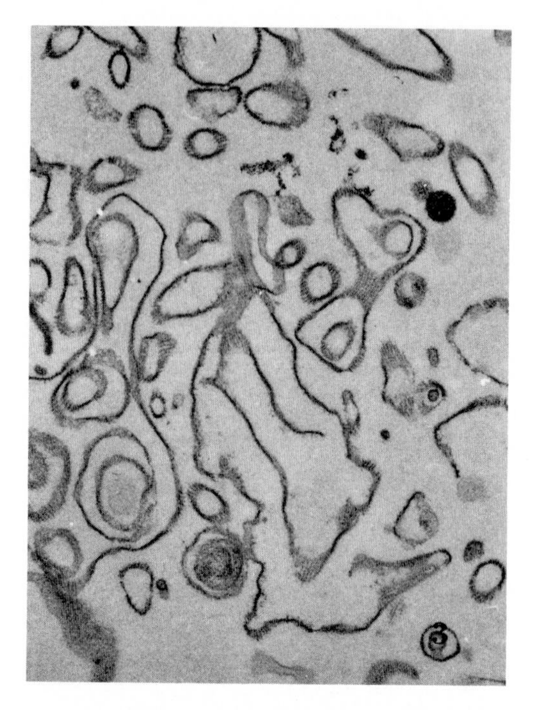

FIG. 3. Thin section of the isolated plasma membrane pellet. Transverse and tangential (*en face*) views show particles attached in rows to one of the membrane surfaces. Little nonmembranous material is evident in these pellets. ×33,750.

are associated with the isolated differentiated plasma membranes: alkaline phosphatase (a likely contaminant), β-galactosidase (β-GAL), and *n*-acetyl-β-glucosaminidase (NAG).

ISOLATION AND CHARACTERIZATION OF THE PARTICULATE COAT

The particles were released from the membranes by 10 mM $CaCl_2$ to give long, individual strips of particles which retain the 14.5 nm periodicity (Fig. 4) (17,23). We refer to these strips of particles as "decorated strips." In addition to the decorated strips, coated membranes were present in this fraction. Since we presumed that the strips were protein it seemed reasonable to expect that they would have very different densities from the membranes which are protein-like complexes. Therefore, we used isopycnic centrifugation in an effort to isolate the particles from the membranes. Two populations of membranes were resolved: one that had retained a lattice and banded at 1.1560ρ and another consisting of membranes which were free of particles banded at 1.0900ρ. We were surprised to find that in addition to the denuded membranes in the lighter fraction, there were also

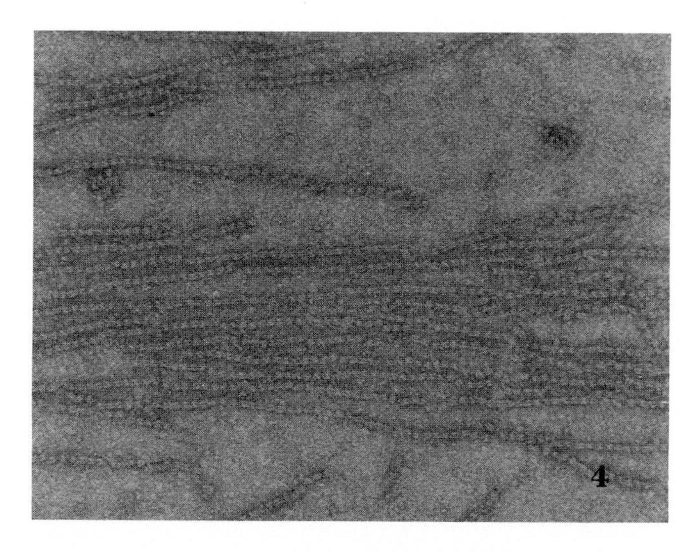

FIG. 4. Decorated strips released from the membrane surface by $CaCl_2$ and negatively stained with 1% PTA containing bacitracin. Each decorated strip consists of individual ~7.5-nm diameter particles joined together with a center-to-center separation of ~14.5 nm. ×87,750.

numerous short decorated strips. The lighter fraction contained predominantly NAG activity with a small amount of β-GAL activity and no alkaline phosphatase activity (Fig. 5). Most of the β-GAL and all of the alkaline phosphatase activities were in the heavier fraction (Fig. 5). In this heavier fraction (1.1560ρ) no recognizable decorated strips were found by negative stain electron microscopy.

Since isopycnic centrifugation did not separate the decorated strips from the denuded membranes, rate centrifugation was employed. The decorated strips were dialyzed against 0.5 mM EGTA, which decreased their sedimentation coefficients. They were then separated by rate centrifugation from the denuded and particulate-coated membranes. The supernatant was further resolved into two protein components by gel filtration through a Sephadex

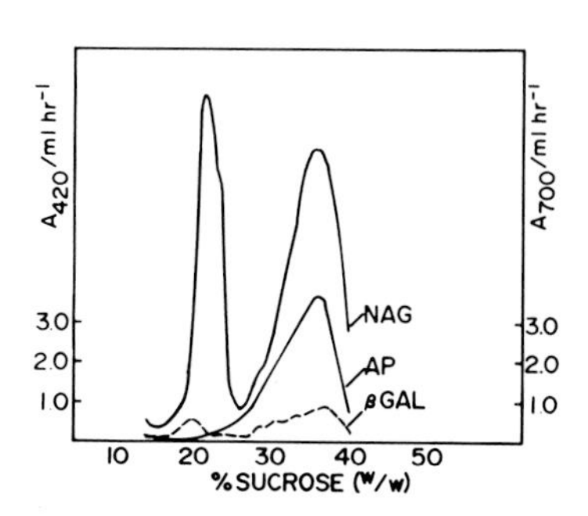

FIG. 5. Sedimentation profile of $CaCl_2$-treated membranes after an isopycnic centrifugation on a linear 15–30% (w/w) sucrose gradient. Left ordinate is the specific activity of NAG and β-GAL expressed as A_{420} units/ml sample per hr. Right ordinate is the specific activity of alkaline phosphatase (AP) expressed as A_{700} units/ml sample per hr.

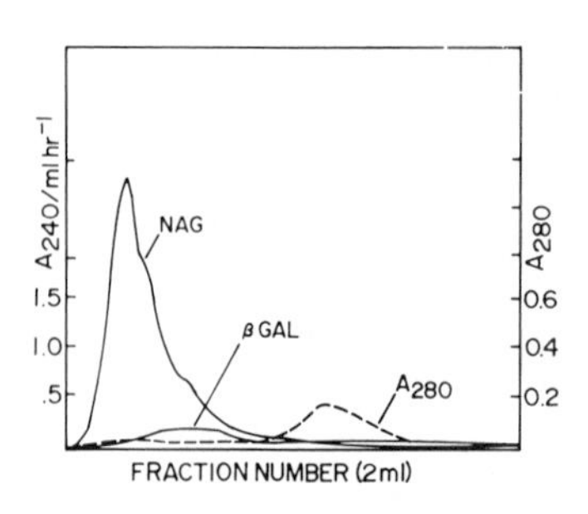

FIG. 6. Elution profile from a Sephadex G200 separation of the protein components of the decorated strips. *Left ordinate,* Specific activity of NAG and β-GAL expressed as A_{420} units/ml sample per hr. *Right ordinate,* Protein concentration expressed as A_{280} units. Essentially two proteins are resolved: one excluded from the Sephadex G200 bed, NAG, and one retarded by the Sephadex G200 bed shown only by its A_{280} peak. The β-GAL activity is consistently present, but always as a minor enzymic activity, and therefore is considered to be a contamination.

G200 column (17,23). A high specific activity of NAG was excluded from the Sephadex bed (Fig. 6). When the gel filtration was performed in a phosphate buffer, negative stain electron microscopy demonstrated that the fractions with NAG activity contained square particles ~10.5 nm in diameter (Fig. 7) whose appearance was reminiscent of the ~7.5 nm particles seen edge on while still attached to the membrane surface (4) (Fig. 7, see insert). The NAG fraction from the Sephadex G200 column consistently migrated

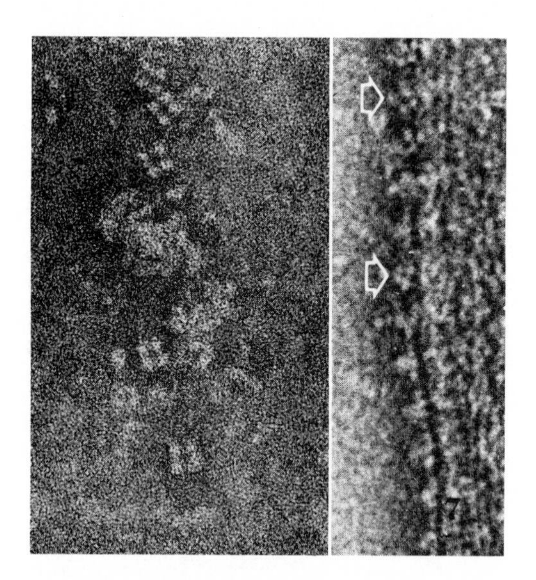

FIG. 7. Negative stain preparation of pooled NAG fractions voided from the Sephadex G200 column (see Fig. 6). The squares are 10.5 nm in diameter. The particles appear to consist of four globular subunits which make up the four corners of the square, but this appearance may result from a pool of stain located in the center of the particle. *Inset,* An edge-on view of the particles still attached to the membrane surface in a crude membrane fraction. In most instances these particles (~7.5 nm in diameter) appear to be triangular; however, occasionally a diamond shape is seen (*arrows*). ×362,000. Insert: ×484,000. [Inset from Robertson et al. (4).]

as a doublet with apparent molecular weights of 110–115,000 and 100–105,000 on SDS polyacrylamide gels (17,23) (Fig. 8). Both bands were PAS-positive. Also present in the NAG fraction were two minor proteins with apparent molecular weights of 80,000 and 70,000 which were not PAS-positive. The doublet of 100,000 and 110,000 daltons for NAG from these membranes was highly reproducible with a constant ratio of 1:1. This suggests that two NAGs are present with one slightly modified either in charge or in carbohydrate content. Usually, NAG is a soluble enzyme sequestered in the lysosomes of the cell (24). However, in the human spleen, two forms of NAG occur: only form A is PAS-positive and this enzyme is thought to be membrane-bound (24).

The second protein resolved by gel filtration was retarded by the Sephadex bed. Negative stain preparations of this fraction did not show any recognizable structure. However, after 10 mM $CaCl_2$ was added to this fraction, large numbers of single and tangled bundles of thin (~ 3 nm) filaments of indefinite length were easily found by negative stain electron microscopy (Fig. 9) (17,23). It is thought that this filamentous protein interacts directly with the lipids of these membranes because the filament protein clearly

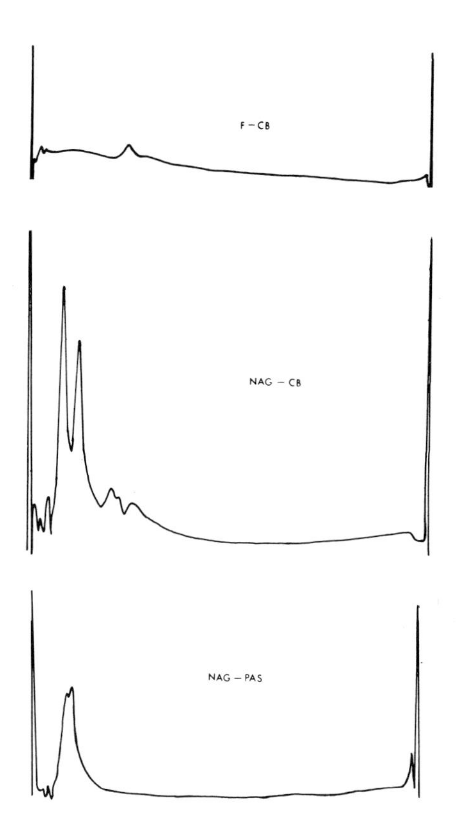

FIG. 8. Densitometer traces of SDS polyacrylamide gels of the filamentous protein and NAG. The uppermost profile shows the A_{550} scan of the filament protein (60 μg) stained with Coomassie blue (F-CB). The apparent molecular weight of this protein is 80,000. The middle and lower densitometer traces show the migration of NAG (30 μg) in SDS polyacrylamide gels stained either with Coomassie blue (A_{550} NAG-CB) or Schiff's reagent (A_{560} NAG-PAS). NAG migrates as a doublet with apparent molecular weights of 110–115,000 and 100–105,000. Both NAG bands are PAS-positive. Two minor polypeptide chains are also seen in the NAG gels stained with Coomassie blue. Of these two bands with molecular weights of 80,000 and 70,000, only the 80,000 molecular weight polypeptide is consistently present.

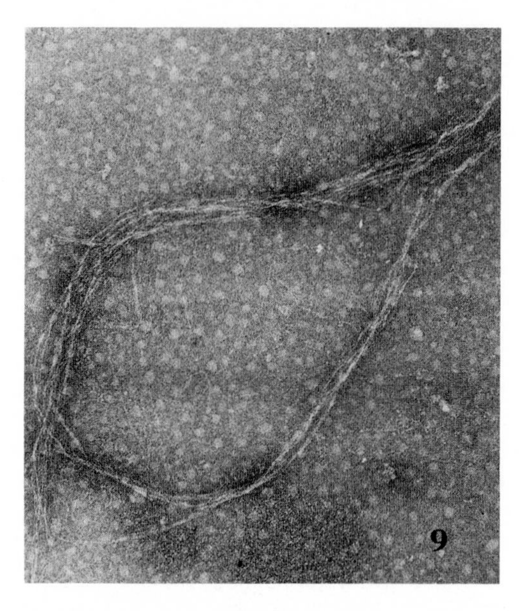

FIG. 9. Negative stain preparation (1% PTA with bacitracin) of the filament fraction after chromatography through a Sephadex G200 column and the readdition of $CaCl_2$. The filaments are 3 nm in diameter and of varying lengths. They are seen in tangled bundles as well as individual filaments. ×114,750.

must have been released from the membrane by $CaCl_2$ as parts of the decorated strips. The flotation density of the decorated strips, being the same as that of the denuded membranes, shows that they contain lipids and these have been identified as phospholipids and cholesterol. The attachment of NAG to the filament is by an ionic interaction because the particles separate from the decorated strips at the isoelectric point of NAG (17). Molecular weight estimates for the subunit of this polymer by SDS polyacrylamide gel electrophoresis indicates an apparent molecular weight of 80,000 (Fig. 8). However, gel filtration of this protein through a 6 M guanidine-HCl column suggests that its electrophoretic migration in SDS is aberrant. The filamentous protein does not stain well with Coomassie blue and is not PAS-positive. However, the protein is a glycoprotein having mannosamine and galactosamine residues which may explain its aberrant migration in SDS. On the basis of its polymerization properties, apparent molecular weight and inability to decorate with heavy meromyosin, the filament is not actin (25,26), spectrin (27,28), or nectin (29,30). Functionally, it is similar to nectin, a protein isolated from *Streptococcus faecalis* which binds ATPase to the plasma membrane (29,30), and we believe that its function is to hold the particles in the decorated strips together and mediate their binding to the membranes. We call this filamentous protein ligatin.

SUMMARY

The particulate array covering the invaginated plasma membranes of the ileal epithelial cells in the suckling rat has been found to consist of two proteins: ligatin, a filamentous protein, and NAG. Ligatin, which polymerizes in the presence of $CaCl_2$, binds NAG to the membrane surface and possibly other enzymes not usually bound to membranes. The attachment of the decorated strips to the membrane surface is disrupted by $CaCl_2$, but this is not a competitive interaction of the cation because concentrations of $CaCl_2$ greater than 10 mM do not cause a further (>40%) release of the particles from the membranes (17,23). NAG, which presumably is the 7.5 nm particles, cleaves the β-linked *n*-acetylglucosamine groups from glycoproteins, glycolipids, and mucopolysaccharides (24). The membranes of this invaginated complex therefore function as sites for the extracellular digestion of carbohydrate in the maternal milk. The intimate attachment of the enzyme to the external surface of the membrane constitutes an ideal arrangement to facilitate immediate absorption of the reaction products through the membrane into the cytoplasm of the cell.

REFERENCES

1. Wissig, S. L., and Graney, D. O. (1968): *J. Cell Biol.,* 39:564–579.
2. Knutton, S., Limrick, A. R., and Robertson, J. D. (1974): *J. Cell Biol.,* 62:679–694.
3. Clark, S. L. (1959): *J. Biophys. Biochem. Cytol.,* 5:41–49.
4. Robertson, J. D., Knutton, S., and Limrick, A. R. (1975): *J. Cell Biol.* (*in press*).
5. Porter, K. R., Kenyon, K., Badenhauser, S. (1967): *Protoplasm,* 63:262–274.
6. Cornell, R., and Padykula, H. A. (1969): *Am. J. Anat.,* 125:291–316.
7. Henning, S. J., and Kretchner, N. (1973): *Enzyme,* 15:3–23.
8. Koldovsky, O. (1969): *Development of the Functions of the Small Intestine in Mammals and Man.* Karger, Basel.
9. Koldovsky, O., and Herbst, J. (1971): *Biol. Neonate,* 17:1–9.
10. Koldovsky, O., Heringova, A., Hoskova, J., Jirsova, V., Noack, R., Friedrich, M., Schenck, G. (1965): *Biol. Neonate,* 9:33–43.
11. Lebenthal, E., Sunshine, P., and Kretchner, N. (1972): *J. Clin. Invest.,* 51:1244–1250.
12. Koldovsky, O., and Palmieri, M. (1971): *Biochem. J.,* 125:697–701.
13. Koldovsky, O., and Sunshine, P. (1970): *Biochem. J.,* 117:467–471.
14. Koldovsky, O., and Herbst, J. J. (1973): *Gastroenterology,* 64:1142–1149.
15. Boass, A., and Wilson, T. H. (1963): *Am. J. Physiol.,* 204:101–104.
16. Robberecht, P., Deschodt-Lanckman, M., Camus, J., Bruylands, J., and Christophe, J. (1971): *Am. J. Physiol.,* 221:376–381.
17. Jakoi, E. R., Zampighi, G., and Robertson, J. D. *J. Cell Biol.* (*in press*).
18. Forstner, G. G., Sabesin, S. M., and Isselbacher, K. J. (1968): *Biochem. J.,* 106:381–390.
19. Eichholz, A. (1967): *Biochim. Biophys. Acta,* 135:475–482.
20. Nordstrom, C., and Dahlqvist, A. (1973): *Scand. J. Gastroenterol.,* 8:407–416.
21. Gianetto, R., and deDuve, C. (1955): *Biochem. J.,* 59:433–438.
22. deDuve, C., and Wattiaux, R. (1966): *Annu. Rev. Physiol.,* 28:435–492.
23. Jakoi, E. R., Zampighi, G., and Robertson, J. D. (1975): *Biophys. J.,* 15:218a.
24. Robinson, D., and Stirling, J. L. (1968): *Biochem. J.,* 107:321–327.
25. Moore, P. B., Huxley, H. E., deRosier, D. J. (1970): *J. Mol. Biol.,* 50:279–295.

26. Spudich, J. A., and Watt, S. (1971): *J. Biol. Chem.,* 246:4866–4871.
27. Clark, M. (1971): *BBRC 45,* 1063–1070.
28. Fuller, G. M., Boughter, J. M., and Morazzani, M. (1974): *Biochemistry,* 13:3036–3041.
29. Baron, C., and Abrams, A. (1971): *J. Biol. Chem.,* 246:1542–1544.
30. Abrams, A., Baron, C., and Schebli, H. P. (1974): *Methods Enzymol.,* 32:428–439.

Membranes and Disease, edited by L. Bolis, J. F. Hoffman, and A. Leaf. Raven Press, New York, © 1976.

Studies on the Mechanism of Action of Cholera Toxin on Adenylate Cyclase

Jorge Flores and Geoffrey W. G. Sharp

Biochemical Pharmacology Unit, Medical Services, Massachusetts General Hospital, Boston, Massachusetts 02114

It is now well known that the diarrhea of cholera is mediated by stimulation of adenylate cyclase by a toxin produced by *Vibrio cholerae* (1). The increased secretion of an isotonic fluid throughout the length of the small intestine, in excess of the reabsorptive capacity, appears to be a function of increased intracellular concentrations of cyclic AMP in the mucosal epithelial cells. The fecal fluid has raised levels of bicarbonate and potassium relative to plasma, contains little protein, and remains essentially isotonic. In the duodenal fluid the bicarbonate concentration is lower than the plasma, whereas in the ileal fluid the bicarbonate concentration is higher (2–4). No manifest damage occurs to the mucosa which can be judged intact by both light- and electron-microscopic studies and by studies on the permeability of the mucosa to large marker molecules (5–7). The most significant change which can be detected by microscopic examination is closure of the intercellular spaces, a change which would be in agreement with a dominant blood to intestinal lumen active ion transport process (8). It appears now that in cholera, the secretory transport processes are stimulated to such an extent that they overwhelm the intestinal reabsorptive mechanisms and diarrhea results.

Purification and crystallization of cholera toxin has been achieved (9–13), and considerable information is available on the mechanism of action of the toxin. The toxin is a protein with a molecular weight estimated at 84,000. It is heat- and acid-labile and consists of two different species of subunits. Estimates of the subunit molecular weights vary, but the toxin may consist of four subunits of approximately 15,000 and one subunit of 25,000 (14). It seems clear that the four subunits of 15,000 MW are responsible for the binding of the toxin to cell membranes and subsequent uptake into the cells, and that the activity of cholera toxin to cause diarrhea lies in the large (25,000 MW) subunit (15). The small subunits interact with ganglioside GM_1 in a highly specific manner to suggest that the membrane ganglioside GM_1 may be the natural receptor utilized by cholera toxin (14,16–18). Binding occurs rapidly, and only brief exposure of the intestinal cells to cholera toxin is sufficient to cause fluid secretion (19).

An important observation was that cyclic AMP stimulates active secretion of chloride and possibly bicarbonate in rabbit ileum (20). In this study, segments of distal ileum were stripped of the outer mesothelium and the two principal muscle layers so that the remaining mucosal epithelial layer could be mounted as a membrane between the two halves of a lucite chamber. Both surfaces were bathed in oxygenated Ringer's solution and short-circuit current was measured by the technique of Ussing and Zerahn (21) as an indicator of net ion flux. Addition of cyclic AMP to the serosal bathing medium increased the short-circuit current. Thus, either cations were stimulated to move from the mucosal medium to the serosal medium or anions were stimulated to move from serosa to mucosa. By the use of sodium-free, chloride-free, or bicarbonate-free Ringer's solutions it was possible to show that increases in current, although smaller than previously, occurred in the absence of sodium and in the absence of chloride, but not in the absence of both chloride and bicarbonate. This study indicated a possible action of cholera toxin to stimulate anion secretion into the lumen of the intestine in excess of reabsorptive capacity. When a crude preparation of cholera toxin was added to the luminal surface of the stripped ileum, the short-circuit current started to increase after 30–45 min and reached a peak plateau 90 min later (22). After elevation by cholera toxin, the short-circuit current was only slightly affected by the further addition of cyclic AMP. Other studies also showed an association of the action of cholera toxin with the action of cyclic AMP. Pierce et al. (23) demonstrated that prostaglandins and theophylline caused fluid production in the small intestine of dogs when infused into the mesenteric artery. Similarly, lipolysis in fat cells (24,25) and glycogenolysis in platelets (26), both of which are stimulated by cyclic AMP, were both stimulated by treatment with cholera toxin. Direct evidence of a relationship between intracellular cyclic AMP and cholera toxin in the intestine came from the work of Schafer, Lust, Sircar, and Goldberg (27). Working on dog intestine, they found little effect of the cholera toxin during the first hour after administration. In the second hour, however, cyclic AMP concentrations in the mucosal epithelial cells began to rise and reached levels 3–4 times higher than control values at 4–5 hr. Reports that adenylate cyclase activity was elevated by cholera toxin in rabbits, guinea pigs, and man, confirmed the hypothesis and identified the key enzyme involved (28–31). Cholera toxin increased adenylate cyclase activity in the lower ileum, the jejunum, and the duodenum, a finding in agreement with an effect of the toxin on the whole of the small intestine. Increased enzyme activity was detected only 15 min after the application of toxin, and the activity then increased progressively, a sequence of events in general accord with the time course of effect of cholera toxin on electrolyte flux changes (28). In a more precise study on the temporal relationship of adenylate cyclase activity and flux changes, dogs given a 10-min pulsed application of cholera toxin showed a close relationship between sodium and water movement across the intestine and the stimulation of adenylate cyclase by the toxin (19). It was found that 90 min after application of cholera toxin the normal reabsorptive activity of the intestine had changed to a secretory function, which corresponded with an increase in adenylate cyclase activity in the mucosal epithelial cells. Adenylate cyclase activity and the secretion of fluid increased to maximum values at about 3 hr. Cyclase activity and fluid production were still elevated at 24 hr, and had returned to normal by 48 hr. Thus, a close correlation between adenylate cyclase activity and fluid production was found.

The main feature of the effect of cholera toxin is a large increase of basal enzyme activity without a proportional increase in the level of fluoride activation. Fluoride universally stimulates adenylate cyclase from mammalian tissues to near maximal levels. The lack of a proportional increase with fluoride strongly suggests that no synthesis of adenylate cyclase occurs in toxin-treated tissue because fluoride activation should increase in direct proportion to the increase in amount of enzyme. The possibility that an activating factor is released into the cytoplasm, or the idea that the cholera toxin is itself a soluble adenylate cyclase, are both unlikely because the increased enzyme activity after exposure to cholera toxin can be observed in well-washed membrane preparations. In fact, the latter suggests that a relatively permanent change in the enzyme has occurred. This may be due to the binding of toxin or a fragment of toxin by the enzyme. It is also possible that the toxin is itself an enzyme which brings about a permanent change in membrane structure, causing increased activity of adenylate cyclase.

Conditions for the assay of adenylate cyclase activity in mucosal epithelial cells of rabbit small intestine have been defined in the basal state and after the enzyme has been activated by prior exposure of the tissue to cholera toxin (32). Optimal conditions for activity were the same with regard to magnesium concentration, pH, and ATP concentration. Examination of the activity of adenylate cyclase at different ATP concentrations revealed that cholera toxin stimulated the activity by an increase in the V_{max} without any change in K_m. No evidence of an allosteric confirmational change was detected. Calcium was inhibitory; manganese could replace magnesium at low concentration but was inhibitory at high concentration. Again, no difference was observed between the effects of these agents on the basal enzyme activity and on the enzyme activated by cholera toxin. Differences noted after activation of adenylate cyclase by cholera toxin were the increased V_{max} (32) and increased responsiveness to prostaglandin E_1 (28) and beta-adrenergic stimulation (33).

Studies on adenylate cyclase activity of mucosal epithelial cells of human jejunum, taken by biopsy during acute choleraic diarrhea, have confirmed many of the observations made on the enzyme from animal models. In these patient studies, biopsy material was taken within 24 hr of the onset of cholera when the average water loss was 11 liters per day, and again after cessation of diarrhea during convalescence. During the diarrheal stage, adenylate cyclase activity was increased to more than double that during the convalescent stage (29). The activated human jejunal enzyme had similar characteristics to those described thus far for rabbit ileum (30).

A complicating factor in the sequence of events between uptake of cholera toxin by the brush-border membrane of the intestinal epithelial cells is that adenylate cyclase does not appear to be present in the brush-border membranes. In experiments on the localization of the enzyme, membrane fractions were separated and purified by techniques of differential homogenization and centrifugation (34). The brush borders are characterized by high alkaline phosphatase activity, low Na, K-ATPase and by electron microscopic appearance. Basal and lateral plasma membranes were characterized by low alkaline phosphatase activity, high Na, K-ATPase activity and also by electron-microscopic appearance. Adenylate cyclase activity is very low in the brush border, so low that the activity could be accounted for by contamination from other membrane fractions, e.g., the lateral tags which can be seen adhering to the brush-border edges. Thus, it appears that adenylate cyclase, like Na, K-ATPase may be absent from the brush border membranes. If this is the case, then cholera toxin, which is present initially at the luminal brush border membrane, cannot interact with adenylate cyclase in that membrane. Rather it affects the enzyme in the lateral and basal membranes. One reservation to this conclusion is that the cells in the crypts do not have brush borders. Therefore, luminal membranes from the crypt cells can be expected to be present in the fraction labeled lateral and basal membranes, and could possibly contain adenylate cyclase.

In trying to explain the mechanism of action of cholera toxin, several factors need to be fitted into the schema. Only brief exposure of the intestinal mucosa to toxin is required for a large diarrheal response; a latent period of variable length occurs before the secretory response begins; the effect of cholera toxin is essentially irreversible and cannot be removed by washing (either the intestine or the membranes); the effect is not direct, because, while the toxin has its initial interaction with brush borders, adenylate cyclase, the target enzyme, is present in the basal and lateral membranes.

Several possibilities have been suggested to explain the toxin action. For example, the toxin could: (i) stimulate adenylate cyclase directly, much in the manner of a hormone acting through its receptors, except that the toxin effect would have to be irreversible; (ii) cause the synthesis or release of an activator; and (iii) remove a normal physiological inhibitor of the enzyme.

In a recent study we examined the action of cholera toxin on adenylate cyclase in liver cell membrane, and solubilized the enzyme by treatment with the nonionic detergent, lubrol (33). After such treatment the enzyme is "freed" from the membranes and is solubilized in the operational sense that it does not sediment when centrifuged at 200,000 $\times$ g and can be passed through 0.22 μm Millipore filters without loss of activity. The solubilized enzyme was still activated by prior treatment with cholera toxin, suggesting that the toxin directly affects the adenylate cyclase complex. That is, it so changes the enzyme by stimulation or removal of an inhibitor that permanent stimulation of the enzyme results. Cholera toxin could do this either by binding, perhaps just the 25,000 subunit, which has alone been shown to stimulate adenylate cyclase (35), or by enzymatically altering the adenylate cyclase—that is to say, that cholera toxin is also an enzyme.

It has been suggested recently that cholera toxin is an enzyme and that nicotinamide adenine dinucleotide (NAD) is an essential cofactor. For example, it has been demonstrated that cholera toxin can stimulate adenylate cyclase in tissue homogenates provided that adequate NAD is present in the homogenate (36–38). This stimulatory effect can be obtained with concentrations of the toxin as low as 10^{-10} M in the presence of 0.3–1.0 mM NAD (38). The characteristics of the adenylate cyclase stimulated under these conditions were similar to those of the enzyme stimulated *in vivo,* i.e., the stimulation was persistent, the action of hormones was enhanced, and the enzyme was solubilized in the activated state. These findings are important because the ability to use an *in vitro* model for the study of cholera toxin should facilitate efforts to understand the detailed mechanism of its action. In this respect, there are interesting parallels between the action of cholera toxin and the action of the analogs of guanosine triphosphate such as 5′-guanylylimidodiphosphate (Gpp(NH)p) to stimulate adenylate cyclase. For instance, Gpp(NH)p, like the toxin, stimulates adenylate cyclase irreversibly, enhances the action of hormones, and the enzyme after treatment with either agent can be solubilized in the activated state (39). These similarities between the action of cholera toxin and guanylylimidodiphosphate prompted an examination of the interrelationship of the two agents and their effects (40). Using the recently developed procedure to stimulate adenylate cyclase in broken cell preparations, it was possible to perform sequential incubations with the two agents in the study of the interaction. It was found that, after stimulation of the enzyme by cholera toxin administered either *in vivo* or *in vitro* to a broken cell preparation in the presence of NAD, the addition of Gpp(NH)p failed to cause stimulation. In fact, Gpp(NH)p, which alone stimulates adenylate cyclase, partially inhibited the toxin-stimulated enzyme. Thus, interference by Gpp(NH)p in the stimulatory effect of cholera toxin was observed. In a further series of experiments, the order of treatment by these two agents was reversed so that the tissues were first treated with Gpp(NH)p and subsequently with cholera toxin and NAD, both incubations being carried out at room temperature. Dose–response studies had shown that 10^{-3} M Gpp(NH)p was a maximally stimulating concentration under these conditions. It was found that 10^{-3} M Gpp(NH)p stimulated adenylate cyclase and completely blocked the effect of cholera toxin and NAD. Thus, the results showed blockade of the stimulatory effect of cholera toxin by prior treatment with Gpp(NH)p and interference with the stimulation of cholera toxin when added subsequently (40).

The mutual interference of the action of these two agents becomes of considerable interest when taken in conjunction with their similarity of effects on adenylate cyclase. As noted earlier, both cholera toxin and Gpp(NH)p produce marked stimulation of adenylate cyclase, the stimulation is essentially irreversible, hormone action is enhanced and the enzyme can be solubilized still in the activated state. Thus, the possibility exists that cholera toxin and Gpp(NH)p share a common target in the adenylate cyclase complex—the so-called guanyl nucleotide binding protein. Thus, one can postulate that cholera toxin or its active subunit acts through a reaction involving NAD on the site on the adenylate cyclase complex, which is sensitive to the action of guanyl nucleotides, to produce an irreversible stimulation of the enzyme. Such a reaction between cholera toxin and NAD may consist of the incorporation of a NAD metabolite onto the guanyl nucleotide binding site. This type of reaction takes place in diphtheria when the diphtheria toxin acts by incorporating the adenosine diphosphoribose moiety from NAD onto elongation factor 2, a ribosomal protein (41).

Several recent studies have focused on the effects of guanyl nucleotides on adenylate cyclase (39,42–44). Pfeuffer and Helmreich (39) have detected a guanyl nucleotide binding protein which they suggested might regulate the activity of adenylate cyclase. Thus interaction of guanyl nucleotides with this membrane-bound component of the adenylate cyclase complex would essentially remove an inhibitory restraint on the enzyme activity. Similarly, a model has been developed by Schramm and Rodbell (42) for the effect of Gpp(NH)p on adenylate cyclase from frog erythrocyte membranes. In this model the guanyl nucleotides produce different activation states of the enzyme in the presence and absence of isoproterenol. They suggest that the hormone sensitizes the enzyme to the stimulatory effects of the nucleotide to produce a highly activated state from which the hormone dissociates. Readdition of high concentrations of isoproterenol can partially inhibit this activity by setting up a new equilibrium position for the various activated states of the enzyme. A similar effect of isoproterenol has been observed after stimulation by cholera toxin, pointing to yet another similarity between the effects of Gpp(NH)p and cholera toxin.

In conclusion, it is suggested that cholera toxin and NAD act to stimulate adenylate cyclase by interaction with the same regulatory protein on which guanyl nucleotides exert their effects.

REFERENCES

1. De, S. N. (1959): *Nature,* 183:1533.
2. Banwell, J. G., Pierce, N. F., Mitra, R. C., et al. (1970): *J. Clin. Invest.,* 49:183.
3. Carpenter, C. C. J., Sack, R. B., Feeley, J. C., and Steenberg, R. W. (1968): *J. Clin. Invest.,* 47:1210.

4. Sack, R. B., Carpenter, C. C. J., Steenberg, R. W., and Pierce, N. F. (1966): a canine model. *Lancet,* 2:206.
5. Gangerosa, E. J., Beisel, W. R., Benyajati, C., Sprinz, H., and Piyaratn, P. (1960): *Am. J. Trop. Med. Hyg.,* 9:125.
6. Elliott, H. L., Carpenter, C. C. J., Sack, R. B., and Yardley, J. H. (1970): *Lab. Invest.* 22:112.
7. Norris, H. T., and Jajho, G. (1968): *Am. J. Pathol.,* 53:263.
8. DiBona, D. R., Chen, L. C., and Sharp, G. W. G. (1970): *J. Clin. Invest.,* 53:1300.
9. Finkelstein, R. A., and LoSpalluto, J. J. (1969): *J. Exp. Med.,* 130:185.
10. Finkelstein, R. A., and LoSpalluto, J. J. (1970): *J. Infect. Dis.,* 121:563.
11. Richardson, S. H., and Noftle, K. A. (1970): *J. Infect. Dis.,* 121:573.
12. Richardson, S. H., Evans, D. G., and Feeley, J. C. (1970): *Infec. Immunol.,* 1:546.
13. Finkelstein, R. A., and LoSpalluto, J. J. (1972): *Science,* 175:530.
14. Van Heyningen, S. (1974): *Science,* 183:656.
15. Lonnroth, I., and Holmgren, J. (1973): *J. Gen. Microbiol.,* 76:417.
16. King, C. A., and van Heyningen, W. E. (1973): *J. Infect. Dis.,* 127:639.
17. Cuatrecasas, P. (1973): *Biochemistry,* 12:3558.
18. Holmgren, J., Lonnroth, I., and Svennerholm, L. (1973): *Infect. Immun.,* 8:208.
19. Guerrant, R. L., Chen, L. C., and Sharp, G. W. G. (1972): *J. Infect. Dis.,* 25:377.
20. Field, M., Plotkin, G. R., and Silen, W. (1968): *Nature,* 217:469.
21. Ussing, U. H., and Zerahn, K. (1951): *Acta Physiol. Scand.,* 23:110.
22. Field, M., Fromm, D., Al-Awqati, Q., and Greenough, W. B. (1972): *J. Clin. Invest.,* 51:796.
23. Pierce, N. F., Carpenter, C. C. J., Elliott, H. L., and Greenough, W. B. (1971): *Gastroenterology,* 60:22.
24. Greenough, W. B., Pierce, N. F., and Vaughan, M. (1970): *J. Infect. Dis.,* 121:5111.
25. Vaughan, M., Pierce, N. F., and Greenough, W. B. (1970): *Nature,* 225:568.
26. Zieve, P. D., Pierce, N. F., and Greenough, W. B. (1970): *Clin. Res.,* 18:890.
27. Schafer, D. E., Lust, W. D., Sircar, B., and Goldberg, N. D. (1970): *Proc. Natl. Acad. Sci. USA,* 67:851.
28. Sharp, G. W. G., and Hynie, S. (1971): *Nature,* 229:266.
29. Chen, L. C., Rohde, J. E., and Sharp, G. W. G. (1971): *Lancet,* 1:939.
30. Chen, L. C., Rohde, J. E., and Sharp, G. W. G. (1972): *J. Clin. Invest.,* 51:731.
31. Kimberg, D. V., Field, M., Johnson, J., Henderson, A., and Gershon, E. (1971): *J. Clin. Invest.,* 50:1218.
32. Sharp, G. W. G., Hynie, S., Ebel, H., Parkinson, D. K., and Witkum, P. A. (1973): *Biochim. Biophys. Acta,* 309:339.
33. Beckman, B., Flores, J., Witkum, P. A., and Sharp, G. W. G. (1974): *J. Clin. Invest.,* 53:1202.
34. Parkinson, D. K., Ebel, H., DiBona, D. R., and Sharp, G. W. G. (1972): *J. Clin. Invest.,* 51:2292.
35. Van Heyningen, S., and King, C. A. (1975): *Biochem. J.,* 146:269.
36. Gill, D. M., and King, C. A. (1975): *J. Biol. Chem.* 250:6424.
37. Gill, D. M. (1975): *Proc. Natl. Acad. Sci. USA,* 72:2064.
38. Flores, J., Witkum, P. A., and Sharp, G. W. G. (1975): *J. Clin. Invest.,* 57:450.
39. Pfeiffer, T., and Helmreich, E. J. M. (1975): *J. Biol. Chem.,* 250:867.
40. Flores, J., and Sharp, G. W. G. (1975): *J. Clin. Invest.,* 56:1345.
41. Gill, D. M., Pappenheimer, A. M., and Unhida, T. (1973): *Fed. Proc.,* 32:1508.
42. Schramm, M., and Rodbell, M. (1975): *J. Biol. Chem.,* 250:2232.
43. Spiegel, A. M., and Aurbach, G. D. (1974): *J. Biol. Chem.,* 249:7630.
44. Lefkowitz, R. J. (1975): *J. Biol. Chem.,* 250:1006.

Membranes and Disease, edited by L. Bolis, J. F. Hoffman, and A. Leaf. Raven Press, New York, © 1976.

Alterations in the Cell-Surface Membrane of the Intestinal Epithelial Cell During Mitosis and Differentiation, and After Neoplastic Transformation

Milton M. Weiser

Department of Medicine, Harvard Medical School and the Gastrointestinal Unit, Massachusetts General Hospital, Boston, Massachusetts 02114

Recent studies, primarily with cells in tissue culture, have demonstrated changes in the cell surface that occur during mitosis, in early growth phase, and after viral transformation (1–3). Many of these changes appear to be similar, and the implication has been that these cell surface alterations are necessary and serve to control and to determine cell growth and behavior (4). Some of these apparent membrane-dependent changes which occur in association with malignancy and viral transformation of cells (as well as during normal mitosis) are loss of cell adhesion, contact inhibition of movement, density-dependent inhibition of growth, changes in cell-surface charge density, and the appearance of variably specific surface antigens. The alterations in the social behavior of the cell are accompanied by changes in cell-surface glycolipids and glycoproteins, but may also involve alterations of the membrane-associated enzymes active in the biosynthesis of these membrane constituents, the glycosyltransferases (5).

Differences in surface glycoproteins have been described for chick embryo and 3T3 mouse embryo cells in tissue culture that are related to changes in cell growth (6,7). Buck et al. have described a fucose-containing glycoprotein which is increased in virally transformed cells (7) and Warren, Critchley, and MacPherson (8) have described an increase in fucose-containing glycoproteins on the surface membranes of chick embryo fibroblasts transformed with a temperature-sensitive mutant of Rous sarcoma virus. Mallucci, Poste, and Wells (9) presented data showing that transformed cells have an increase in the rate of cell-coat synthesis. The concept that surface-membrane glycoproteins and glycolipids may serve as general control factors in cell growth patterns and cell turnover has been strengthened by the work of Oppenheimer et al. (10), who have shown that the conversion of nonadhesive cells to adhesive cells requires the biosynthesis of membrane glycoproteins and glycolipids.

Roth, McGuire, and Roseman (11) have demonstrated cell-surface glycosyltransferases on embryonic chicken neural retinal cells in tissue culture. Autoradiographic data have confirmed the surface membrane location of galactosyltransferase on established tissue culture lines (mouse 3T3 and 3T12) (5) and Bosmann (12) confirmed the presence of these enzymes on 3T3 cell lines and also showed that this activity increased after virus transformation of these cells.

Recent evidence has suggested that some of the physical, biochemical, and antigenic changes found in malignant cells may be similar to those also seen in the fetal state. In the gastrointestinal tract, certain surface-membrane antigens of colon and liver cells have been shown to disappear during the process of malignant transformation (13,14). In the colon this has been associated with the appearance of a tumor-specific surface antigen, which seems to be identical to an embryonic colon membrane antigen, CEA (15). Many of the reported enzyme (and isoenzyme) changes and the increased glycolysis that have been considered to be tumor-specific have also been shown to be characteristic of fetal cells.

One of the most mitotically active cell populations in animals is the small intestinal epithelium. With this tissue we have been able to obtain data that suggest that the cell surface changes previously noted in tissue culture systems also occur in the intestine and that these changes appear to be correlated with cellular differentiation, mitosis, and neoplastic transformation. Specifically, we have been able to show that the fetal intestinal epithelial cell, the mitotically undifferentiated crypt cell and the intestinal tumor cell appear to have on their cell surface glycosyltransferases as well as incomplete glycoproteins which can serve as endogenous acceptors for these surface enzymes (16,17). These factors are largely absent in the mature villus epithelial cell-surface membrane. These "immature" crypt cells also showed increased susceptibility to agglutination by the plant lectin concanavalin A (18).

SOME GENERAL FEATURES OF THE SMALL INTESTINE

Intestinal cells consist of two major types: (i) *crypt cells,* which are undifferentiated and the site of active mitosis; and (ii) the more distal *villus cells,* which are differentiated, mature, and actively engaged in nutrient uptake and transport (Fig. 1). The mature villus cell also demonstrates a functional polarity which is reflected in the properties of the two major areas of its surface membrane. The luminal end forms extensive microvilli on which carbohydrate-rich material has been shown to be an integral part of its structure. This outermost component has been termed the glycocalyx and appears to be the first barrier to absorption and the site of the hydrolytic enzymes of the microvillus membrane. This contrasts with the lateral-basal membrane part of the epithelial cell, which is rich in Na^+, K^+-ATPase and which appears to have a totally different protein composition (19).

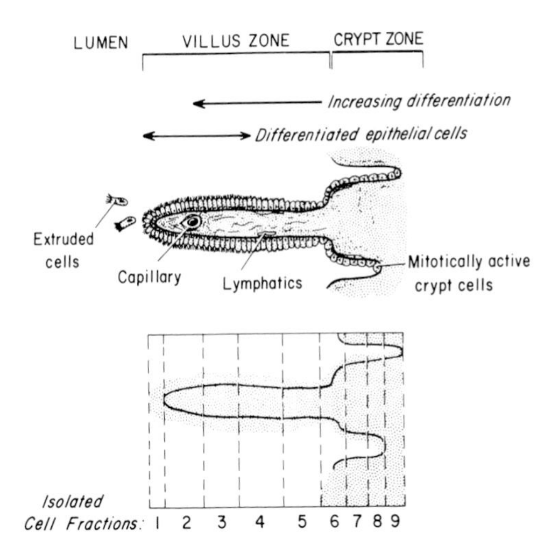

FIG. 1. Schematic representation of an intestinal villus demonstrating the polarity of cell differentiation and the source of isolated intestinal cell fractions.

VILLUS–CRYPT GRADIENTS OF ISOLATED INTESTINAL EPITHELIAL CELLS

Recent advances in methodology have made possible the isolation and identification of the different glycoproteins and glycolipids present in the plasma membrane of the intestinal epithelial cell. In addition to a method for isolating the plasma membrane of the microvilli (20), a technique was recently developed for the isolation of the plasma membrane of the basal and lateral portion of the epithelial cell (19). We have also developed a method for making isolated rat intestinal epithelial cell preparations that segregate crypt cells from villus cells (21). This has permitted us to study factors influencing cell differentiation along the villus. The isolated cell preparation has also permitted us to establish techniques for the interaction of these isolated epithelial cell preparations with phytoagglutinins.

In Fig. 1 the vertical lines represent an approximation of the area of the different isolated cell fractions that were obtained by the technique we have developed. It should be emphasized that this results in an isolated epithelial cell preparation. At the end of the procedure, it was observed that the villus substructure remained intact with vascular elements still present behind the collagenous basement membrane on which the cells were originally attached.

INTESTINAL GLYCOPROTEIN SYNTHESIS

The development of the above method provided us with an opportunity to investigate whether glycoprotein synthesis might parallel the growth of

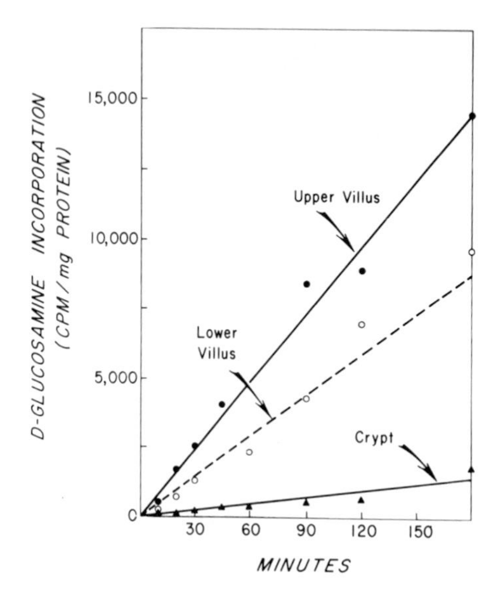

FIG. 2. Villus–crypt gradients of glycoprotein synthesis (*in vitro*).

the glycocalyx-rich microvilli as the cell moves from crypt to villus regions. Data by Forstner (22), which we confirmed, indicated that the highest rate of D-glucosamine and L-fucose incorporation occurred in the microsomal pellet at 1 hr and in microvillus preparations at 3 hr (21). To further evaluate surface-membrane glycoprotein synthesis, various precursors were injected intraperitoneally into rats and their rate of incorporation measured 3 hr later. There were sharp gradients of incorporation for D-glucosamine, L-fucose, and D-galactosamine in a pattern that suggested that the upper villus cells had the highest rate of incorporation.

It could also be demonstrated that this gradient of glycoprotein synthesis was maintained when isolated cells were incubated with D-glucosamine *in vitro* (Fig. 2). This suggested that the high rate of glycoprotein synthesis by villus cells and the low rate exhibited by crypt cells were intrinsic properties of these cells. This experiment also represented an important control for the data described below, which show that the crypt cell surface is characterized by incomplete oligosaccharides on the glycoproteins as well as the presence of glycosyltransferases.

GLYCOSYLTRANSFERASES AND INCOMPLETE GLYCOPROTEINS OF THE CRYPT CELL SURFACE MEMBRANE

The incompleteness of surface membrane glycoproteins in relationship with membrane-associated glycosyltransferases has been invoked by Roseman (23) as a mechanism for intercellular adhesion, cell agglutination and contact inhibition. In addition, Roth et al. (5) demonstrated galactosyltransferase activity on cell surface membranes of chick neural retina cells in tissue culture and they have again suggested that these enzymes may determine cell behavior. Webb and Roth (24) recently demonstrated that 3T3 cells in mitosis have cell surface galactosyltransferase activity characteristics similar to transformed 3T12 cells with high endogenous acceptor activity. We have evaluated isolated intestinal villus and crypt cells, fetal intestinal cells, and intestinal tumor cells for the presence of glycosyltransferase activities and corresponding glycoprotein acceptors on their cell surface.

As shown in Fig. 3, incubation of isolated epithelial cells with the nucleotide sugar UDP-*N*-acetylglucosamine-^{14}C resulted in a tenfold greater incorporation of label onto crypt cells as compared to villus cells. This was also found to be true for a series of nucleotide sugars tested (UDP-glucose, GDP-fucose, and GDP-mannose). The only important exception was CMP-sialic acid. With CMP-sialic acid, sialic acid incorporation into an endogenous membrane acceptor was found to be highest in the villus rather than the crypt zone. Both sialyltransferase activity and the galactosyltransferase activity were active with exogenous acceptors as detected with *intact* cells. This indicated that the enzymes were exposed (or had been exposed) and remained active on the outer aspect of the cell surface membrane.

These data suggested not only that glycosyltransferases were characteristic of the crypt cell surface, but that there were endogenous acceptors that could be glycosylated. This appeared to be an indication of what has been termed "incomplete" glycoproteins. Further evidence that these ac-

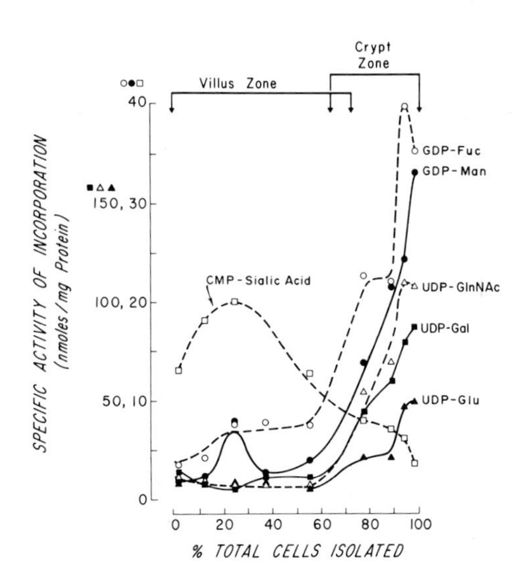

FIG. 3. Villus-crypt gradients of cell-surface glycosyltransferase : endogenous acceptor activities.

TABLE 1. *Incomplete cell surface glycoproteins: Enhancement of galactosyltransferase:endogenous acceptor activity by preincubation with UDP-N-acetylglucosamine*

Substrate	Ratio of Gal to acGln
Each nucleotide alone, UDP-(1-^{3}H)Gal or UDP-ac(1-^{14}C)Gln	0.73
UDP-(1-^{3}H)Gal first, then UDP-ac(1-^{14}C)Gln	1.00
UDP-ac(1-^{14}C)Gln first, then UDP-(1-^{3}H)Gal	2.27

ceptors did represent incomplete oligosaccharide chains of glycoproteins is shown in Table 1. When crypt cells were first incubated with UDP-(1-^{3}H)galactose and then incubated with UDP-*N*-acetyl-(1-^{14}C)glucosamine no significant increase of incorporation was observed over that seen with each nucleotide alone. However, prior incubation of cells with UDP-*N*-acetyl-(1-^{14}C)glucosamine markedly increased the subsequent incorporation of (1-^{3}H)galactose from UDP-(1-^{3}H)galactose. This suggested that more acceptor sites became available for galactose incorporation with the prior insertion onto the surface glycoproteins of *N*-acetylglucosamine. Other data indicated that the crypt cell glycosyltransferase reactions involved one crypt cell with another (16). That is, the glycosyltransferase of one cell was glycosylating an incomplete glycoprotein of another crypt cell rather than the same cell. Roth (5) has termed this transglycosylation by cell-surface glycosyltransferases.

GLYCOSYLTRANSFERASES AND INCOMPLETE GLYCOPROTEINS OF THE FETAL INTESTINAL CELL AND INTESTINAL TUMOR CELLS

We also examined the cell surface characteristics of human fetal cells and rat intestinal tumor cells. Dr. J. T. LaMont examined rat intestinal tumors which were induced in rats by 1,2-dimethylhydrazine, a proven carcinogen. After extirpation of the tumors, isolated cell preparations were made of both tumor cells and adjacent normal tissue cells. These cells were then evaluated for surface-membrane glycosyltransferase : endogenous acceptor activities and for agglutination by lectins (see below) (17). Similarly, human fetal intestinal cells were examined as an isolated cell preparation (16,18). Table 2 compares the cell-surface glycosyltransferase activities of rat mature villus cells, undifferentiated crypt cells, and "dedifferentiated" tumor cells. The data would suggest that 1,2-dimethylhydrazine-induced rat small intestinal tumors had cell-surface characteristics more like those of the undifferentiated crypt cell than of the differentiated villus cell. Human fetal cells also demonstrated high galactosyltransferase: endogenous acceptor activities (16).

TABLE 2. *Glycosyltransferase: endogenous acceptor activities of rat intestinal cell-surface membranes*

Origin of cell membranes	Glycosyltransferases (CPM/mg protein)		
	Sialyl-	*N*-acetyl-glucosaminyl-	Galactosyl-
Villus	12,420	1,200	3,600
Crypt	2,230	20,640	50,100
Tumors (dimethylhydrazine)	2,620	13,550	51,400

CONCANAVALIN A AGGLUTINATION OF CRYPT, FETAL, AND TUMOR CELLS

A major technique used to study changes in surface membrane glycoproteins and glycolipids involves certain plant derived proteins (phytoagglutinins or lectins) which react with specific sites on the cell surface. These agglutinins have been used to identify and isolate membrane-receptor sites. All evidence to date indicates that these membrane sites are specific carbohydrate portions of membrane glycolipids or glycoproteins. Increased agglutinability of cells appears to be associated with viral transformation, decreased contact inhibition, increased mitotic rate, malignancy, and fetal cell lines (3,4,18,25).

The phytoagglutinin concanavalin A (Con A) has been found to agglutinate transformed tissue culture cells, leukemic cells, and implantable tumor cells but not the derivative normal cell lines (26). Con A is a purified protein derived from the jack bean. Its agglutinating properties appear to be partly due to specific binding to α-glycosidic linkages in polysaccharides as carefully worked out by Goldstein (27).

The basis of the agglutinability of malignant cells by lectins is still unclear, but may involve (i) exposure of deeper membrane sites, (ii) a concentration of exposed binding sites as a result of a decrease in cell size, or (iii) a rearrangement and clustering of exposed sites (28). We have evidence that the surface membrane galactosyltransferases may be an important binding site for Con A agglutination of some cells (29). However, whatever its molecular mechanism, agglutination requires a distinct membrane alteration that is clearly associated with transformation and malignancy. In fact, there is evidence that agglutination by Con A of transformed cells is manifested only after virus replication has occurred (30), and that the alteration in the membrane surface is under the control of the viral genome (31).

Moscona showed that chick embryonal retinal and liver cells are preferentially agglutinated by Con A compared to their normal, adult tissue cell counterpart (3). We recently demonstrated that human fetal intestinal cells

TABLE 3. *Surface membrane properties of differentiated and undifferentiated intestinal cells*

Property	Cell type			
	Villus	Crypt	Fetal	Tumor (small intestine)
Surface membrane glycoproteins	Complete	Incomplete	Incomplete	Incomplete
Surface membrane glycosyltransferases				
Sialyltransferase	+++	0	0	0
Galactosyltransferase (and most other transferases)	+/−	+++	+++	+++
Con A agglutination	+/−	+++	+++	+++

are also agglutinated by Con A, but not human adult intestinal cells (18). This important finding raises the question of the relationship of the fetal cell surface Con A reactive sites to fetal antigens such as CEA which are found to be associated with human cancer (15). Dr. LaMont (17) has also examined rat small intestinal tumor cells and found them to be preferentially agglutinated by Con A. Table 3 summarizes these data, showing the correlations among crypt, fetal, and intestinal tumor cells.

Table 3 also demonstrates that there appears to be a correlation between incomplete glycoprotein on the cell surface, the presence of glycosyltransferase activity, and the phenomenon of Con A agglutination, and that these properties of the cell surface are more indicative of the undifferentiated mitotically active or dedifferentiated cell. Many investigators believe these changes are important in the control of cell processes directly related to the malignant state and the ability for a cell to metastasize (23,32).

Most studies in the past have been on tissue culture cells which, although advantageous in that the experimental design can be more easily manipulated, are disadvantageous in that their pertinence to intact animal systems cannot be assured. Our studies with intestinal epithelium have demonstrated some of the same cell-surface alterations noted in tissue culture studies; in a sense demonstrating again that studies in tissue culture are indeed applicable to intact organ systems.

SPECULATIONS

The cell-surface characteristics noted for the normal crypt cell as contrasted with the villus cell have been interpreted to be evidence of new cell membranes. Since the crypt cell is undergoing mitosis, one might assume that such a cell may require new cell-surface membrane. The new undifferentiated cell-surface membrane might then exhibit some of the characteristics of the internal membrane from which it may have been derived. Golgi membrane is characterized by glycosyltransferase activities. It is possible that one explanation for finding glycosyltransferase activities on cell surface membranes is that the "early" cell surface membrane is, in fact, externalized Golgi membrane. Such factors on the cell-surface membrane could then serve as signal or control factors or, as differentiation occurs, assume membrane structural or cell adhesive properties.

The clinical implications of these studies may appear, at first, to be remote. However, these cell-surface changes accompanying differentiation and neoplastic transformation may prove to be important in understanding the pathophysiology of some disease processes and even helpful in diagnosis. For example, in some intestinal bacterial infections and in gluten-sensitive enteropathy, the intestine appears to have a higher rate of turnover and exhibit a less mature epithelial cell. We have preliminary data that show that untreated patients with gluten-sensitive enteropathy have cell-surface glycosyltransferase activities characteristic of the crypt cell. This supports the histological and organ culture data of Trier (33) showing that these patients have a true increase in their mitotic index with a resultant higher percentage of undifferentiated cells. This higher percentage of undifferentiated, crypt-like cells appears to be part of the explanation for the malabsorption seen in this disease.

Cell-surface changes occurring with neoplastic transformation may also prove to be indicative of cell membrane factors that could, on occasion, be released into the serum of patients, just as they have been shown to be shed into the tissue culture medium (34). The parallel with CEA is obvious, and indeed D. K. Podolsky, working in our laboratory, has recently obtained evidence of the presence of an unusual isoenzyme of galactosyltransferase in the sera of patients with cancer (35). We are investigating the possibility that this isoenzyme is being released (or shed) from the surface of these tumor cells.

ACKNOWLEDGMENTS

This work was supported by a grant from USPHS (CA14294) and American Cancer Society Grant (BC-93).

REFERENCES

1. Fox, T. O., Shepard, J. R., and Burger, M. M. (1971): *Proc. Natl. Acad. Sci. USA*, 68:244.
2. Gahmberg, C. G., and Hakamori, S. (1974): *Biochem. Biophys. Res. Commun.*, 59:283.
3. Moscona, A. A. (1971): *Science*, 171:905.
3. Burger, M. M. (1971): *Curr. Top. Cell. Regul.*, 3:135–193.
5. Roth, S., and White, D. (1972): *Proc. Natl. Acad. Sci. USA*, 69:485.
6. Wu, H. C., Meezan, E., Black, P. H., and Robbins, P. W. (1969): *Biochemistry*, 8:2509.
7. Buck, C. A., Glick, M. C., and Warren, L. (1970): *Biochemistry*, 9:4567.
8. Warren, L., Critchley, D., and MacPherson, I. (1972): *Nature*, 235:275.

9. Malluci, L., Poste, G. H., and Wells, V. (1972): *Nature* [*New Biol.*], 235:222.
10. Oppenheimer, S. B., Edidin, M., Orr, C. W., and Rosemen, S. (1969): *Proc. Natl. Acad. Sci. USA,* 50:613.
11. Roth, S., McGuire, E. J., and Roseman, S. (1971): *J. Cell Biol.,* 51:536.
12. Bosmann, H. B. (1972): *Biochem. Biophys. Res. Commun.,* 49:1256.
13. Nairn, R. C., Richmond, H. G., and Forthergill, F. (1960): *Br. Med. J.,* 2:1341.
14. Baldwin, R. W., and Glaves, D. (1972): *Int. J. Cancer,* 9:76.
15. Gold, P., and Freedman, S. O. (1965): *J. Exp. Med.,* 121:439.
16. Weiser, M. M. (1973): *J. Biol. Chem.,* 248:2542.
17. LaMont, J. T., Weiser, M. M., and Isselbacher, K. J. (1974): *Cancer Res.,* 34:3225.
18. Weiser, M. M. (1972): *Science,* 177:525.
19. Fujita, M., Kawai, K., Asano, S., and Nakao, M. (1973): *Biochim. Biophys. Acta,* 307:141.
20. Forstner, G. G., Sabesin, S., and Isselbacher, K. J. (1968): *Biochem. J.,* 106:381.
21. Weiser, M. M. (1973): *J. Biol. Chem.,* 248:2536.
22. Forstner, G. G. (1970): *J. Biol. Chem.,* 245:3584.
23. Roseman, S. (1970): *Chem. Phys. Lipids,* 5:270.
24. Webb, G. C., and Roth, S. (1974): *J. Cell Biol.,* 63:796.
25. Pollack, R. E., and Burger, M. M. (1969): *Proc. Natl. Acad. Sci. USA,* 62:1074.
26. Inbar, M., and Sachs, L. (1969): *Proc. Natl. Acad. Sci. USA,* 63:1418.
27. Goldstein, I. J., Hollerman, C. E., and Smith, E. E. (1965): *Biochemistry,* 4:876.
28. Inbar, M., Ben-Bassat, H., and Sachs, L. (1972): *Nature* [*New Biol.*], 236:3.
29. Podolsky, D. K., and Weiser, M. M. (1975): *Biochem. J.,* 146:213.
30. Ben-Bassat, H., Inbar, M., and Sachs, L. (1970): *Virology,* 40:854.
31. Eckhart, W., Dulbecco, R., and Burger, M. M. (1971): *Proc. Natl. Acad. Sci. USA,* 68:283.
32. Abercrombie, M., and Ambrose, E. S. (1962): *Cancer Res.,* 22:525.
33. Trier, J. S., and Browning, T. H. (1970): *N. Engl. J. Med.,* 283:1245.
34. Tompkins, W. A. F., Watrach, A. M., Schmale, J. D., Schultz, R. M., and Harris, J. A. (1974): *J. Nat. Cancer Inst.,* 52:1101.
35. Podolsky, D. K., and Weiser, M. M. (1975): *Biochem. Biophys. Res. Commun.* 65:545.

Membranes and Disease, edited by L. Bolis, J. F. Hoffman, and A. Leaf. Raven Press, New York, © 1976.

Hormone Receptors in the Kidney

G. D. Aurbach

Metabolic Diseases Branch, National Institute of Arthritis, Metabolism, and Digestive Diseases, National Institutes of Health, Bethesda, Maryland 20014

The initial event in the action of many polypeptide and amine hormones involves binding to the plasma membrane of the cell and then activation of the enzyme adenylate cyclase with consequent generation of cellular cyclic AMP. The first physiological evidence *in vivo* that this system was involved in the action of hormones on the kidney derived from physiological experiments with parathyroid hormone or vasopressin. Experiments of Takahaski et al. (1) suggested that vasopressin influenced urinary excretion of cyclic AMP, and Chase and Aurbach (2) showed that intravenous injection of parathyroid hormone in rats caused a prompt increase in the urinary excretion of cyclic AMP, which preceded the classically recognized phosphaturic affect of the hormone. Other experiments indicated that the cyclic AMP excreted in response to parathyroid hormone was elaborated directly into the urine by the kidney itself. This observation suggested that specific receptors exist within the kidney and that the corresponding cells contain adenylate cyclase enzyme activated by the hormone.

The experiments described above prompted a search for hormone-sensitive adenylate cyclase in kidney cell membrane fractions. It was found that parathyroid hormone causes direct activation of adenylate cyclase in plasma membrane preparations from the kidney. Vasopressin, which had been shown to activate the cyclic AMP–adenylate cyclase system in the kidney (3) as well as in the toad bladder *in vitro* (4), also was found to activate adenylate cyclase in the crude homogenates that contain receptors for parathyroid hormone. However, further analysis showed that the receptor–adenylate cyclase complex sensitive to vasopressin in the kidney is localized in an anatomical region clearly distinct from that responding to parathyroid hormone (5). These early experiments established the existence of specific hormone-sensitive adenylate cyclases within the kidney that explain the increased elaboration of cyclic AMP into the urine upon injection of parathyroid hormone or vasopressin (the response to vasopressin is small; even pharmacological amounts of vasopressin cause only a twofold increase in urinary excretion of cyclic AMP).

It is known now that receptors for a number of hormones, as well as prostaglandins, exist in the kidney. Parathyroid hormone, vasopressin, glucagon, calcitonin, VIP, epinephrine, and prostaglandins all have been shown to cause an increase in renal adenylate cyclase or in cyclic 3′,5′-AMP content of kidney tissue *in vitro*. However, of the several polypeptide hormones, only parathyroid hormone, glucagon, and vasopressin have been found to affect urinary excretion of cyclic AMP, and it is only the effect of parathyroid hormone that has been shown definitively to cause renal elaboration of cyclic AMP directly into the urine. [Glucagon causes a significant increase in urinary cyclic AMP, but in this instance the response reflects hepatic contribution of cyclic AMP to the plasma compartment and subsequent clearance of cyclic AMP from the plasma compartment into the urine via glomerular filtration. Vasopressin increases urinary excretion of cyclic 3′,5′-AMP slightly, and only when pharmacological amounts of hormone are given (2,6).]

LOCALIZATION OF SPECIFIC HORMONE RECEPTORS WITHIN THE KIDNEY

Much experimental evidence shows that renal receptors for the several hormones exist at discrete and unique locations within the nephron. Evidence for this has been obtained through one of four experimental approaches: (i) direct determination of hormone–receptor interaction utilizing high affinity radioiodinated hormone ligand; (ii) determination of adenylate cyclase responses in various segments or fractions of the kidney; (iii) determination of ion fluxes or cyclic AMP in renal tubular luminal fluid obtained through micropuncture or stop-flow techniques; and (iv) direct immunofluorescent analysis for cyclic AMP in sections of renal tissue. So far, the first approach has not been applied in detail to study this problem. Indeed the only valid radioreceptor interaction determined to date in kidney tissue is that described for calcitonin (7–9). Attempts to develop radioreceptor assays for parathyroid hormone have so far yielded results that are not clearly physiologically valid (10). The second approach was applied initially with parathyroid hormone and vasopressin. Gross separation of rat into cortex and medulla showed that the receptors for parathyroid hormone and vasopressin must be separate (5). Further recent studies have shown that the receptors for calcitonin, parathyroid hormone, and lysine vasopressin are each located in relatively distinct regions of the kidney. Parathyroid hormone causes the greatest relative activation of adenylate cyclase in the cortex. Calcitonin causes the greatest relative effect in the outer medulla region and vasopressin produces the greatest effect in the renal papillary area (9). This is illustrated in Fig. 1. Other studies (11) with isolated renal tubules show that receptors for parathyroid hormone, calcitonin, and vasopressin are located within tubular elements of the cortex, corticomedullary junction, and medulla, respectively. The above experiments suggest that receptors for parathyroid hormone are located in the

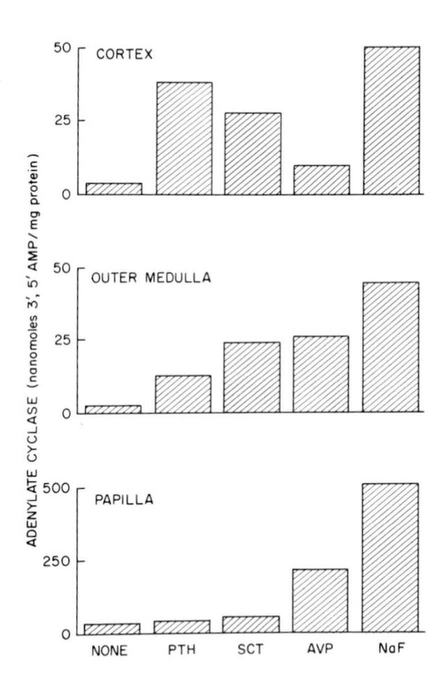

FIG. 1. Adenylate cyclase in membranes from three zones of the rat kidney. The kidneys were dissected into cortical, red (outer) medullary, and white medullary (papilla) zones. Adenylate cyclase was assayed at 22°C for 30 min in 50 mM Tris, 0.013% bovine serum albumin, 30 mM KCl, 4.5 mM $MgCl_2$, 1.1 mM ATP, 4 mM creatinine phosphate, 10 μg creatine phosphokinase, and 9 mM theophylline, at pH 7.5; the total volume was 70 μl. Final concentrations were: parathyroid hormone (PTH) (1500 USP units/mg), 20 μg/ml; salmon calcitonin (SCT) (4200 MRCU/mg), 200 ng/ml; arginine vasopressin (AVP) (60 IU/mg), 800 ng/ml; sodium fluoride, 7 mM. The points represent the mean $\pm$1 SE of three determinations for the cortex and outer medulla, or the average and range of two determinations for the papilla. All three hormones and fluoride caused significant ($p <$ 0.001) stimulation of the enzyme prepared from the cortex or outer medulla. In membranes from the outer medulla SCT caused greater activation than PTH, whereas the latter caused greater hormonal activation in the cortical preparation. (From ref. 9, with permission.)

proximal renal tubule. The experiments of Agus et al. (12) can be taken as supportive for this thesis; they found by micropuncture studies that parathyroid hormone influenced ion transport predominantly in the proximal tubule. Further evidence to support this thesis was provided by the work of Scurry and Pauk (13), who utilized the stop-flow technique in rats. The results of their experiments suggested that cyclic AMP is elaborated into the urine by proximal tubular cells in a region similar to or identical with the locus for inhibition of phosphate reabsorption. Two novel but distinct approaches used by Charbardès et al. (14) and by Chase (*personal communication*) lend still further proof to the proximal tubular locus for parathyroid hormone receptors. Charbardès et al. (14) succeeded in preparing single isolated tubules from each of several regions of the nephron and determined adenylate cyclase in each. They found parathyroid hormone-sensitive adenylate cyclase confined to the early proximal tubule and an entirely distinct locus, although less active, in the distal tubule (Fig. 2). Chase (*personal communication*), in preliminary experiments, has utilized the indirect Coombs immunofluorescent technique with anti-cyclic AMP antibody to show that parathyroid hormone influences predominantly proximal tubular cells. He injected rats with either normal saline as a control or parathyroid hormone. Two minutes after injection was concluded the kidney

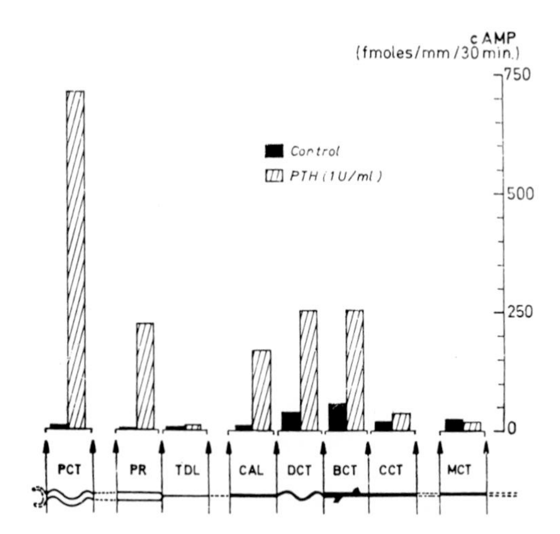

FIG. 2. Profile of control and PTH-sensitive adenylate cyclase activities along the rabbit nephron. The different segments of the nephron are schematically represented in abscissa. The size of each bar corresponds to the mean value of three to four replicate adenylate cyclase measurements for both control and PTH-treated structures. PCT: proximal convoluted tubule (initial portion). PR: pars recta (terminal portion). TDL: thin descending limb (from outer medulla). MAL: thick ascending limb (from outer medulla). CAL: thick ascending limb (terminal cortical portion). DCT: distal convoluted tubule. BCT: collecting tubule (first, branched portion). CCT: collecting tubule (cortical portion, distally connected to BCT). MCT: collecting tubule (from outer medulla). (From ref. 14, with permission.)

was perfused with formaldehyde and then sectioned. He found a strong immunofluorescent localization of cyclic AMP in the luminal border of proximal tubular cells. The control sections showed diffuse fluorescence, but in several experiments there was no intense band of fluorescence localized as in the parathyroid-treated preparations. Distal convoluted tubules in the same sections showed very little fluorescence and no discernible effect of parathyroid hormone. Thus, evidence gathered through several different biological techniques suggests that parathyroid hormone regulates renal function predominantly through interaction with receptors in proximal convoluted tubules. There also appears to be some effect of parathyroid hormone on the distal convoluted tubule. There is a less diverse body of evidence localizing receptors for calcitonin to the corticomedullary junction and those for vasopressin to medullary tubular sites.

HORMONE, CYCLIC AMP, AND ION TRANSPORT

The influence of vasopressin on sodium transport across the toad bladder is well recognized as the consequence of the studies of Orloff and Handler (15), Leaf (16), and Edelman et al. (17). These studies have shown also that cyclic AMP mediates the action of vasopressin on ion transport and that addition of the cyclic nucleotide directly to the medium can mimic the effect of the hormone on water and sodium transport. Catecholamines represent another class of hormonal agonist that influence sodium transport as shown by Morel and Jard with the frog skin (18) and Gardner et al. (19) with the turkey erythrocyte. Recent studies indicate also that polypeptide and amine hormones can influence sodium transport in the kidney. The studies of Agus et al. (12) show that parathyroid hormone inhibits sodium reabsorption in the proximal tubule. Dibutyryl cyclic AMP mimics this effect of parathyroid hormone. Gill and Casper (20) showed that dibutyryl cyclic AMP inhibits proximal sodium reabsorption in hypophysectomized dogs. Catecholamines produced a similar effect. Although both parathyroid hormone and catecholamines influence sodium transport in the proximal tubule, they do not cause significant natriuresis. It appears that the sodium rejected at the proximal tubule is reabsorbed in the distal tubule leaving a net increase in excretion of water, potassium, and phosphate. Calcitonin (8,21) as well as vasopressin (22), albeit in very high concentrations, also produce significant increases in natriuresis. The latter effect of calcitonin has been observed in human beings, as well as dogs and rats (21). It is tempting to propose that, although catecholamines, parathyroid hormone, and calcitonin all can influence sodium transport in the kidney, the ultimate expression of these effects differ because the sites of action within the renal tubule for these two hormones are distinct. Thus it is possible that influences on sodium reabsorption in the early proximal tubule (parathyroid hormone) are reflected by increased free-water clearance, whereas effects on sodium transport lower in the nephron, for example the corticomedullary junction region (calcitonin), is reflected in natriuresis. It is perhaps of significance that this same region of the kidney (the corticomedullary junction) shows the highest concentration of the enzyme sodium-potassium-ATPase (23). The potential significance of this is discussed further.

Hormone-induced cyclic AMP-mediated ion transport has been clearly established in recent studies on the avian erythrocyte (19,24–28). This system represents a single uniform cell type which shows a rapid response to one hormone class only, the beta-adrenergic catecholamines. Within a few seconds after adding isoproterenol to these cells, there is a significant increase in the rate of sodium and potassium transport. These effects can be mimicked completely by exogenous cyclic 3′,5′-AMP. There is now considerable evidence indicating that the following sequence of events is involved in catecholamine-stimulated ion transport: (i) interaction of catecholamine with specific receptor on cell surface; (ii) activation of adenylate cyclase; (iii) generation of cyclic AMP; (iv) activation on cyclic AMP of a cyclic nucleotide-sensitive protein kinase; (v) kinase-catalyzed phosphorylation of a protein specifically involved in ion transport. All of these steps seem to take place within, or intimately involve, the plasma membrane of the cells. The localization of the receptor–adenylate cyclase complex to the plasma membrane in avian erythrocytes was first shown by Davoren and Sutherland (29). More recently, Rudolph and Greengard (30) have shown that a unique plasma membrane protein in turkey erythrocytes is phosphorylated at a rate parallel to the increase in sodium transport caused when isoproterenol is added to these cells. We (Gardner and Aurbach, *unpublished*) had observed earlier that plasma membrane of these cells contained cyclic AMP binding protein and a kinase system. Thus a series of diverse experiments suggest that the above sequence is correct. The link between phosphorylation of a specific protein in the plasma membrane and activation of ion transport, however, remains to be established.

The avian erythrocyte system may be a valid model for hormone-regulated ion transport in specific segments of the kidney. We have already pointed out that parathyroid hormone, calcitonin, catecholamines, and vasopressin all have been shown to influence sodium transport in the kidney. All of these hormones act on kidney cells through the intermediation of cyclic AMP, and the cyclic AMP receptor–protein kinase system has been identified in kidney (31–33). There is suggestive evidence that in renal cells, in contrast to erythrocytes, there is a distinct difference in polarity for location of hormone receptor–adenylate cyclase on the one hand, and cyclic AMP receptor–protein kinase on the other. Wilfong and Neville (34) prepared isolated brush borders of renal tubular cells and found that specific activity for the enzyme was no higher in the purified preparation than in the crude homogenate. They concluded thus that the hormone-sensitive adenylate cyclase was most likely located at the plasma front or basilar portion of the cell. This was supported further by Marx et al. (7) who found that the hormone-sensitive enzyme was enriched in a membrane fraction enriched also with Na/K ATPase activity, an enzyme marker for the lateral-basal surface of the cell. Conversely, the observations of Kinne et al. (35) and Insel et al. (33) suggest that the cyclic AMP receptor–kinase system is located at the luminal surface of the tubular cell. This conclusion is supported further by the observations of Chase (*personal communication*) who utilized the indirect Coombs' fluorescent-labeled antibody technique to show that cyclic AMP accumulates at the luminal border of proximal tubular cells following injection of parathyroid hormone intravenously into rats.

In the turkey erythrocyte sodium transport is mediated by a mechanism related to, but distinguishable from, the classical Na/K ATPase system. Ouabain inhibits basal transport in these cells but potentiates (under certain circumstances) catecholamine-stimulated sodium transport. The latter, but seemingly not basal sodium, transport is mediated by cyclic 3′,5′-AMP. After prolonged incubations in the presence of isoproterenol and potassium, there is a loss of responsiveness to effects of cyclic AMP on sodium transport. This loss of responsiveness can be prevented by including ouabain in the medium (24). Since ouabain seems to inhibit hydrolysis of the second phosphorylated intermediate of the Na/K ATPase reaction [36], it is possible that this is a common point in pathways for cyclic AMP-dependent as well as cyclic nucleotide-independent ion transport. The facts that in the kidney the corticomedullary junction contains the highest concentration of Na/K ATPase, that this is the region richest in calcitonin receptors, that calcitonin causes increased accumulation of cyclic AMP by stimulating adenylate cyclase, and that calcitonin causes natriuresis (8) raise the possibility that a sequence analogous to the avian erythrocyte cyclic AMP-mediated ion transport system described above operates in response to calcitonin, parathyroid hormone, vasopressin, and catecholamines, each at specific loci within the nephron.

HORMONE-REGULATED CALCIUM TRANSPORT IN THE KIDNEY

The principal physiological function of parathyroid hormone is regulation of calcium concentration in the extracellular fluid. This is effected in part through control of calcium reabsorption from the glomerular filtrate. Physiological studies (37) suggest that calcium and sodium compete within the kidney for similar transport systems. Is it possible that cyclic AMP-mediated sodium transport can regulate calcium transport as well? We have meager information about this. There are no direct physiological experiments, for example, micropuncture studies, to show that parathyroid hormone directly influences calcium transport at a specific locus within the kidney. There are suggestions that parathyroid hormone could influence calcium reabsorption at the distal nephron (38), as compared to the experiments of Jamison et al. (39) indicating that the bulk of calcium reabsorption takes place in the descending limb of the loop of Henle (39). It is of interest that the golden hamster displays one of the most significant renal regulatory mechanisms in controlling calcium metabolism. Parathyroidectomy of the golden hamster causes a striking increase in calcium excretion into the urine (40). In the first few hours after parathyroidectomy, loss of calcium into the urine seems to account completely for the fall in calcium concentration in the extracellular fluids (40). Thus, the hamster kidney seems to be extremely sensitive to the parathyroid hormone-regulated mechanism influencing calcium transport in the nephron. The hamster, being a desert-type rodent, should have a high percentage of long-looped nephrons, and the average length of the loop of Henle should be high. If it is true that the bulk of calcium absorption takes place in the loop of Henle, there might be a significant correlation between the length and number of loops and the degree of sensitivity to parathyroid hormone-controlled renal tubular transport of calcium. Further comparisons between various desert versus domestic rodents might shed some light on this question.

On the other hand, there is no direct proof that the parathyroid hormone influence on calcium reabsorption from the glomerular filtrate takes place at either the corticomedullary junction or in the loop of Henle. As indicated above, the region rich in parathyroid hormone receptors does not seem to be the area of importance for calcium transport. Indeed the area for calcium transport may overlap with the corticomedullary junction, the region containing the highest concentrations of calcitonin receptors (also containing the highest concentration of Na/K ATPase). Also as noted above, some evidence suggests that the parathyroid hormone-regulated calcium transport site is located in the distal nephron (38).

Another possibility is that a calcium-binding protein is involved in tubular reabsorption of calcium under the influence of parathyroid hormone. A calcium-binding protein similar to that found in the intestine (where synthesis of the protein is regulated by vitamin D) has been found in the kidney (see ref. 41 for review). This protein appears to be located in the renal cortex. However, the protein in the kidney has not been adequately studied, and there is no definitive evidence that parathyroid hormone-regulated calcium transport in the kidney is mediated by a mechanism involving this protein. Resolution of the question of exactly how parathyroid hormone regulates calcium transport in the kidney awaits further discovery. Whether calcium transport is dependent upon cyclic AMP-mediated sodium transport in the kidney also remains to be established.

REFERENCES

1. Takahashi, K., Kamimura, M., Shinko, T., and Tsuji, S. (1966): *Lancet,* 2:967.
2. Chase, L. R., and Aurbach, G. D. (1967): *Proc. Natl. Acad. Sci. USA,* 58:518.
3. Brown, E., Clarke, D. L., Roux, V., and Sherman, G. H. (1963): *J. Biol. Chem.,* 238: PC852.
4. Bar, H., Hechter, O., Schwartz, I. L., and Walter, R. (1970): *Proc. Natl. Acad. Sci. USA,* 67:7.
5. Chase, L. R., and Aurbach, G. D. (1968): *Science,* 159:545.
6. Kaminsky, N. I., Broadus, A. E., Hardman, J. G., Jones, Jr., D. J., Ball, J. H., Sutherland, E. W., and Liddle, G. W. (1970): *J. Clin. Invest.,* 49:2387.
7. Marx, S. J., Fedak, S. A., and Aurbach, G. D. (1972): *J. Biol. Chem.,* 247:6913.
8. Marx, S. J., Woodard, C. J., Aurbach, G. D., Glossmann, H., and Keutmann, H. T. (1973): *J. Biol. Chem.,* 248:4797.
9. Marx, S. J., Woodard, C. J., and Aurbach, G. D. (1972): *Science,* 163:999.
10. Heath, D. A., and Aurbach, G. D. (1975): Calcium Regulating Hormones. *Proc. 5th Parathyroid Conf., Amsterdam 1975,* 159–162.
11. Melson, G. L., Chase, L. R., and Aurbach, G. D. (1970): *Endocrinology,* 86:511.
12. Agus, Z. S., Puschett, J. B., Senesky, D., and Goldberg, M. (1971): *J. Clin. Invest.,* 50:617.
13. Scurry, M. T., and Pauk, G. L. (1974): *Acta Endocrinol.,* 77:282.
14. Chabardès, D., Imbert, M., Clique, A., Montégut, M., and Morel, F. (1975): *Pfluegers Arch.,* 354:229.
15. Orloff, J., and Handler, J. (1967): *Am. J. Med.,* 42:757.
16. Leaf, A. (1967): *Am. J. Med.,* 42:745.
17. Edelman, I. S., Petersen, M. J., and Gulyassy, P. F. (1964): *J. Clin. Invest.,* 43:2185.
18. Morel, F., and Jard, S. (1971): *Ann. NY Acad. Sci.,* 185:351.
19. Gardner, J. D., Klaeveman, H. L., Bilezikian, J. P., and Aurbach, G. D. (1973): *J. Biol. Chem.,* 248:5590.

20. Gill, Jr., G. N., and Casper, A. G. T. (1971): *J. Clin. Invest.,* 50:112.
21. Marx, S. J., and Aurbach, G. D. (1975): Calcium Regulating Hormones, *Proc. 5th Parathyroid Conf., Amsterdam 1975,* 163–171.
22. Kurtzman, N. A., Rogers, P. W., Boonjarern, S., and Arruda, J. A. L. (1975): *Am. J. Physiol.,* 228:890.
23. Hendler, E. D., Torretti, J., and Epstein, F. H. (1971): *J. Clin. Invest.,* 50:1329.
24. Gardner, J. D., Klaeveman, H. L., Bilezikian, J. P., and Aurbach, G. D. (1974): *J. Biol. Chem.,* 249:516.
25. Gardner, J. D., Klaeveman, H. L., Bilezikian, J. P., and Aurbach, G. D. (1974): *Endocrinology,* 96:499.
26. Gardner, J. D., Mensh, R. S., Kiino, D. R., and Aurbach, G. D. (1975): *J. Biol. Chem.,* 250:1155.
27. Gardner, J. D., Kiino, D. R., Jow, N., and Aurbach, G. D. (1975): *J. Biol. Chem.,* 250: 1164.
28. Gardner, J. D., Jow, N., and Kiino, D. R. (1975): *J. Biol. Chem.,* 250:1176.
29. Davoren, P. R., and Sutherland, E. W. (1963): *J. Biol. Chem.,* 238:3009.
30. Rudolph, S. A., and Greengard, P. (1974): *J. Biol. Chem.,* 249:5684.
31. Dousa, T. P., Sands, H., and Hechter, O. (1972): *Endocrinology,* 91:757.
32. Winickoff, R., and Aurbach, G. D. (1970): Program of the 52nd Meeting, Endocrine Society, 45, *Abstract No. 17.*
33. Insel, P., Balakir, R., and Sacktor, B. (1975): *J. Cyclic Nucleotide Res.,* 1:107.
34. Wilfong, R. F., and Neville, Jr., D. M. (1970): *J. Biol. Chem.,* 245:6106.
35. Kinne, R., Schlatz, L. J., Kinne-Saffran, E., and Schwartz, I. L. (1973): *Proc. 9th Int. Cong. Biochem. 1973,* 258, *Abstract No. 5b.*
36. Dahl, J. L., and Hokin, L. E. (1974): *Annu. Rev. Biochem.,* 43:327.
37. Walser, M. (1973): *Handbook of Physiology, Renal Physiology,* p. 555. American Physiological Society, Washington, D.C.
38. Widrow, S. H., and Levinsky, N. G. (1962): *J. Clin. Invest.,* 41:2151.
39. Jamison, R. L., Frey, N. R., and Lacy, F. B. (1974): *Am. J. Physiol.,* 227:745.
40. Biddulph, D. M. (1972): *Endocrinology,* 90:1113.
41. Taylor, A. N., and Wasserman, R. H. (1972): *Am. J. Physiol.,* 223:110.

Membranes and Disease, edited by L. Bolis, J. F. Hoffman, and A. Leaf. Raven Press, New York, © 1976.

Some Aspects of the Cellular Action of Antidiuretic Hormone in Normal Mammalian Kidney and in Nephrogenic Diabetes Insipidus

Thomas P. Dousa

Mayo Clinic and Foundation, and Mayo Medical School, Rochester, Minnesota 55901

Our understanding of how antidiuretic hormone (ADH) regulates the water permeability in the distal segments of mammalian nephron progressed considerably in recent years. It has been established, and it is generally accepted, that cyclic AMP is the intracellular mediator of the ADH action in mammalian nephron (1,2). Initial steps in the cyclic AMP-mediated action of ADH involve hormone-dependent cyclic AMP formation within the cell (3). ADH receptor–adenylate cyclase complex appears to be localized in the plasma membranes, while the cyclic AMP degrading enzymes, cyclic AMP phosphodiesterases, have the highest specific activity in a fraction of soluble proteins (4). Using a large number of chemical analogues of ADH, Jard and associates (5,6) found a close correlation between ability of the hormone to bind on the plasma membrane fraction from renal medulla and ability to stimulate the adenylate cyclase. Comparing a number of various mammalian species, the ADH-sensitive renal medullary adenylate cyclase appears to be most sensitive to the type of hormone molecule naturally present in the respective species (3,5).

Many aspects of the ADH-dependent cyclic AMP metabolism were elucidated in recent studies, but further extensive investigations are required to clarify these initial steps in the ADH action. However, it appears that the frontier of investigations on the cellular action of ADH is moving toward the question of how cyclic AMP formed under ADH stimulation regulates the water permeability of epithelial cells of distal nephron, namely, permeability of the luminal plasma membrane, which appears to be a limiting barrier for the water movement (2). Two such major mechanisms and their components are under active experimental investigation at the present time: (i) cyclic AMP-dependent protein phosphorylation, and (ii) role of cytoplasmic microtubules.

CYCLIC AMP-DEPENDENT PROTEIN PHOSPHORYLATION IN ADH ACTION

Cyclic AMP-dependent protein kinase was found in numerous tissues, and its role in cellular mechanism of some hormones was established (7).

In relation to the action of ADH on kidney, it was proposed several years ago that cyclic AMP-dependent phosphorylation of specific protein in luminal plasma membranes of renal tubular cells may be involved in the water permeability control (8). Implicit is the assumption that such specific phosphoprotein located within the membrane is critical for the structural and functional changes of the membrane associated with water permeability and that reversible cyclic AMP-dependent phosphorylation of this protein is associated with membrane permeability function (8). Basic elements of cyclic AMP-dependent phosphorylation of membrane protein appears to be present in the renal medulla. Cyclic AMP-dependent protein kinase partially purified from this tissue was shown to be able to catalyze phosphorylation of the proteins of plasma membrane fraction from the bovine renal medulla (8) and the phosphorylated plasma membrane can be dephosphorylated by the protein phosphatase which is present also in the renal medullary extract (4,8). Protein phosphatase from mammalian renal medulla was never found to be effected by cyclic AMP or other cyclic 3′,5′-nucleotide tested (4,8). Thus, a cyclic AMP-dependent component of the proposed phosphoprotein metabolism appears to be the protein kinase. In a study, attempting to isolate the contraluminal from luminal portions of the renal medullary plasma membranes, the cyclic AMP-dependent protein kinase activity was described in the membrane fraction, considered to be luminal plasma membrane fragments (9). It was proposed that this membrane-bound cyclic AMP-dependent protein kinase is involved in the protein membrane phosphorylation (9); cyclic AMP-stimulated phosphorylation of renal medullary plasma membranes was also reported from another laboratory (10).

In our study on the subcellular distribution of the protein kinase in the renal medulla we found that cyclic AMP-dependent protein kinases were found almost exclusively in the cytosol and only minute activities of this enzyme were apparently associated with membrane fractions; it appears possible that certain portions of the cytosolic enzyme were unspecifically absorbed on the membrane structures (4). Moreover, no stimulation of the protein kinase activity in plasma membrane fraction by cyclic AMP was found. Likewise, most of the protein phosphatase activity (using several protein substrates) was mostly localized in the soluble fraction of the renal medullary homogenate (4). All the above-mentioned studies on the cyclic AMP-dependent protein phosphorylations and on the activation of protein kinase by cyclic AMP were conducted in isolated cell-free systems from renal medulla.

In our current studies we addressed ourselves to the question of whether cyclic AMP-dependent protein kinase is indeed activated by ADH (and subsequently through cyclic AMP) in the intact, unbroken cell. Another appealing question was whether protein kinase, which is apparently localized (predominantly or almost exclusively) in the cytosol, is the enzyme involved in the phosphorylation of the plasma membrane proteins. In such a

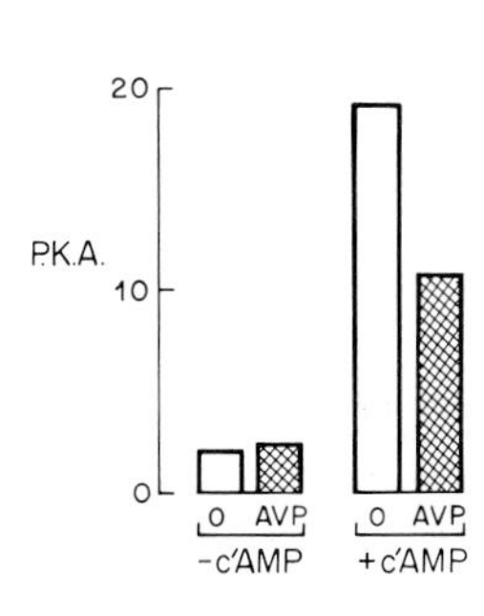

FIG. 1. Protein kinase activity (PKA) in 100,000 × *g* supernate of homogenate from the renal medullary slices incubated without (0) or with 2.5×10^{-7} M [8-arginine]-vasopressin (AVP). **Left:** PKA assayed without addition of cyclic AMP. **Right:** PKA after addition of 5×10^{-6} M cyclic AMP. The protein kinase activation is expressed as a ratio of PKA measured with or without addition of cyclic AMP (−cyclic AMP/+cyclic AMP) (11,13). This ratio was 0.111 in control and increased to 0.237 in presence of AVP. There was marked decrease in the total PKA (after maximal stimulation with cyclic AMP) in preparation from slices incubated with AVP.

case, protein kinase would be activated by cyclic AMP in the cytosol, and the catalytic subunit would attach to the specific protein substrate in the luminal membrane.

We explored these questions using preparation of renal medullary slices from bovine kidney (11,12). Slices were first incubated with ADH in modified Krebs-Ringer solution and then quickly homogenized and centrifuged at high speed to separate bulk of membranes from the soluble protein. Protein kinase was assayed in the supernate of the extract and the extent of the activation of the protein kinase was assessed from the ratio of protein kinase activity in the absence of added cyclic AMP to the activity assayed after addition of the maximal stimulatory concentration of cyclic AMP (13). We found that in renal medullary slices exposed to ADH there was a significant increase in the protein kinase activity ratio (13) indicating that protein kinase was stimulated in the intact cells by ADH (Fig. 1) (11). The increase in protein kinase activity ratio was specific for ADH: inactive chemical analog of ADH, or hormones which do not act on renal medullary structures, did not produce significant protein kinase activation (11). Also, a close parallelity was found between the ADH concentration, tissue levels of cyclic AMP and the extent of protein kinase activation (12). Another interesting feature was the finding that specific activity of protein kinase under the maximal stimulation by cyclic AMP added *in vitro* (reflecting the total cyclic AMP-dependent protein kinase) was significantly decreased in the supernates from slices exposed to ADH (12) (Figs. 1,2). There was a smaller but significant increase in the activity of protein kinase assayed without addition of exogenous cyclic AMP *in vitro* in the slices exposed to ADH (12). The net decrease in the total protein kinase activity may reflect a translocation of catalytic subunits of cytosolic protein kinase to the particular material contained in the sediment of the homogenate (12). The possibility that the above-mentioned decrease in the specific protein kinase activity indeed represents such translocation to particulate material is supported by the

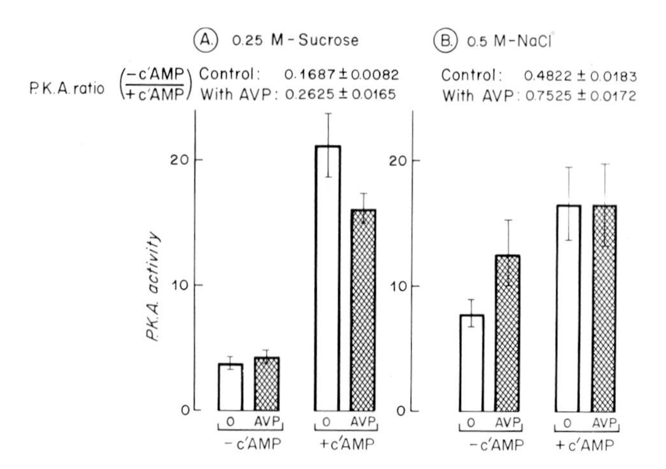

FIG. 2. Experimental conditions and symbols are similar to that described in legend to Fig. 1 with the exception that PKA was measured in 30,000 × *g* supernate. **A:** Slices homogenized in buffered isotonic sucrose. **B:** Slices homogenized on 0.5 M sodium chloride. Using both medium an increase in PKA ratio was observed; however, in the presence of 0.5 M NaCl no decrease in specific PKA was detected in supernate from slices incubated with AVP.

finding that if extracts are prepared in the presence of high concentrations of salts, the protein kinase activation was also detected (Fig. 2) (12), but no decrease in the total protein kinase activity in the supernate occurred. Since it was reported recently that salts can reverse the translocation of catalytic subunits to particulate fraction of homogenates in perfused heart (14), it

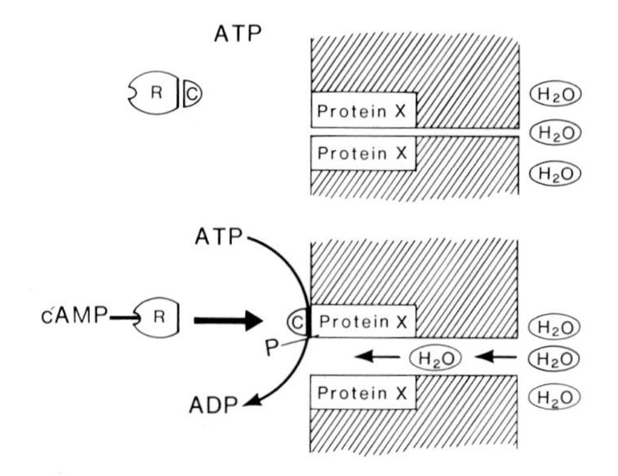

FIG. 3. Hypothetical phosphorylation of specific protein (*protein x*) controlling water permeability of luminal plasma membrane by cytosolic protein kinase. **Top:** Resting state, membrane impermeable to water. **Bottom:** Cyclic AMP formed under influence of ADH binds to regulatory subunit (R) of cytosolic protein kinase holoenzyme and catalytic subunit (C) binds to membrane *protein x* and catalyzes its phosphorylation. Phosphorylation of *protein x* causes the increase in osmotic water permeability of luminal plasma membrane.

appears probable that high salt concentration in our experiments also caused the detachment of catalytic subunits of the protein kinase from particulate material to soluble proteins fraction. Therefore, the present results are apparently in support of the view that ADH stimulates cytosolic protein kinase in the intact cells through the action of cyclic AMP, and that a sizeable portion of catalytic subunits of activated cytosolic protein kinase translocates to the particular material (which includes also plasma membranes). It is tempting to speculate that catalytic subunits of cytosolic protein kinase may attach and subsequently catalyze phosphorylation of the postulated specific luminal plasma membrane protein, involved in the control of water permeability. Such proposed mechanism is outlined in Fig. 3.

ROLE OF CYTOPLASMIC MICROTUBULES

Several lines of recent experimental evidence suggest that integrity of the cytoplasmic microtubules is required for the cellular action of ADH (15). This evidence is based mainly on both *in vivo* and *in vitro* experiments, which showed that variety of compounds which interfere with the integrity of microtubules, such as microtubule-disrupting alkaloids colchicine and vinblastine, block the action of ADH at the cellular level, probably in the steps subsequent to the cyclic AMP generation (15). The questions arise whether microtubules constitute entirely passive cytoskeleton structure which keeps elements (enzymes, modulators) of cyclic AMP-dependent mechanism acting on the luminal plasma membrane in the proper cytotopic position, or whether cyclic AMP acts directly or indirectly on microtubules and influences their function or structure. A prominent feature of cytoplasmic microtubules is the fact that microtubules are in dynamic equilibrium between free, soluble monomeric protein tubulin and polymerized microtubules. In our recent studies, we have developed an *in vitro* system which assesses indirectly the assembly of free tubulin into microtubules in the tissue slices of the renal medulla (16). The polymerization can be detected by incubating tissue slices under a variety of conditions and then homogenizing the slices in a medium which preserves polymerized microtubules. Polymerized microtubules are separated for soluble tubulin by ultracentrifugation of homogenate (16). The amount of microtubules is indirectly detected by colchicine-binding assay (15). Microtubule polymerization in this system was inhibited by low concentration of vinblastine, supporting the previous findings in the intact animals and in cell-free systems which show that this alkoloid blocks the action of ADH through interference with the microtubule assembly (17). The potential effect of ADH on the extent of microtubule polymerization was explored in this system. Incubation of renal medullary slices with ADH caused a marked increase in the tissue level of cyclic AMP, but no significant change in extent of microtubule polymerization was detected (Fig. 4) (16,18). These observations would favor

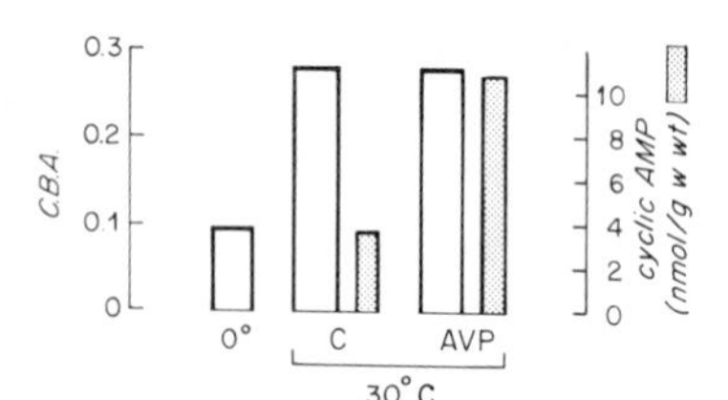

FIG. 4. Effect of ADH on the microtubule assembly in renal medullary tissue slices. C = control incubation, AVP = with 2.5 × 10^{-7} M [8-arginine]-vasopressin. Although AVP increased the tissue level of cyclic AMP, there was no effect on microtubule assembly as measured (16) by colchicine binding activity (CBA).

the view that cyclic AMP does not act directly or indirectly on the microtubules, or at least it does not appear to influence the ability of tubulin to assemble into microtubular structures. It cannot be excluded, however, that ADH and, subsequently, cyclic AMP influences structure and function of microtubules in some other way such as to change their charge (e.g., through phosphorylation) or orientation within the cell. Microtubules modified in their properties by cyclic AMP conceivably may (directly or indirectly) interact with luminal plasma membrane and influence its water permeability.

NEPHROGENIC DIABETES INSIPIDUS

Some aspects of the cellular action of ADH in the kidney were studied in several animal models of the nephrogenic diabetes insipidus (19). These models include congenital nephrogenic diabetes insipidus in mice (20), decrease in renal concentrating ability induced by drugs such as BAX 349, lithium, or declomycin (19), or in other states characterized by the unresponsiveness to ADH such as hypercalcemia (21) or a subnormal concentrating ability in a long-standing hypothalamic diabetes insipidus in Brattleboro strain rats (22). Investigations in these models were so far focused mainly on the possible impairment in the ADH-dependent cyclic AMP metabolism. It is noteworthy that in many, if not in all, of those above-mentioned models of urinary concentrating defects, resistant to ADH, there is a decreased ability to form the cyclic AMP under stimulation with ADH in cell-free systems and/or decreased accumulation of the cyclic AMP in response to ADH in the renal medulla was observed (19). When the adenylate cyclase prepared from animals with above-mentioned different models of nephrogenic insipidus was examined *in vitro* there was always a similar pattern of the decreased response of this enzyme to stimulation by ADH. In general, the affinity of adenylate cyclase complex to ADH, reflected as the dose required for only half-maximal stimulation of adenylate cyclase (3), was not changed; the total capacity to form cyclic AMP (maximal activity or intrinsic activity) (3) was diminished, but never totally absent. These findings pose an interesting question of whether partial decrease in the intrinsic activity of the ADH-sensitive adenylate cyclase can be a primary cause of decreased functional response to the hormone, providing that the steps subsequent to the cyclic AMP generation are intact. It appears that basically three extreme possibilities (grossly oversimplified for sake of argument) can be considered as outlined in Fig. 5. First (Fig. 5A), it is possible that all cyclic AMP formed under influence of ADH is available for the effect on systems controlling water permeability in the steps subsequent to cyclic AMP generation. In such a case, functional defect in the cyclic AMP formation would be proportionally related to the decrease in functional response to hormone. The other possibility (Fig. 5B) is that a certain fraction of the cyclic AMP which is formed within the cells under influence of ADH is not available for the control of water permeability, due to the breakdown through action of cyclic AMP phosphodiesterases, diffusion out of the cells or sequestration by binding on the proteins (different from regulatory subunit of cyclic AMP-sensitive protein kinase). In such a case, even partial decrease in the capacity to form cyclic AMP may result in the profound decrease in the ultimate functional response (in terms of increased water permeability and urine concentration). Still, the other extreme possibility portrayed in

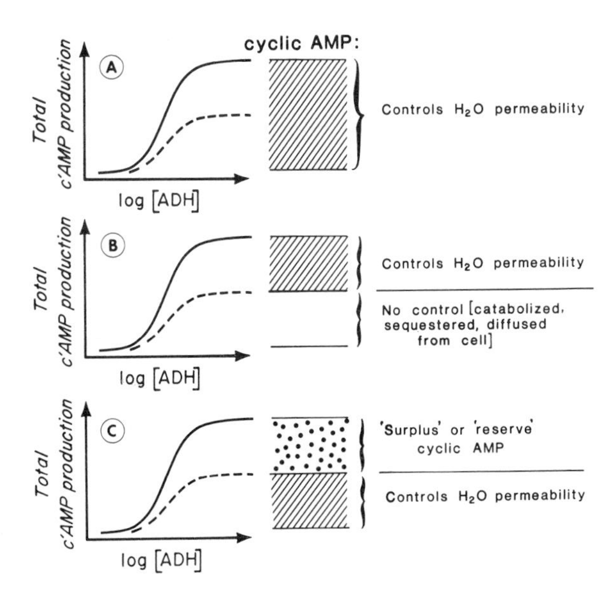

FIG. 5. Three model possibilities of relationship between ADH-dependent cyclic AMP production and pool of cyclic AMP available for the control of water permeability in the steps subsequent to cyclic AMP generation. Solid line: Dose–response curve of renal medullary adenylate cyclase to vasopressin (AVP) in normal state. Dashed line: Decreased dose–response curve of the adenylate cyclase to vasopressin as found in the several models of nephrogenic diabetes insipidus (19). **A:** All cyclic AMP produced is utilized for the control of water permeability: functional defect would be directly proportional to decrease in cyclic AMP formation. **B:** Large portion of cyclic AMP formed is unavailable for water permeability control: even partial decrease in the cyclic AMP formation may result in profound defect in the renal concentrating ability. **C:** Cyclic AMP in presence of high concentration of ADH can be formed in amounts which are in excess to that which is required for permeability control: partial decrease in cyclic AMP formation would have little or no effect on the final functional response.

Fig. 5C, is that not all the cyclic AMP formed under maximal stimulation with ADH is utilized for control of water permeability, and there exists a fraction of cyclic AMP referred to as a "reserve" or "surplus" cyclic AMP. In the last case, partial decrease in the capacity for forming cyclic AMP under influence of high levels of ADH would have minimal or no effect on the final functional response. Experimental observations compatible with several of the above-mentioned models were described. The view that supramaximal capacity to form cyclic AMP or the "reserve" cyclic AMP exists was suggested (23), or is at least compatible mostly with the observations made on the amphibian toad bladder. In this useful model of ADH-responsive epithelial membrane, where the functional response in terms of osmotic water flow and cyclic AMP accumulation can be monitored simultaneously, some observations show that by increasing concentration of ADH to very high levels more of cyclic AMP is formed than is required for the maximal hydroosmotic response to the hormone (24,25). Observations on toad bladder are to a certain degree complicated by the presence of two ADH-responsive systems, both mediated by cyclic AMP: water permeability system and active sodium transport; it is not often possible to discriminate the cyclic AMP related to water control from that related to the regulation of the active sodium transport. In mammalian kidney it is much more difficult to measure simultaneously ADH-dependent cyclic AMP metabolism and a functional response. Several observations, both made on the mammalian kidney *in vitro,* as well as some observations in toad bladder, seem to support the view that not all cyclic AMP formed under the influence of ADH is used for the control of water permeability. Very active cyclic AMP phosphodiesterases are present in renal tissue (4), and this itself suggests that an appreciable portion of cyclic AMP formed is quickly destroyed by these enzymes. This is compatible with the observation that the cyclic AMP accumulation in renal medullary tissue *in vitro,* after addition of ADH, is greatly potentiated by inclusion of cyclic AMP phosphodiesterase inhibitors (T. P. Dousa and L. D. Barnes, *unpublished results*). Similar observations were made on the toad bladder system (25). Here also the inhibition of cyclic AMP phosphodiesterase through the action of aldosterone appears to greatly potentiate not only accumulation of cyclic AMP, but also the hydroosmotic response of toad bladder to ADH (26). In our recent studies in mammalian kidney mentioned above we found a very close correlation between the increase in cyclic AMP in the renal medulla by increasing doses of ADH and *in situ* activation of the protein kinase; the dose of ADH required for maximal accumulation of cyclic AMP is identical with the dose which produced the maximal activation of the protein kinase (12). Accepting the assumption that the protein kinase is an obligatory step in the cellular action of ADH, this observation would also be against the concept that surplus cyclic AMP is produced under maximal stimulation with ADH, at least in the mammalian kidney. Thus, the latter observations appear to be compatible with the possibility that even partial decrease in the intrinsic activity of ADH-responsive adenylate cyclase complex in the renal nephron might cause complete or partial functional unresponsiveness to ADH as occurs in many of the above-mentioned animal models of this syndrome. Solution of this question would be important also for considering the rational basis for therapeutical approach to ADH-resistant concentrating defects. Suitable cyclic AMP derivative, which would accumulate preferentially in renal medulla, would substitute for the insufficient cyclic AMP formation and may correct renal concentrating defects of this type, or at least it may be used for testing in differential diagnosis of renal concentrating defects.

The present discussion deals only with cyclic AMP metabolism. There are at least indications that in some concentrating defects such as those induced by declomycin, lithium, hypercalcemia, or potassium depletion, an impairment may occur also in steps subsequent to the cyclic AMP generation (19).

CONCLUSION AND PROSPECTIVES

Research on the cellular action of ADH in mammalian kidney and related epithelial tissues progressed considerably over several recent years, namely, the ADH-dependent cyclic AMP formation was elucidated to a great degree. Identification of cyclic AMP-sensitive phosphoprotein metabolism and microtubule and microfilament systems as potential components in the cellular ADH action distal to the cyclic AMP generation serve as a useful working hypotheses for future investigations. Studies on the animal models of nephrogenic diabetes insipidus provided the first insights to the possible molecular basis in pathogenesis of this disease.

Major experimental challenges for the future investigations appear to be the search for the natural substrate for cyclic AMP-dependent protein kinase in the ADH-responsive epithelial cells, intracellular localization of site of ADH-stimulated cyclic AMP accumulation, site of breakdown of cyclic AMP as well as cytotopic localization of the phosphoprotein metabolism. Elucidation of the role of cytoplasmic microtubules in the water permeability control is another major problem, and relationship between protein phosphorylations, microtubular function, and functional state of luminal membrane are exciting new areas of experimental exploration. Judicious utilization of the animal models of ADH-resistant concentrating defects will help both to elucidate the importance of individual components involved in the cellular action of ADH and elucidate the molecular basis of pathogenesis of nephrogenic diabetes insipidus.

ACKNOWLEDGMENTS

Recent studies in author's laboratory were conducted in close collaboration with Dr. Larry D. Barnes with expert technical assistance of Mrs.

Yvonne S. F. Hui and Mr. Christopher Wilson. Mrs. Joyce Wellik provided expert secretarial assistance. Research in the author's laboratory was supported by USPHS research grant AM 16105 from National Institute of Arthritis, Metabolic and Digestive Diseases, by Grant-in-Aid from the American Heart Association, with funds partially contributed by the Minnesota Heart Association, by grant from the Kidney Foundation of Upper Midwest and by the Mayo Foundation. The author is Established Investigator of the American Heart Association.

REFERENCES

1. Dousa, T. P. (1973): *Life Sci.,* 13:1033.
2. Grantham, J. J. (1974): In: *Kidney and Urinary Tract Physiology,* edited by K. Thurau, Chap. 8, University Park Press, Baltimore, Md.
3. Dousa, T. P., Walter, R., Schwartz, I. L., and Hechter, O. (1972): *Adv. Cyclic Nucleotide Res.,* 1:121.
4. Barnes, L. D., Hui, Y. S. F., Frohnert, P. P., and Dousa, T. P. (1975): *Endocrinology,* 96:118.
5. Roy, C., Barth, T., and Jard, S. (1975): *J. Biol. Chem.,* 250:3149.
6. Roy, C., Barth, T., and Jard, S. (1975): *J. Biol. Chem.,* 250:3157.
7. Walsh, D., and Ashby, D. (1973): *Rec. Prog. Horm. Res.,* 29:329.
8. Dousa, T. P., Sands, H., and Hechter, O. (1972): *Endocrinology,* 91:757.
9. Schwartz, I. L., Schlatz, L. J., Kimme-Saffran, E., and Kinne, R. (1974): *Proc. Natl. Acad. Sci. USA,* 71:595.
10. Czoka, F. C., and Ettinger, M. T. (1975): *Fed. Proc.,* 34:543.
11. Dousa, T. P., Hui, Y. S. F., and Barnes, L. D. (1975): *Physiologist,* 18:194.
12. Dousa, T. P., Hui, Y. S. F., and Barnes, L. D. (1975): *Clin. Res.,* 23:542A.
13. Corbin, J. D., Keely, S. L., Soderling, T. R., and Park, C. R. (1975): *Adv. Cyclic Nucleotide Res.,* 5:265.
14. Keely, S. L., Corbin, J. D., and Park, C. R. (1975): *Adv. Cyclic Nucleotide Res.,* 5:829.
15. Dousa, T. P., and Barnes, L. D. (1974): *J. Clin. Invest.,* 54:252.
16. Barnes, L. D., Hui, Y. S. F., and Dousa, T. P. (1975): *Physiologist,* 18 (*in press*).
17. Dousa, T. P., and Barnes, L. D. (1975): *Abstr. 6th Int. Cong. Nephrology, Florence Italy, 1975,* Abstr. 162.
18. Barnes, L. D., Hui, Y. S. F., and Dousa, T. P. (1975): *Clin. Res.,* 23 (*in press*).
19. Dousa, T. P. *Mayo Clin. Proc.,* 49:188.
20. Dousa, T. P., and Valtin, H. (1974): *J. Clin. Invest.,* 54:753.
21. Beck, N., Singha, M., Reed, S. W., Murdaugh, H. V., and Davis, B. B. (1974): *J. Clin. Invest.,* 54:1049.
22. Dousa, T. P., Hui, Y. S. F., and Barnes, L. D. (1975): *Endocrinology,* 97:802.
23. Eggena, P., Schwartz, I. L., and Walter, R. (1970): *J. Gen. Physiol.,* 56:250.
24. Flores, J., Witkum, P., Beckman, B., and Sharp, G. W. G. (1974): *Biochem. Biophys. Acta,* 362:501.
25. Omachi, R. S., Robbie, D. E., Handler, J. S., and Orloff, J. (1974): *Am. J. Physiol.,* 226: 1152.
26. Stoff, J. S., Handler, J. S., and Orloff, J. (1972): *Proc. Natl. Acad. Sci. USA,* 69:805.

Membranes and Disease, edited by L. Bolis, J. F. Hoffman, and A. Leaf. Raven Press, New York, © 1976.

Localization and Function of Na-K-ATPase Activity in Various Structures of the Nephron

Udo Schmidt and Hans-Alfred Habicht*

Enzyme Laboratory, Medical Policlinic, Department Internal Medicine, University of Basel, Switzerland, and Institute of Pathology, University of Tübingen, West Germany

The energy for the active transport of sodium in renal tubule is derived from ATP. This has been proved at least for the proximal convoluted tubule, where the bulk of net sodium reabsorption takes place (1). The membrane-bound Na-K-ATPase catalyzes the hydrolysis of ATP and causes thus the active transport of sodium (2). Yet we still do not know how the chemical energy, released from ATP hydrolysis, is transferred to sodium.

In the very large number of publications of the last few years, dealing with the role of Na-K-ATPase in transmembrane sodium transport, four results are of particular importance:

1. Antibodies to Na-K-ATPase abolished completely the ouabain-sensitive efflux of sodium and simultaneously Na-K-ATPase which was incorporated into human red cell ghosts (3).
2. A purified Na-K-ATPase preparation, incorporated into an artificial membrane, developed a net short-circuit current flow which could be inhibited by the addition of ouabain (4,5).
3. In vesicles prepared from purified Na-K-ATPase from the rectal gland an actual ATP-dependent accumulation of sodium could be demonstrated, which proves that the Na-K-ATPase is the molecular machine for sodium transport (6).
4. First evidence of a sodium binding to the enzyme preparation was provided (7).

These new data suggest strongly that the Na-K-ATPase is the molecular machine that effects sodium and potassium transport. The major function of the kidney, which is a rich source of Na-K-ATPase, is sodium reabsorption (8). For the past seven years we have been working at the localization and regulation of Na-K-ATPase in the nephron. The distribution of the enzyme within the nephron may provide some further knowledge about its functional properties as to sodium transport. Therefore we studied the distribution of Na-K-ATPase along the nephron of various mammals in-

* With technical assistance of B. Funk, E. Engeler, and A. Mall.

cluding man. Furthermore we investigated the velocity of changes in the ouabain-sensitive activity of single nephron segments, when sodium transport was increased or decreased.

Because of the anatomical and functional complexity of the kidney we have developed an analytical procedure for quantitating Na-K-ATPase in single anatomically well-defined portions of the nephron of 20–100 μm length and 5–50 ng dry weight. This method enables us to evaluate Na-K-ATPase per millimeter tubular length and to compare the activity value with the flux rate of sodium.

METHODOLOGICAL CONSIDERATION

The method for measuring Na-K-ATPase activity in single structures of the nephron was described in detail previously (9–11). It is based on techniques developed by Lowry and associates (12), and includes various steps:

1. Kidney tissue is quickly frozen in liquid nitrogen.
2. Sections of at least 10-μm thickness are cut in a cryostat and dried under vacuum.
3. When the lyophilized sections have been rewarmed the desired tubular structures are identified and dissected under a stereomicroscope.
4. Length and dry weight of each sample are measured.
5. Na-K-ATPase activity in each tubular structure is quantitated with the aid of the oil-well technique combined with an enzymic cycling system.

For the ATPase reaction, an incubation time of 5 min is used, although the plot of liberation of P_i against time exhibits linearity during 30 min (Fig. 1). The reason for choosing such a short incubation time is that under the stereomicroscope the tubular structure, brought into the aqueous assay system, remains visible during these 5 min.

During this short incubation only 3–7% of the excess ATP are split according to the structure and dry weight, i.e., the ATP concentration decreases from 2.28 mM to 2.15 mM. Thus an optimal substrate concentration is maintained throughout the reaction time. As demonstrated in Fig. 2, the optimal substrate concentration is 1 mM ATP, regardless of whether 1 or 2 mM Mg is used. Our concentration of 2 mM is lower than the ATP concentrations of 3–6 mM, which were used by almost all investigators working with purified Na-K-ATPase preparation (13). However, the ATP content of the tubular cell is probably not more than 2 mM (*unpublished data*).

In membrane preparations from mammalian tissues Na-K-ATPase activity is manifest only in part, a substantial (6-fold) increase in activity can be observed after incubation with detergents like deoxycholate (14). Almost all investigators use deoxycholate for unmasking the catalytic activity of the Na-K-ATPase preparation. In our assay conditions deoxycholate in a concentration of 0.6 mg/ml increased Na-K-ATPase activity only slightly

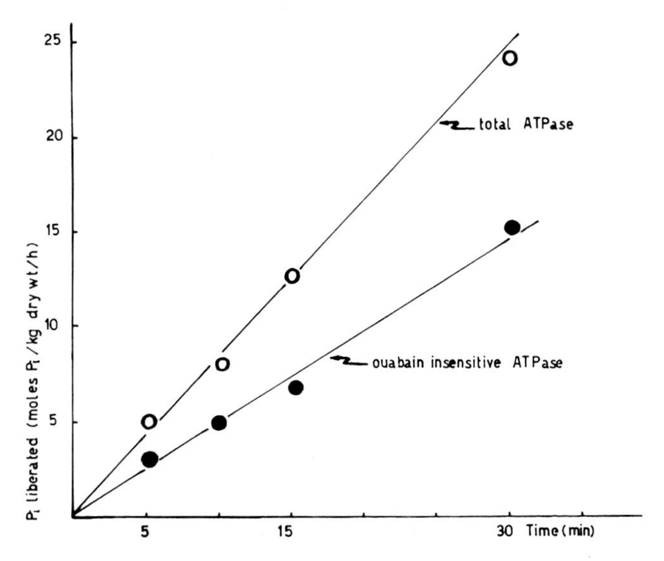

FIG. 1. Time course of release of P_i by proximal convoluted tubule in the presence and absence of 2 mM ouabain, 55 mM Na, 5 mM K, 2 mM Mg, and 2 mM ATP (means of three experiments). (○) Complete assay; (●) 2 mM ouabain added and K omitted.

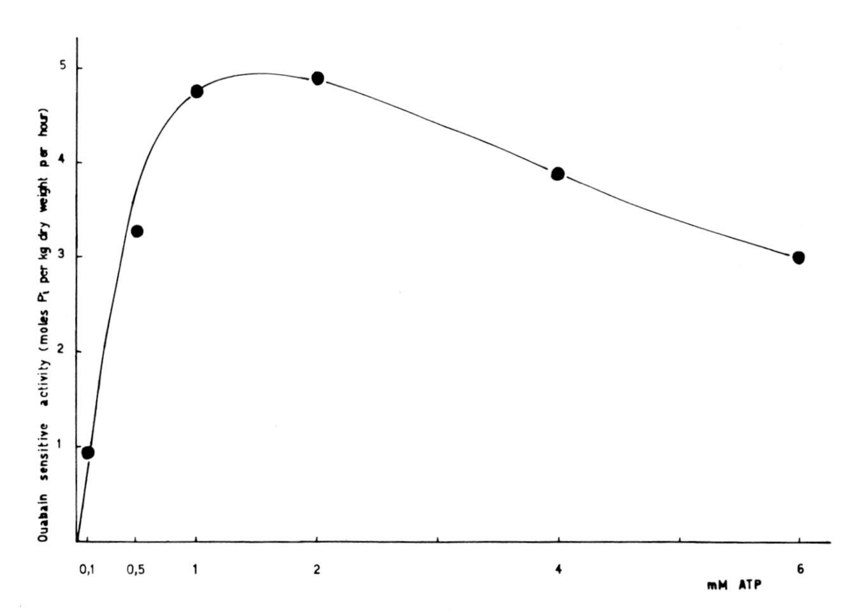

FIG. 2. Effect of increasing ATP concentration on the release of inorganic phosphate by rat renal Na-K-ATPase (means of five studies). Aliquots of 10 μl of the homogenate, containing 500 ng dry wt, were incubated at 37°C with 2 mM $MgCl_2$, 55 mM Na, 5 mM K, 100 mM Tris pH 7.4 for total ATPase and with 2 mM ouabain for ouabain-insensitive Mg ATPase. The reaction was stopped by boiling for 2 min, and P_i was measured with an enzymic reaction (11). The difference in activity with and without ouabain is taken as Na-K-ATPase.

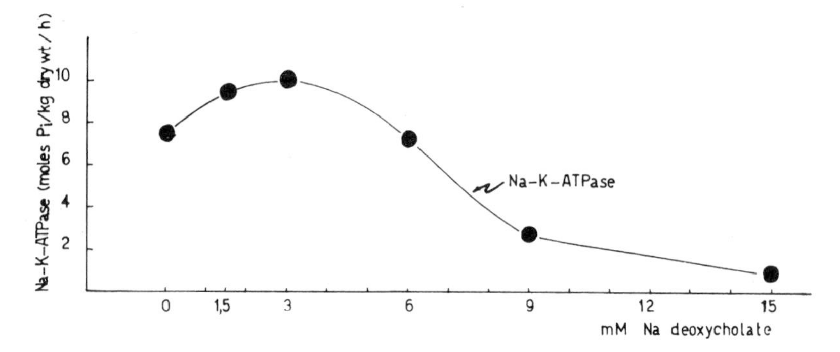

FIG. 3. Effect of an increasing deoxycholate concentration on the release of inorganic phosphate by rat renal Na-K-ATPase at 20°C. Lyophilized homogenate corresponding to 400 μg dry wt was incubated in 1 ml with the indicated concentrations of deoxycholate, 2 mM EDTA and 50 mM imidazole pH 6.9 for 30 min. Five μl aliquots were transferred to test tubes (100 μl) containing the reagents for total ATPase and for ouabain-insensitive Mg ATPase (Experimental conditions in Fig. 2). The substrate concentration was 2 mM ATP (Na-salt).

(Fig. 3); therefore, in the following studies we omitted the detergent. The sodium requirements are shown in Fig. 4. Maximal activity occurred within a range of 40–50 mM sodium (K_m 8 mM). Higher concentrations of sodium became inhibitory. Increasing potassium concentrations from 3 to 5 mM resulted in a progressive increase in Na-K-ATPase activity; above

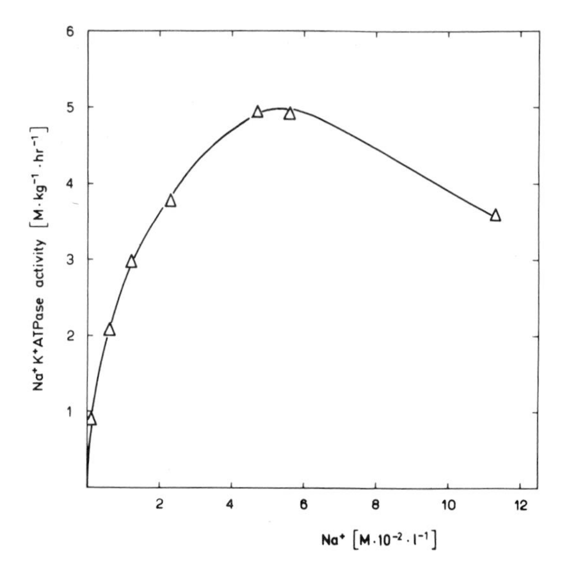

FIG. 4. Effect of increasing sodium concentration on the release of inorganic phosphate by rat renal Na-K-ATPase (means of five studies). An amount of lyophilized homogenate corresponding to 450 ng dry wt was incubated. Experimental conditions as in Fig. 2.

this range a significant decrease in activity was not observed up to a concentration of 31 mM (Fig. 5).

The sodium and potassium concentrations used by investigators who work with purified enzyme preparation are much higher than ours, e.g., 120 mM for sodium and 20–30 mM for potassium (15). For the conditions in the living cell it can be assumed that on the sodium site of the enzyme no more than 30 mM of sodium are active, as to the potassium site no more than 5 mM of potassium. In our assay system we therefore chose cation concentrations as they probably occur in the living cell. Moreover, no difference in the order of magnitude of Na-K-ATPase activity was detected when comparing a Na:K concentration of 50:5 and 120:30, as demonstrated in Fig. 6.

Because rat renal tissue is rather insensitive to ouabain, we used this specific inhibitor for Na-K-ATPase in a 10^{-2} M concentration (Fig. 7). The dry weight and length of the various tubular structures from the rat nephron amounted for the proximal convoluted tubule (PC) to 26 ng/100 μm; for the thick ascending limb of Henle's loop (ALH) to 16 ng/100 μm, and for the distal convoluted tubule (DC) to 11 ng/100 μm. We found no difference in the order of magnitude of Na-K-ATPase activity, using one dissected PC for both ATPases (i.e., one-half for the total ATPase and the other one for the ouabain-insensitive ATPase) or two samples, one for each ATPase.

The rate of inorganic phosphate formation was linear with respect to the concentration of tissue in the range of 5–50 ng of proximal and distal tubule, both in the presence and absence of ouabain, as shown in Fig. 8. In the PC the amount of inorganic phosphate was 1.2 pmoles P_i per 20 ng dry tissue.

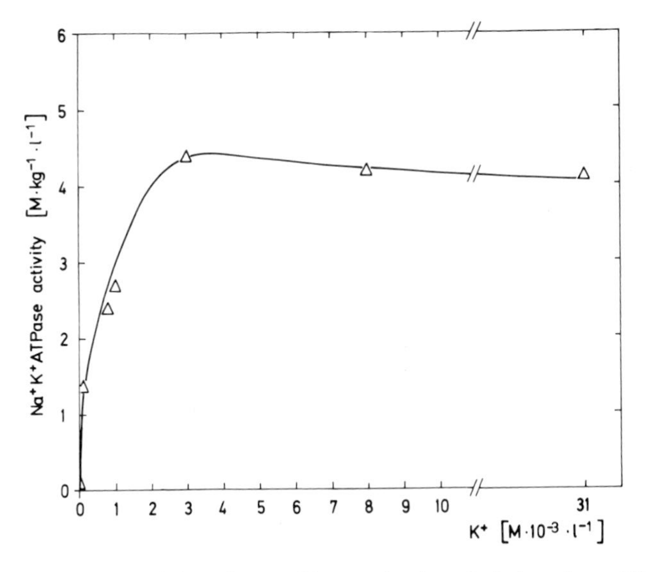

FIG. 5. K concentration and the release of inorganic phosphate by rat renal Na-K-ATPase (means of five experiments). Experimental conditions as in Fig. 2.

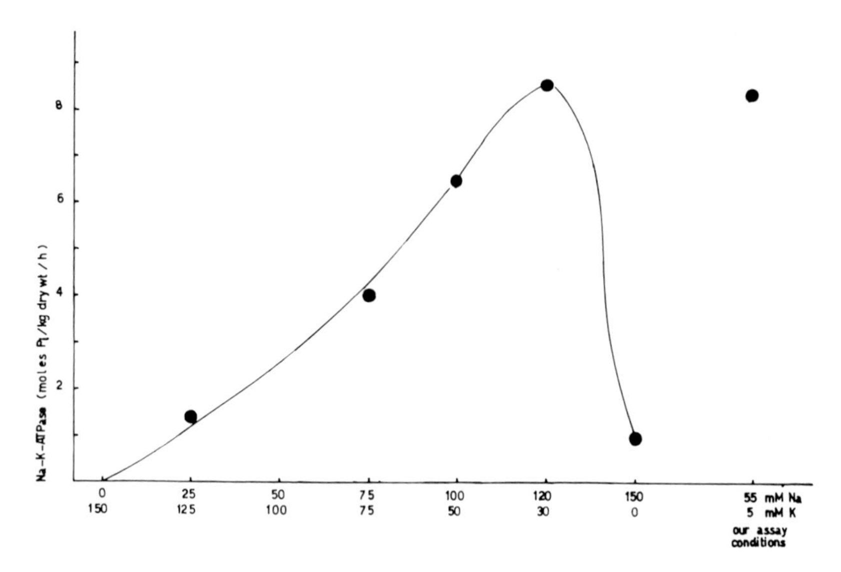

FIG. 6. Activity of Na-K-ATPase as function of sodium and potassium. The enzyme was prepared from the outer medulla of male rats. The ratio of sodium and potassium in the test tubes was varied at a constant ionic strength. ATP 2 mM, Mg 2 mM, Tris 100 mM pH 7.4, incubation time 15 min.

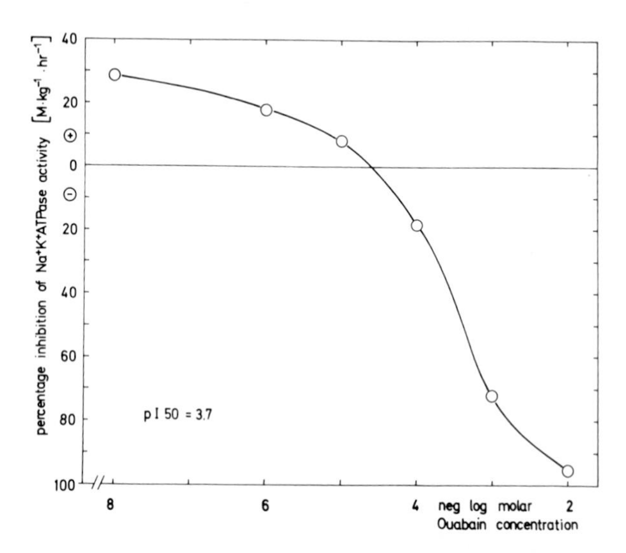

FIG. 7. The effect of ouabain concentration on the release of inorganic phosphate by rat renal cortex. Aliquots containing 500 ng dry wt were incubated at 37°C in 10 μl with concentrations of ouabain as indicated on the abscissa, 2 mM ATP, 55 mM Na, 2 mM $MgCl_2$, and 100 mM Tris pH 7.4. After 15 min the reaction was stopped by boiling and P_i measured enzymatically. The absolute activity was 4.6 moles P_i/kg dry wt/hr.

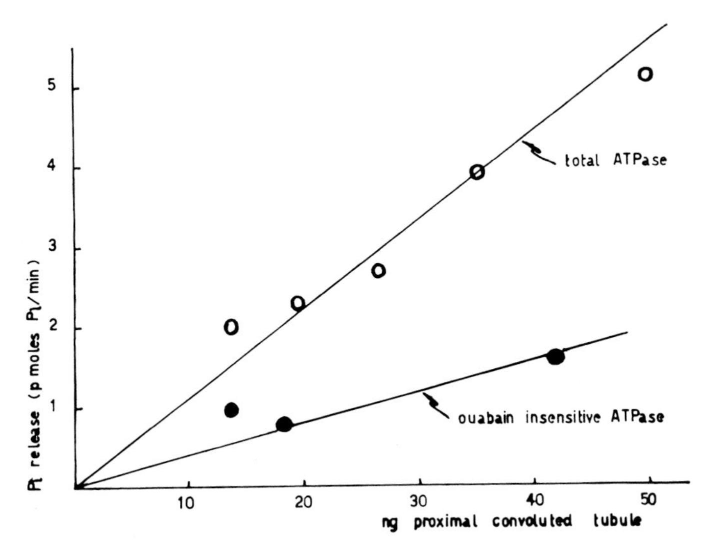

FIG. 8. Proportionality of the rate of ATP hydrolysis to the amount of dry weight of proximal convoluted tubule as indicated on the abscissa. A single tubular portion was incubated in 600 nl containing 2 mM ATP, 55 mM Na, 5 mM K, 2 mM $MgCl_2$, 100 mM Tris pH 7.4. For ouabain-insensitive ATPase K was omitted and 2 mM ouabain added. After an incubation time of 5 min at 37°C the reaction was stopped by boiling, for P_i measurement aliquots of 200 nl were transferred to 1 μl with imidazole 50 mM pH 6.9, Glycogen 0.08%, NADP 300 μM, 5′-AMP 10 μM, G6PDH 1 μg/ml, phosphoglucomutase 3 μg/ml, phosphorylase 25 μg/ml, EDTA 1 mM, Mg 500 μM. After 30 min incubation at 37°C, 400 nl of the reagent were added to 250 μl 0.04 N NaOH to destroy excess NADP. Aliquots of 6 μl were pipetted into 25 μl cycling reagent, containing: Tris 100 mM pH 8.0, oxoglutarate 5 mM, glucose-6-P 1 mM, ADP 300 mM, NH_4-acetate 25 mM, Glu DH 150 μg/ml, G6PDH 35 μg/ml. After 1 hr at 37°C the reaction was stopped by boiling and aliquots of 20 μl were transferred to 200 μl with 20 mM Tris-pH 8.0, NADP 100 μM, 6-phosphogluconate dehydrogenase 100 μg/ml, EDTA 100 μM, Mg 3 mM.

The ATPase activity, however, amounted to 13.4 pmoles P_i/20 ng dry wt/5 min at 37°C. Thus 9% of the liberated P_i were not formed by ATP-hydrolysis. For the distal tubule, including ALH and DC, we found a tissue concentration of inorganic phosphate of 0.25 pmoles P_i/10 ng dry wt and a P_i release of 11 pmoles/10 ng dry wt/5 min; 2% of P_i were therefore not derived from ATP hydrolysis by the enzyme. The activity values in the following results are slightly higher, which is due to tissue P_i concentration.

Identical conditions are found with ouabain-insensitive ATPase. The small amount of tissue phosphate is neglected in the calculation of Na-K-ATPase as the result is obtained from the difference between total ATPase and ouabain-insensitive ATPase. Our method for the Na-K-ATPase assay differs from the Na-K-ATPase determination with purified enzyme preparations so far as the single tubular structures are put directly into the reagent, containing effectors and substrate in concentrations as they occur in the living cell. This means that no steps of disintegration of the tissue are in the procedure. This may be one of the reasons why in our assay conditions the Na-K-ATPase changes the activity acutely when sodium reabsorption is elevated or depressed in some way.

RESULTS AND DISCUSSION

Distribution Along the Nephron of Various Species

In an early experiment we could demonstrate that Na-K-ATPase activity in the *rat* nephron is much more concentrated in the distal than in the proximal tubule [1–2 MKH (=moles P_i/kg dry wt/hr) for the PC, 6–13 MKH for the ALH, 8–10 MKH for the DC, and 4–8 MKH for the CD (16)]. The ratio of proximal to distal tubule as to Na-K-ATPase activity amounts to ~1:8, i.e., the distale tubule reveals an 8 times higher ouabain-sensitive activity than the proximal tubule.

In the meantime, several investigators have established the pattern of Na-K-ATPase distribution throughout the kidney, working with purified enzyme preparations of the various kidney zones from rat, rabbit, dog, and man. According to their results, the outer medulla, where the ALH are predominant, reveals a 2–6 times higher ouabain-sensitive activity than the cortex. The nephron from *human* kidney showed a Na-K-ATPase distribution which corresponds intimately to the pattern, demonstrated with the rat nephron [2 MKH for the PC, 13 MKH for the ALH, 8 MKH for the DC and 2 MKH for the CD (*unpublished results*)].

The Na-K-ATPase distribution along the nephron as shown in rat and man does not seem to be a general pattern for all mammals. In the *rabbit* nephron we found that single portions of the ALH as well as from DC, which are adjacent to the glomerulus, had no Na-K-ATPase, as shown in Fig. 9. The PC, however, revealed a significant ouabain-sensitive activity which corresponds in order of magnitude to the PC of rat and man. This is an intriguing finding because we know from the work of Jørgensen and Skou (18) that the purified membrane fraction from the outer medulla of the rabbit kidney possesses the highest specific Na-K-ATPase activity (934 μmoles P_i/mg protein-hr) ever found in renal tissue. This means that in our assay conditions all Na-K-ATPase sites in the single segments of the distal part of the rabbit nephron must be latent.

In membrane preparations from mammalian tissues only a part of the Na-K-ATPase activity is manifest, a substantial increase in activity may be observed during incubation with detergents like deoxycholate (14). Deoxycholate acts by removing extraneous protein and lipid from the membrane; however, it is not attached to the membrane. We treated single portions of the distal part of the rabbit nephron with deoxycholate (5 ng DOC/400 nl and 15 ng protein) before assaying Na-K-ATPase and found an increase of the ouabain-sensitive activity from 0 to 4–9 MKH.

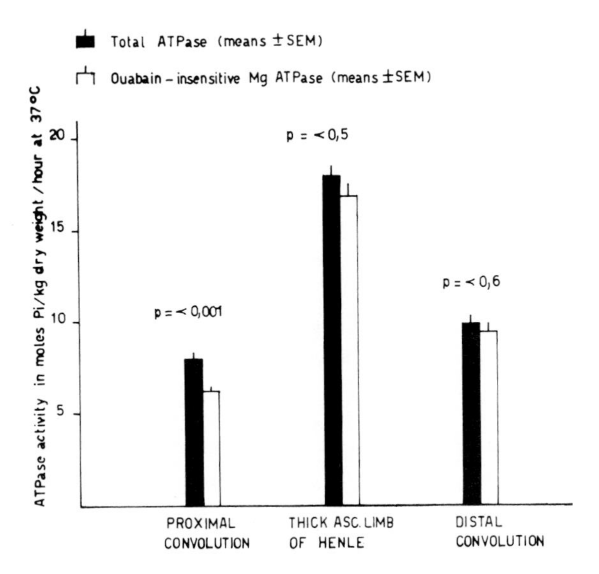

FIG. 9. Distribution of total ATPase and ouabain-insensitive Mg ATPase along the rabbit nephron. Experimental conditions as described in Fig. 8.

We do not know if the latency of the Na-K-ATPase in the distal part of the rabbit nephron has any functional significance or not. Burg et al. (19) have postulated that the ALH of the rabbit nephron has no active sodium transport, but possesses an active chloride pump. They found, however, an active sodium reabsorption in the DC, where in our assay conditions Na-K-ATPase was also zero. This discrepancy may be explained by the inhomogeneity of the distal tubule of rabbit kidney; the DC adjacent to the glomerulus, which we dissected for Na-K-ATPase measurements, belongs probably to the ALH rather than to the DC.

Functional Significance of the Na-K-ATPase Distribution Along the Nephron

The fact that the Na-K-ATPase activity in the PC is relatively low compared to the ALH or DC was interpreted to the effect that in the PC there are two kinds of sodium transport mechanisms, one dependent and the other independent on Na-K-ATPase (20,22). It was suggested that the Na-K-ATPase-dependent or ouabain-sensitive pathway predominates in the ALH and DC, while an ouabain-insensitive pathway is prominent in the proximal tubule (17,20).

Yet we have no satisfying explanation for the difference between proximal and distal tubule as to their Na-K-ATPase activities. It seems very likely that sodium is transported differently in the proximal and distal tubule.

Micropuncture studies have shown that in the PC from rat kidney only 30% of the sodium net transport are mediated actively and 70% passively (21). This active component of net sodium flux was calculated to be 1.3 pmoles Na mm^{-1} sec^{-1} (21). The Na-K-ATPase activity in the PC of rat kidney is 0.13–0.6 pmoles split ATP mm^{-1} sec^{-1}. Thus at least 50% of the active sodium transport may be maintained by the ouabain-sensitive activity, depending on which ratio of moles transported sodium to moles ATP is used. From experiments with tissue slices and with perfused kidney the existence of two sodium pumps was postulated, one ouabain-sensitive and the other ouabain-insensitive (22,23). It is interesting that only 50% of the sodium net reabsorption in the perfused rat kidney were found to be ouabain-sensitive (23).

In connection with this discussion a further experiment with dogs should be mentioned: Digoxin, a specific inhibitor of Na-K-ATPase, infused into one renal artery in doses ranging from 0.4 to 4 μg/kg-min resulted in a significant natriuresis which was accompanied by an inhibition of Na-K-ATPase in cortex and medulla of the infused kidney. In spite of an enzyme inhibition of 90%, at least 80% of the filtered sodium was reabsorbed. Only the changes in medullary Na-K-ATPase activity indicated a direct relationship to alterations in fractional solute-free water reabsorption. The inhibition of cortical enzyme activity alone was not associated with naturiuresis (46).

There is a further important point which has been taken into consideration when discussing the functional significance of the high Na-K-ATPase concentration in the distal tubule, namely, the fact that the distal tubule is the main site for the regulation of potassium excretion (44). It was suggested that one role of Na-K-ATPase may be the mediation of the potassium excretion into urine (19). It could be shown indeed that Na-K-ATPase increases in an adaptive way when the requirements of potassium excretion are augmented, i.e., in a homogenate from rat kidney, Na-K-ATPase increased when the dietary intake of potassium was raised, which was detected first in the outer medulla where it was most prominent (45). However, the suggestion of a direct relationship between Na-K-ATPase and potassium excretion in the distal tubule is not convincing as long as there is no evidence for an acute change in ouabain-sensitive activity when potassium excretion is increased or decreased.

The Site of Na-K-ATPase Within Renal Tubular Cell

As shown in Table 1, there is a strong parallelism between Na-K-ATPase activity and amount of basal infoldings per cell surface in the two segments from the proximal part of the rat nephron (20). In the distal convolution the ratio of Na-K-ATPase to the amount of basal infoldings is 3 times higher, which indicates that there are probably more Na-K-ATPase sites per unit membrane area in the cell of the distal tubule than in the cell of the proximal

TABLE 1. *Na-K-ATPase activity vs. the amount of basal infoldings per cell* (M) *in the various structures from the rat nephron*

	Tubule		
	Proximal convoluted	Proximal straight	Distal convoluted
Na-K-ATPase	1.6[a]	0.3[a]	13
M	10–40[b]	2–10	86–95
ATPase/M	0.05	0.05	0.14

[a] Moles split ATP/kg dry wt/hr at 37°C.
[b] Percent of total cell surface (from ref. 20).

tubule. The assumption of a high density of Na-K-ATPase or Na pump sites per unit membrane area in the distal tubule cell was strongly supported by the extensive work of Jørgensen (47). The parallelism between ouabain-sensitive activity and amount of basal infoldings per cell along the proximal tubule is a strong evidence of the assumption that the Na-K-ATPase is localized in the basal infoldings. We could prove this by dividing the proximal tubular cell in a basal and luminal area (9). It was found that only the basal area contained a Na-K-ATPase activity, whereas the brush border fragments revealed no ouabain-sensitive activity.

The polarity of the proximal tubule cell with regard to Na-K-ATPase was also established by separating brush border membrane from basal infolding membrane with preparative free-flow electrophoresis (24). The finding that the Na-K-ATPase is restricted to the basal infolding membrane in the proximal tubule cell is consistent with micropuncture studies which resulted in the concept of a ouabain-sensitive sodium pump located in the cell base and a bicarbonate pump situated in the brush border (21).

Velocity of Changes in Ouabain-Sensitive Activity of Single Nephron Structures

One way of looking at the role of Na-K-ATPase in the reabsorptive process is to examine the velocity of change in activity when tubular sodium transport is increased or decreased. A lot of work has been done to elucidate the kind of change in renal Na-K-ATPase activity which is due to altered sodium transport. Models as uninephrectomy, adrenalectomy, and hormone application were the most-used techniques. The general result of the experiments on purified membranes was that Na-K-ATPase changes in an adaptive way when the transport of sodium across the kidney cell membrane is altered. Assaying Na-K-ATPase activity in single structures of the nephron we received quite different results from those obtained with purified enzyme preparation, which indicates that Na-K-ATPase changes the activity at the same time when tubular net sodium transport is elevated or depressed. The following results may support the assumption that Na-K-ATPase is capable of changing the activity acutely. As early as 12 hr after uninephrectomy we could demonstrate in the DC of the remaining kidney the first significant increase in Na-K-ATPase activity. In the PC, however, the activity on the third day signified an increase which coincided with the elevation in tubular net sodium reabsorption after uninephrectomy (48).

An acute fall of Na-K-ATPase activity in PC, ALH, and DC was found in the folate-treated rat when the intratubular pressure in the PC was strongly elevated from 10 to 45 mm Hg (Fig. 10) (41). A single dose of folate (250 mg/kg body wt) impairs markedly the kidney function and results in oliguria of variable duration. Early histologic features are dilated tubular portions and casts which are visible mainly in the collecting ducts and more or less in the ALH. It is interesting that the Na-K-ATPase is zero in the PC when the reabsorptive capacity ($J_{(v)}$)—determined by split droplet experiments—begins to decrease (40). This might indicate that Na-K-ATPase activity is altered *before* tubular net sodium reabsorption is changed and not reversely, as assumed by most of the investigators working on this topic.

In the past few years acute changes of Na-K-ATPase activity were also demonstrated with microsomal membranes from the outer medulla and plasma membranes from perfused kidney. During *in vivo* volume expansion with isotonic saline in the rat, the renal medullary Na-K-ATPase specific

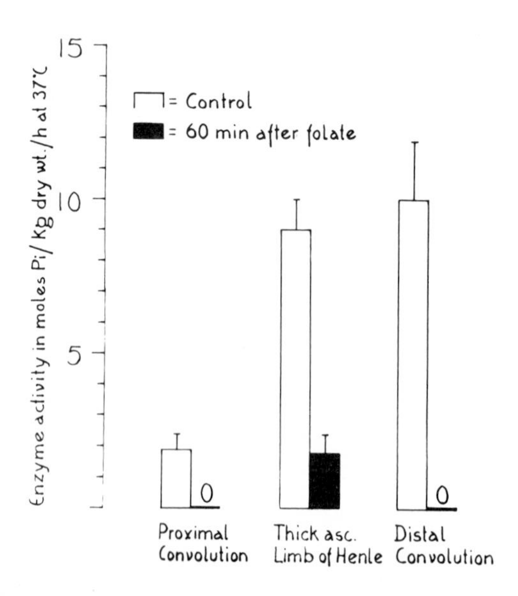

FIG. 10. Na-K-ATPase activity in various structures of the rat nephron 1 hr after injection of folate (41). Experimental conditions as in Fig. 8.

activity increased within 3 hr (42). In a plasma membrane fraction from rat kidney after 1 hr perfusion, Na-K-ATPase activity was significantly higher in the low Na load than in the isotonic Na load experiment (43).

Evidence That Na-K-ATPase Is Regulated by Aldosterone

It has been shown that adrenalectomized rats became natriuretic 5 hr after the maneuver (26). Within 6 hr after adrenal enucleation we found a decrease of Na-K-ATPase activity of 80% below sham-operated controls in the ALH and DC (27). No further activity decrease in the distal tubule could be observed in the following days. This finding is in contrast to the results with purified enzyme preparation, which revealed a significantly slower gradual decline of ouabain-sensitive activity, i.e., only 3 days after adrenalectomy Na-K-ATPase activity was more than 40% below the activity from sham-operated rat kidney (28). The latter result led to the assumption that Na-K-ATPase is not primarily under the control of the adrenals. We questioned this conclusion because we had found that a very low dose of aldosterone (5 μg/100 g rat) was sufficient to normalize Na-K-ATPase in the structures of the proximal and distal nephron parts from adrenalectomized rats within 1 hr after intraperitoneal application (11). This is demonstrated in Fig. 11. Protein synthesis is involved in the aldosterone-induced Na-K-ATPase restoration in the nephron of adrenalectomized rats, as cycloheximide and/or actinomycin D, given at least 1 hr before mineralocorticoid treatment, could inhibit completely this process of steroid-dependent enzyme activation, as shown in Fig. 11. It is well known that the action of aldosterone on sodium transport is mediated by the stimulation of DNA-dependent synthesis of RNA and the subsequent *de novo* synthesis of a specific protein, i.e., the aldosterone-induced protein (AIP). We assume that AIP acts like a detergent, unmasking catalytic sites of Na-K-ATPase which are already present in the tubule cell of the adrenalectomized animal, but are almost completely latent (11). Thus, the synthesis of new enzyme molecules is probably not a prerequisite of the acute aldosterone-induced Na-K-ATPase activation. Very recently, experiments with erythrocytes have shown that proteins immediately enhance the activity of Na-K-ATPase without a *de novo* synthesis of the enzyme (30,31).

Aldosterone has obviously a specific effect on Na-K-ATPase in the proximal and distal tubule of the adrenalectomized rat because spirolactone, a competitive inhibitor of the sodium retaining action of mineralocorticoids, which in a concentration of 10^4:1 displaces aldosterone from its cytoplasmic binding sites, inhibited completely the hormonal-induced activation of the ouabain-sensitive activity in both parts of the nephron (Fig. 12). The results of Fig. 12 agree completely with physiologic studies which showed that the antimineralocorticoid spirolactone reduced the renal cytoplasmic bind-

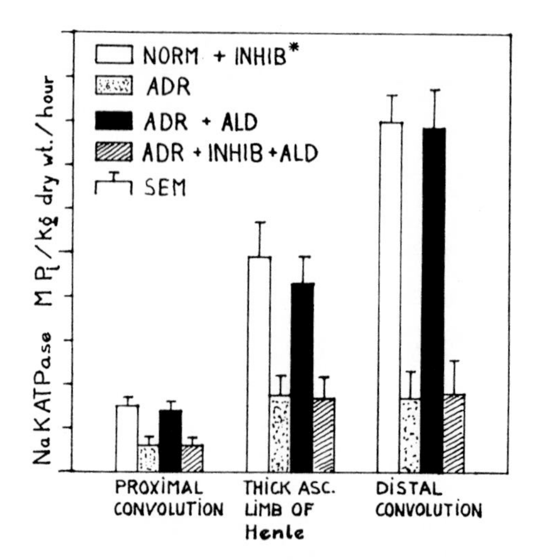

FIG. 11. Acute normalization of Na-K-ATPase in microdissected tubular structures from adrenalectomized rats after a single dose of aldosterone (5 μg aldosterone/100 g rat). The kidneys were removed 1 hr after steroid injection. The first column of each structure represents normal rat with the inhibitors of protein synthesis administered 2 hr before death; the second column represents rats adrenalectomized 10 days before substitution and kept on 0.9% saline. In the third column adrenalectomized aldosterone-treated rats are plotted down, and in the fourth adrenalectomized rats with inhibitor of protein synthesis applicated 2 hr and with aldosterone applicated 1 hr before death. Cycloheximide (4 mg per rat) and actinomycin D (10 μg per rat) were used as inhibitors of protein synthesis. The fourth group of animals was divided into two groups; one group received cycloheximide, the other actinomycin D, before aldosterone was injected.

ing of aldosterone by 70% and produced a maximal inhibition of the steroid-mediated antinatriuresis (33). Further microdissection studies revealed that dexamethasone, administered to adrenalectomized rats 1 hr *before* aldosterone, suppressed also completely the mineralocorticoid-induced Na-K-ATPase activation in the ALH and DC. Dexamethasone has no salt-retaining component.

Micropuncture studies have largely established that the sites of aldosterone action in the nephron are the collecting duct and the DC, and it seems very likely that this mineralocorticoid has an antinatriuretic effect on the PC as well (34). In these physiologic studies a lag period of 30–60 min was described, which corresponds to our experiments. The action of aldosterone on the ALH is obscure. One group of investigators described an effect of aldosterone on free water clearance and the attainment of maximum urinary osmolality (rat) (35). They concluded, however, very early that aldosterone may act on the distal segment rather than on the ALH, whereas cortisol increases the reabsorption of sodium ions in the ALH. Another group was not able to demonstrate any effect of aldosterone on renal concentrating and diluting capacity (dog), hence, they assumed that the mineralocorticoid

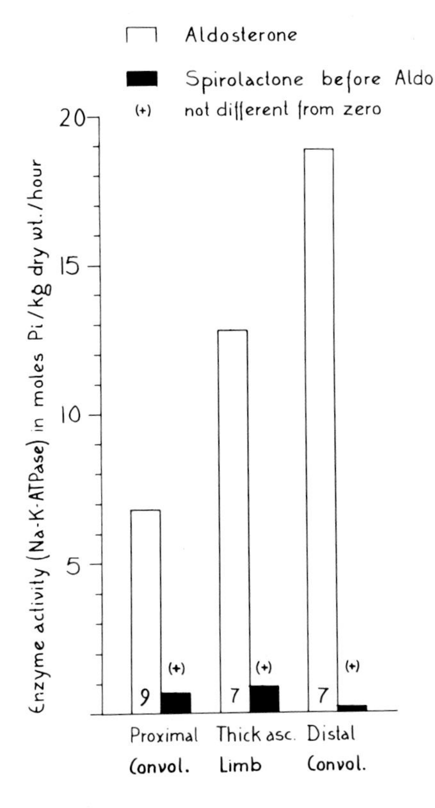

FIG. 12. Inhibition of aldosterone-induced Na-K-ATPase activation in the various structures of the nephron from adrenalectomized rats. Rats were adrenalectomized 10 days before the maneuver and kept on 1% saline. One group of the animals received 50 mg/100 g rat spirolactone 1 hr before aldosterone injection. A second group received aldosterone as described in the legend of Fig. 10. The open bars represent Na-K-ATPase activity after aldosterone substitution, the dark bars the ouabain-sensitive activity after spirolactone before aldosterone administration. Na-K-ATPase was measured with the oil-well technique combined with enzymic cycling reaction, as described in Fig. 8.

has no influence on sodium transport in the ALH (36). It is possible that there exist species differences with regard to the nephron site of hormonal action. As shown in Fig. 11, the activation of Na-K-ATPase in the ALH of adrenalectomized rat by aldosterone is evident.

It is known that sodium deprivation causes a rise in endogenous aldosterone excretion (37). Microdissection studies in salt-depleted rats demonstrated a significant increase of Na-K-ATPase activity in the PC and DC, but not in the ALH, as shown in Fig. 13. We have no valid explanation for the difference between exogenous and endogenous aldosterone action on the ouabain-sensitive activity in the ALH. The results of Fig. 13 are in contrast to earlier studies of the Epstein group (38) who were not able to demonstrate any effect of sodium depletion on medullary or cortical Na-K-ATPase from rat kidney.

Very recently Knox and Sen (39) reported an aldosterone-mediated increase in renal Na-K-ATPase within 3 hr. They performed this study with a purified enzyme preparation from the whole kidney. This may be the reason for the differences in ouabain-sensitive activity between adrenalectomy and hormonal substitution which, however, are relatively small and in our opinion hardly significant ($p < 0.05$).

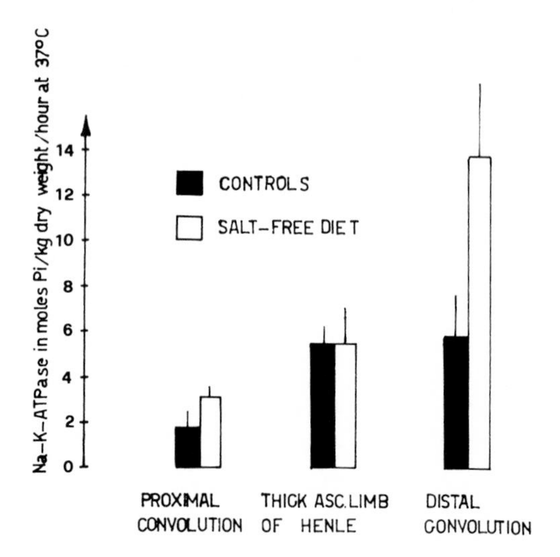

FIG. 13. Increase of Na-K-ATPase in the various structures of the nephron from rats after chronic sodium depletion. Experimental conditions as in Fig. 8.

It may be a very provocative implication when we assume that Na-K-ATPase is a target of aldosterone or of the aldosterone-induced protein (AIP). The rapid restoration of the ouabain-sensitive activity in the various tubular structures from adrenalectomized rats after a very low dose of aldosterone and the inhibition of this hormonal-induced enzyme activation by inhibitors of protein synthesis and by the antimineralocorticoid spirolactone and finally by glucocorticoid in a very high dose furnish striking evidence of the conception that aldosterone acts on Na-K-ATPase in renal tubular cell.

CONCLUSION

With our technique we have tried to answer the questions: does the load of salt to be reabsorbed further the development of more Na-K-ATPase, or are there other regulatory mechanisms which are responsible for a new formation of the enzyme and subsequently for the reabsorption of an increased amount of sodium? Investigators who are working with purified enzyme preparations from kidney tissue have answered these questions to the effect that the enzyme is changed only chronically when a change of sodium transport in the kidney tubule has occurred.

In our studies—using nondisintegrated tubular segments and a reagent with substrate and effector concentrations as in the living cell—we could show that Na-K-ATPase changes in parallel with the tubular sodium transport. A gradual decrease of tubular sodium reabsorption resulted in a loss of ouabain-sensitive activity.

These findings might suggest that an unknown substance—perhaps a protein—regulate the amount of catalytic sites of Na-K-ATPase, which are always present in a masked and/or unmasked form, and that this process is followed by higher or lower sodium reabsorption. Hence, the synthesis of new enzyme molecules is perhaps not a necessary prerequisite for an increase or decrease in ouabain-sensitive activity.

ACKNOWLEDGMENTS

This work was supported by the SNF (project 3.2220.73) and DFG (projects 217/5 and 217/6).

REFERENCES

1. Győry, A. Z., and Kinne, R. (1971): Energy source for transepithelial sodium transport in rat renal proximal tubules. *Pflügers Arch.,* 327:234–260.
2. Askari, A. (editor) (1974): Properties and functions of Na-K-activated adenosinetriphosphatase. *Ann. NY Acad. Sci.,* 242:1–741.
3. Jørgensen, P. L., Hansen, O., Glynn, I. M., and Caviers, J. D. (1973): Antibodies to pig kidney Na-K-ATPase inhibit the Na pump in human red cells provided they have access to the inner surface of the cell membrane. *Biochim. Biophys. Acta,* 291:795–800.
4. Jain, M. K., White, F. P., Strickholm, A., Williams, E., and Cordes, E. H. (1972): Studies concerning the possible reconstitution of an active cation pump across an artificial membrane. *J. Membr. Biol.,* 8:363–388.
5. Shamoo, A. E. (1974): Isolation of a sodium-dependent ionophore from Na-K-ATPase preparations. *Ann. NY Acad. Sci.,* 242:389–405.
6. Hokin, L. E. (1974): Purification and properties of the sodium + potassium-activated adenosinetriphosphatase and reconstitution of sodium transport. *Ann. NY Acad. Sci.,* 242:12–23.
7. Lindenmayer, G. E., Lane, L. K., and Schwartz, A. (1974): Effects of sodium and potassium on the partial reactions of a highly purified Na-K-ATPase: Modulation of the rate of ouabain interaction. *Ann. NY Acad. Sci.,* 242:235–243.
8. Bonting, S. L., Simon, K. A., and Hawkins, N. M. (1961): Studies on Na-K activated adenosinetriphosphatase. I. Quantitative distribution in several tissues of the cat. *Arch. Biochem. Biophys.,* 95:416–422.
9. Schmidt, U., and Dubach, U. C. (1971): Na-K-stimulated adenosine triphosphatase: intracellular localization within the proximal tubules of the rat nephron. *Pflügers Arch.,* 330:265–270.
10. Schmidt, U., Schmid, H., Funk, B., and Dubach, U. C. (1974): The function of Na-K-ATPase in single portions of the rat nephron. *Ann. NY Acad. Sci.,* 242:489–501.
11. Schmidt, U., Schmid, J., Schmid, H., Dubach, U. C. (1975): Sodium- and potassium-activated ATPase. A possible target of aldosterone. *J. Clin. Invest.* 55:655–660.
12. Lowry, O. H., and Passonneau, J. V. (1972): *A Flexible System of Enzymatic Analysis.* Academic Press, New York, London.
13. Epstein, F. H. (1973): Role of sodium-potassium-ATPase in sodium reabsorption by the kidney. In: *Modern Diuretic Therapy in the Treatment of Cardiovascular and Renal Disease,* pp. 188–195. Excerpta Medica, Amsterdam.
14. Jørgensen, P. L., and Skou, J. C. (1971): Purification and characterization of Na-K-ATPase. I. The influence of detergents on the activity of Na-K-ATPase in preparations from the outer medulla of rabbit kidney. *Biochim. Biophys. Acta,* 233:366–380.
15. Skou, J. C. (1975): The Na-K-activated enzyme system and its relationship to transport of sodium and potassium. *Quart. Rev. Biophys.,* 7:401–434.
16. Schmidt, U., and Dubach, U. C. (1969). Activity of Na-K-stimulated adenosine-triphosphatase in the rat nephron. *Pflügers Arch.,* 306:219–226.
17. Hendler, E. D., Torretti, J., and Epstein, F. H. (1971): The distribution of sodium-po-

tassium-activated adenosine triphosphatase in medulla and cortex of the kidney. *J. Clin. Invest.,* 50:1329–1337.
18. Jørgensen, P. L., and Skou, J. C. (1969): Preparation of highly active Na-K-ATPase from the outer medulla of rabbit kidney. *Biochem. Biophys. Res. Commun.,* 37:39–46.
19. Burg, M. B., and Green, N. (1973): Function of the thick ascending limb of the Henle's loop. *Am. J. Physiol.,* 224:659–668.
20. Schmidt, U., and Dubach, U. C. (1971): Na-K-ATPase in the rat nephron related to sodium transport: Results with quantitative histochemistry. In: *Recent Advances in Quantitative Histo- and Cytochemistry,* pp. 320–344, edited by U. C. Dubach and U. Schmidt. Hans Huber Verlag, Bern-Stuttgart-Wien.
21. Frömter, E., Rumrich, G., and Ullrich, K. J. (1973): Phenomenological description of Na, Cl and HCO_3-absorption from proximal tubules of the rat kidney. *Pflugers Arch.,* 343:189–220.
22. Whittenbury, G., and Proverbio, F. (1970): Two modes of Na extrusion in cells from guinea-pig kidney cortex slices. *Pflügers Arch.,* 316:1–25.
23. Ross, B., Leaf, A., Silva, P., and Epstein, F. H. (1974). Na-K-ATPase in sodium transport by the perfused rat kidney. *Am. J. Physiol.,* 226:624–629.
24. Heidrich, H. G., Kinne, R., Kinne-Saffran, E., and Hannig, K. (1972): The polarity of the proximal tubule cell in rat kidney. *J. Cell Biol.,* 54:232–245.
25. Epstein, F. H., and Silva, P. (1974): Role of sodium, potassium-ATPase in renal function. *Ann. NY Acad. Sci.,* 242:519–524.
26. Williamson, H. E. (1963): Mechanism of the antidiuretic action of aldosterone. *Biochem. Pharmacol.,* 12:1449–1501.
27. Schmidt, U., and Dubach, U. C. (1971): Sensitivity of Na-K-ATPase activity in various structures of the rat nephron: Studies with adrenalectomy. *Eur. J. Clin. Invest.,* 1:307–312.
28. Jørgensen, P. L. (1968): Regulation of the Na-K-activated ATP hydrolyzing enzyme system in rat kidney. I. The effect of adrenalectomy and the supply of sodium on the enzyme system. *Biochim. Biophys. Acta,* 151:212–224.
29. Jørgensen, P. L. (1972): The role of aldosterone in the regulation of Na-K-ATPase in rat kidney. *J. Steroid Biochem.,* 3:181–191.
30. Lauf, P. K. (1974): Erythrocyte surface antigens and cation transport. *Ann. NY Acad. Sci.,* 242:324–342.
31. Sachs, J. R. (1974): The kinetics of the Na-K pump in goat red blood cells and the effect of an antibody. *Ann. NY Acad. Sci.,* 242:343–354.
32. Rossier, B. C., Wilce, P. A., and Edelman, I. S. (1974): Kinetics of RNA labeling in toad bladder epithelium: Effects of aldosterone and related steroids. *Proc. Natl. Acad. Sci. USA,* 71:3101–3105.
33. Marver, D., Stewart, J., Funder, J. W., Feldman, D., and Edelman, I. S. (1974): Renal aldosterone receptors: Studies with (^{3}H) aldosterone and the anti-mineralocorticoid (^{3}H) spirolactone (SC 26304). *Proc. Natl. Acad. Sci. USA,* 71:1431–1435.
34. Edelman, I. S., and Fanestil, D. D. (1970): Mineralocorticoids. In: *Biochemical Actions of Hormones,* Vol. 1, pp. 321–364, edited by G. Litwack. Academic Press, New York.
35. Sakai, F., and Murayama, Y. (1971): Effects of aldosterone and cortisol on Henle's loop in the adrenalectomized rat's kidney. *Japan. J. Pharmacol.,* 21:23–31.
36. Rogers, P. W., Flynn, J. J., and Kurtzman, N. A. (1975): The effect of mineralocorticoid deficiency on renal concentrating and diluting capacity. *Proc. Soc. Exp. Biol. Med.,* 148:847–853.
37. Marieb, N. J., and Mulrow, P. J. (1965): Role of the renin angiotensin system in the regulation of aldosterone secretion in the rat. *Endocrinology,* 76:657–664.
38. Hendler, E. D., Torretti, J., Kupor, L., and Epstein, F. H. (1972): Effects of adrenalectomy and hormone replacement on Na-K-ATPase in renal tissue. *Am. J. Physiol.,* 222:754–760.
39. Knox, W. H., and Sen, A. K. (1974): Mechanism of action of aldosterone with particular reference to (Na-K)ATPase. *Ann. NY Acad. Sci.,* 242:471–487.
40. Huguenin, M. E., Birbaumer, A., Thiel, G., Brunner, F. P., Torhorst, J., Schmidt, U., and Dubach, U. C. (1975): Tubular obstruction as the primary cause of acute renal failure due to the folic acid in the rat. A micropuncture study. (*submitted for publication.*)
41. Schmidt, U., and Dubach, U. C. (1975): Acute renal failure in the folate-treated rat:

Early metabolic changes in various structures of the nephron. (*submitted for publication.*)
42. Katz, A. I., and Genant, H. K. (1971): Effect of extracellular volume expansion on renal cortical and medullary Na-K-ATPase. *Pflügers Arch.,* 330:136–148.
43. Franke, H., Mályusz, M., and Weiss, Ch. (1975): Acute changes of the Na-K-ATPase-activity in plasmamembranes of the isolated cell free perfused rat kidney. *Pflügers Arch.,* 353:97–106.
44. Giebisch, G., Boulpaep, E. L., and Whittembury, G. (1971): Electrolyte transport in kidney tubule cells. *Phil. Trans. R. Soc. Lond. Ser. B. Biol. Sci.,* 262:175–195.
45. Silva, P., Hayslett, J. P., and Epstein, F. H. (1973): The role of Na-K-activated adenosine triphosphatase in potassium adaptation. Stimulation of enzymatic activity by potassium loading. *J. Clin. Invest.,* 52:2665–2671.
46. Martinez-Maldonado, M., Allen, J. C., Inagaki, C., Tsaparas, N., and Schwartz, A. (1972): Renal sodium-potassium-activated adenosine triphosphatase and sodium reabsorption. *J. Clin. Invest.,* 51:2544–2551.
47. Jørgensen, P. L. (1975): Isolation and characterization of the components of the sodium pump. *Quart. Rev. Biophys.,* 7:239–274.
48. Schmidt, U., and Dubach, U. C. (1974): Induction of Na-K-ATPase in the proximal and distal convolution of the rat nephron after uninephrectomy. *Pflügers Arch.,* 346:39–48.

Membranes and Disease, edited by L. Bolis, J. F. Hoffman, and A. Leaf. Raven Press, New York, © 1976.

Resolution of the Epithelial Cell Envelope into Luminal and Contraluminal Plasma Membranes as a Tool for the Analysis of Transport Processes and Hormone Action

Rolf Kinne and Irving L. Schwartz

Max Planck Institut für Biophysik, Frankfurt/Main, Germany; and Department of Physiology and Biophysics, Mount Sinai Medical and Graduate Schools of the City University of New York, New York 10029

Studies of isolated whole plasma membrane envelopes of epithelial cells have played a significant role in the provision of general information on molecular correlates of membrane transport processes and hormone action. However, these studies do not reveal the cellular locus or the sequence of the individual reactions in the chain of events that comprises (i) a transepithelial transport process, or (ii) the action of a peptide or amine hormone (such as vasopressin or epinephrine) on an epithelial cell. Therefore, we have sought a higher level of cellular resolution by attempting to separate the luminal plasma membrane and the contraluminal plasma membrane of the transporting and/or hormone-sensitive cell. This separation has been achieved by the application of free-flow electrophoresis to whole plasma membranes derived from renal proximal tubular (1,2) and collecting duct (3) epithelial cells, as well as from epithelial cells derived from the small intestine (4); this approach is currently being extended to other epithelial cells in mammals, amphibians, reptiles, etc.

At the present time the largest experience with isolated luminal and contraluminal plasma membrane preparations has been obtained with mammalian renal epithelial cells. This chapter, therefore, will be concerned primarily with the use of these isolated membrane preparations for the study of transport systems in the proximal tubule and hormone-responsive systems in the collecting duct.

SEPARATION OF LUMINAL AND CONTRALUMINAL PLASMA MEMBRANES BY FREE FLOW ELECTROPHORESIS

The basic features of continuous preparative free-flow electrophoresis for dissociating and isolating luminal and contraluminal plasma membranes are illustrated in Fig. 1. The separation takes place in a vertical chamber formed by 2 glass plates between which a curtain of buffer flows downward to be divided into fractions indicated at the bottom of the chamber. An elec-

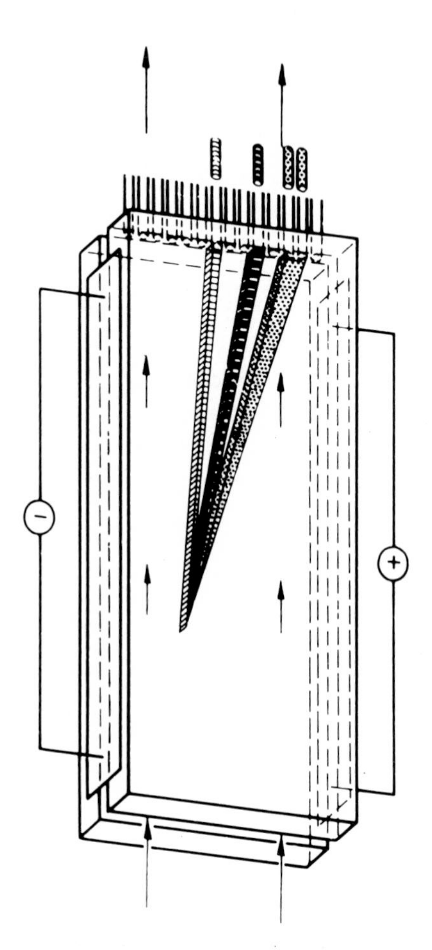

FIG. 1. Sketch of a separation of cells or cell fractions (e.g., luminal and contraluminal plasma membranes) using continuous preparative free-flow electrophoresis. See text.

tric field is applied perpendicularly to the direction of buffer flow as indicated in Fig. 1 by the anodal plus sign at the left and the cathodal minus sign at the right. The plasma membrane homogenate (or any other mixture of biological materials that are to be resolved into components) is injected into a port situated on the right in the upper third of the chamber and the components of the original homogenate are separated in the electric field on the basis of difference in surface charge density, mass, and shape (5,6).

The separation of luminal and contraluminal plasma membranes derived from renal epithelial cells can be evaluated by the use of the following marker enzymes: alkaline phosphatase, trehalase, and maltase (luminal), and Na-K-ATPase and parathyroid hormone-stimulated adenylate cyclase (7) (contraluminal) in the case of the proximal tubule (1); HCO_3-ATPase (8) (luminal) and Ca-ATPase (9) and antidiuretic hormone-stimulated adenylate cyclase (contraluminal) in the case of the collecting duct (3). For example, Fig. 2 illustrates the morphological appearance of the membrane

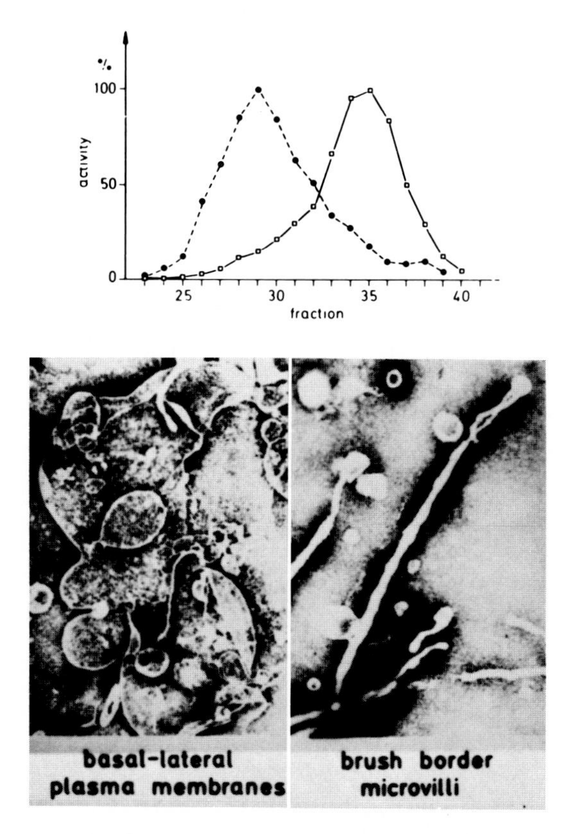

FIG. 2. The upper panel indicates the distribution of Na-K-ATPase (—●—) and alkaline phosphatase (—○—) after separation of luminal (brush border) and contraluminal (basal-lateral) plasma membranes of the rat renal proximal tubule; the values are given as percent of the maximal enzyme activity found in the fractions. The lower panel shows the ultrastructure of the Na-K-ATPase-containing membranes in contrast to the ultrastructure of the alkaline phosphatase-containing membranes.

preparations and the distribution of Na-K-ATPase and alkaline phosphatase following electrophoretic separation of a purified whole membrane homogenate derived from the rat renal proximal tubular epithelial cell. Both luminal and contraluminal membrane fragments migrate to the anode. However, the contraluminal, Na-K-ATPase-containing plasma membranes exhibit greater electrophoretic mobility than the luminal, alkaline phosphatase-containing microvilli. It is noteworthy that these relative electrophoretic mobilities appear to be paradoxical in view of histochemical and other data that indicate that the luminal (brush border) plasma membrane is coated with a glycocalyx, which has a greater net negative charge than the surface of the contraluminal (basal-lateral) plasma membrane. However, it must be kept in mind that geometrical factors (size, shape) may override the factor of surface charge density in dictating the ultimate mobility in the electric field (6). Interestingly, in the case of the collecting duct epithelial cell where there is much less difference in the geometry of luminal and contraluminal plasma membrane fragments, the luminal membranes migrate more rapidly to the anode than the contraluminal membranes, as would be expected when surface charge density is the predominant factor influencing electrophoretic mobility.

STUDIES OF TRANSPORT SYSTEMS IN THE PROXIMAL TUBULE

Although valuable information on the partial reactions of a transport system can be obtained by studying the binding to membranes of transported substrates or specific transport inhibitors (10) and by localization within the cell of "transport enzymes" such as Na-K-ATPase, Ca-ATPase, and HCO_3-ATPase, a more comprehensive, but still microcosmic, analysis of the transport system can be obtained by the study of the vesicles derived from separated luminal and contraluminal plasma membranes. Such vesicle systems are amenable to a broad variety of experimental approaches related to driving forces, orientation of active sites, and the overall symmetry or asymmetry of the transport systems under study, because (i) the intravesicular fluid and extravesicular medium can be varied independently as required by the experimental objectives, and (ii) vesicles can be prepared with "outside-out" or "inside-out" orientation with reference to the *in vivo* orientation of the membranes from which the vesicles were derived. The following are examples of transport functions of the luminal and contraluminal poles of the renal proximal tubular epithelial cell as studied by vesicle systems derived from microvilli and basal-lateral plasma membranes.

Facilitated transport systems have been identified by vesicle preparations from luminal membranes for D-glucose (2), L-phenylalanine (11) and phosphate (12). Figure 3 illustrates the sodium dependence and saturability of the phosphate transport system of luminal membranes; similar phenomena

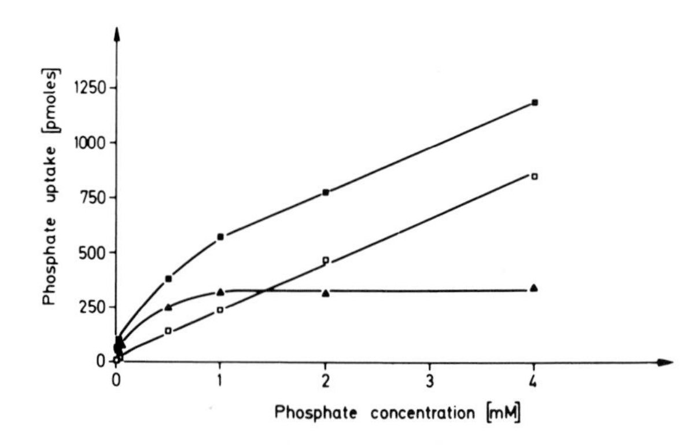

FIG. 3. Relationship between phosphate uptake by isolated renal brush border microvillus vesicles (after 15 sec of incubation) and the phosphate concentration in the extravesicular incubation medium. The latter contained 0.1 M mannitol, 0.02 M Tris-HEPES, 0.1 M NaCl (■) or 0.1 M KCl (□), and the concentration of phosphate is shown on the abscissa. (▲): sodium-dependent part of the uptake. Uptake was determined by a rapid filtration technique (2).

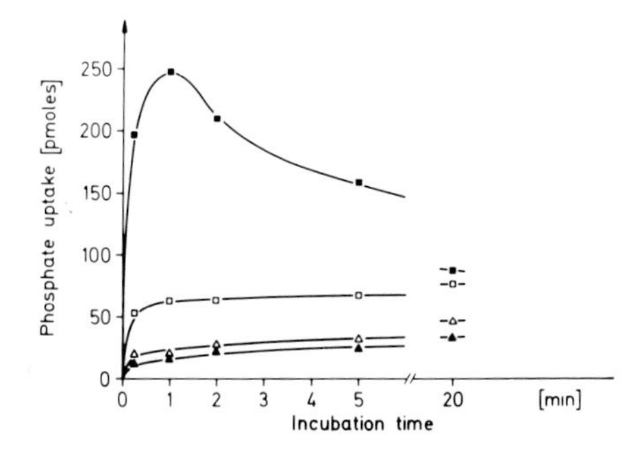

FIG. 4. Comparison of the effect of sodium and potassium on the phosphate uptake by isolated brush border microvillus vesicles under gradient and nongradient conditions. For studies of uptake under cation gradient conditions the membranes were preincubated for 1 hr at 25°C in mannitol Tris-HEPES buffer, and the uptake was determined in a medium containing 0.1 M mannitol, 0.02 M Tris-HEPES, pH 7.4, and 0.1 M NaSCN (□) or 0.1 M KSCN (▲), respectively. For nongradient conditions, the buffer used for the preincubation contained in addition 0.1 M NaSCN (□) or 0.1 M KSCN (△), respectively. Intravesicular space, determined by the amount of D-glucose present in the vesicles after 20 min, was identical under all four incubation conditions.

are observed for the D-glucose and L-phenylalanine transport systems. The effect of sodium on phosphate uptake into microvillus vesicles is shown in more detail in Fig. 4 where the uptake in the presence of a NaCl gradient across the membrane is compared to the uptake in the presence of a KCl gradient and in the absence of either a NaCl or KCl gradient (i.e., with NaCl and KCl equilibrated across the vesicle membrane). It can be seen that there is a transient intravesicular accumulation of phosphate when sodium is distributed asymmetrically across the membrane at the start of the experiment, and that, even in the absence of a gradient, sodium specifically stimulates phosphate uptake, as shown by the difference between phosphate uptake into vesicles preequilibrated with sodium and phosphate uptake into vesicles preequilibrated with potassium. It can therefore be concluded that the overshoot phenomenon and the specific effect of sodium on phosphate transport indicate the presence of a phosphate–sodium cotransport system in the luminal membrane. Similar findings in other studies on microvillus vesicles have uncovered D-glucose–sodium and L-phenylalanine–sodium cotransport systems in the luminal membrane. In addition, as shown in Fig. 5, luminal membrane vesicles exhibit a sodium-dependent ejection of protons, as revealed by the difference in response to the application of extravesicular–intravesicular gradients of sodium and choline (13).

The above-noted phosphate-sodium, L-phenylalanine–sodium, and D-glucose–sodium cotransport systems as well as the sodium–proton exchange system have not been found in studies on contraluminal membrane vesicles; however, D-glucose and L-phenylalanine transport systems which are not influenced by sodium have been observed. Also, as demonstrated in Fig. 6, there is a specific transport system for *p*-aminohippurate (PAH) which is present in the contraluminal membrane, but not in the luminal

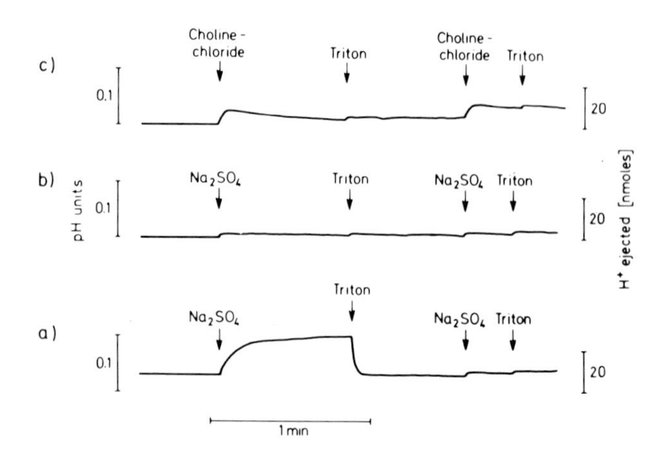

FIG. 5. Proton ejection induced by Na^+. The pH of a rapidly stirred membrane suspension was recorded continuously. The membranes were suspended in 200 μl of 0.15 M k-cyclamate, 0.005 M glycylglycine buffer, pH 6.1. (a) and (c) intact renal brush border membrane vesicles; (b) Triton-solubilized (0.5%) membrane vesicles. Additions: 20 μl 0.5 M Na_2SO_4, 20 μl 1.0 M choline chloride, 20 μl 5% Triton × 100. The membrane suspension contained 1,530 μg protein.

membrane (14). Although PAH uptake by basal-lateral plasma membranes was stimulated by a sodium gradient to a greater degree than by a potassium gradient, this differential stimulation was not observed when sodium and potassium were preequilibrated across the vesicle membranes (Fig. 7). Furthermore, it was demonstrated that valinomycin-enhanced potassium permeability of the vesicle membranes was associated with a PAH uptake which was increased in approximately the same degree as seen in the presence of a sodium gradient (Fig. 8). Presumably, in the presence of a KCl gradient, the intravesicular space is more negative than in the presence of a NaCl gradient, and this negativity is associated with diminished PAH uptake; accordingly the reduction of this negativity following the addition of valinomycin stimulates PAH uptake. The latter findings suggest that the effect of a sodium gradient on the PAH transport system in the contraluminal membrane is related to the membrane potential and not to the existence of a PAH–sodium cotransport system.

Experiments of similar design have been carried out with luminal membrane vesicles to explore the effect of membrane potential on the transport of D-glucose, L-phenylalanine, phosphate, and protons. For example, in the case of L-phenylalanine, when a K gradient directed from the intravesicular to the extravesicular space was established, the L-phenylalanine uptake was stimulated markedly in the presence of valinomycin (Fig. 9). In other words, the valinomycin-mediated increase in potassium permeability and the consequently enhanced intravesicular negativity was associated with increased L-phenylalanine transport, suggesting that the amino acid was moving as a positively charged L-phenylalanine–sodium complex. The same phenomena

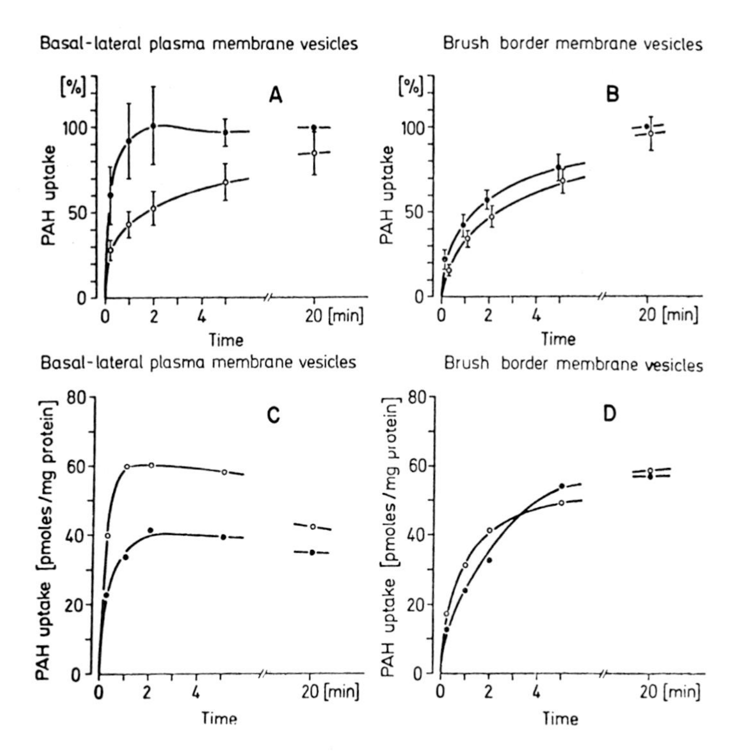

FIG. 6. PAH uptake in basal-lateral plasma membrane vesicles (**A**) and brush border membrane vesicles (**B**) isolated from rat kidney cortex by free flow electrophoresis. The membranes were prepared in 0.1 M mannitol, 0.02 M Tris-HEPES (pH 7.4) and incubated at 25°C in a medium containing 1×10^{-5} M labeled PAH, 0.1 M mannitol, 0.02 M Tris-HEPES (pH 7.4), 0.1 M NaCl in the presence or absence of 5×10^{-4} M probenecid. The mean values $\pm$ SEM of 10 experiments are given in percent of the value reached with 0.1 M NaCl after 20 min of incubation (for basal-lateral plasma membrane vesicles: 22.1 $\pm$ 8.0 pmoles PAH/mg protein; for brush border membrane vesicles: 28.3 $\pm$ 9.6 pmoles PAH/mg protein). The difference of the PAH equilibrium values for basal-lateral plasma membrane vesicles and brush border membrane vesicles reflects a different degree of vesiculation of the membranes. In a separate counterflow experiment, the vesicles of the basal-lateral plasma membranes (**C**) and the brush border membranes (**D**) were preincubated with unlabeled PAH in 6×10^{-5} M PAH, 0.1 M mannitol and 0.02 M Tris-HEPES (pH 7.4) for 1 hr and then incubated at 25°C in a medium containing 2×10^{-5} M labeled PAH (final concentration), 0.1 M mannitol, 0.02 M Tris-HEPES and 0.1 M NaCl. The data given are mean values derived from 3 experiments. Uptake was determined by a rapid filtration technique (2).

have been observed in the case of D-glucose. In the case of phosphate, the establishment of a range of vesicular membrane potentials by using anions (SCN, chloride, cyclamate) with different permeabilities—and kinetic studies on the interaction of sodium with the phosphate transport system—suggested that at pH 6, for example, phosphate is transported predominantly as a positively charged complex of $H_2PO_4^-$ with 2 sodium ions, whereas at pH 7.6 phosphate is transported predominantly as an electroneutral complex of HPO_4^{2-} and 2 sodium ions. Also, in the case of sodium–proton

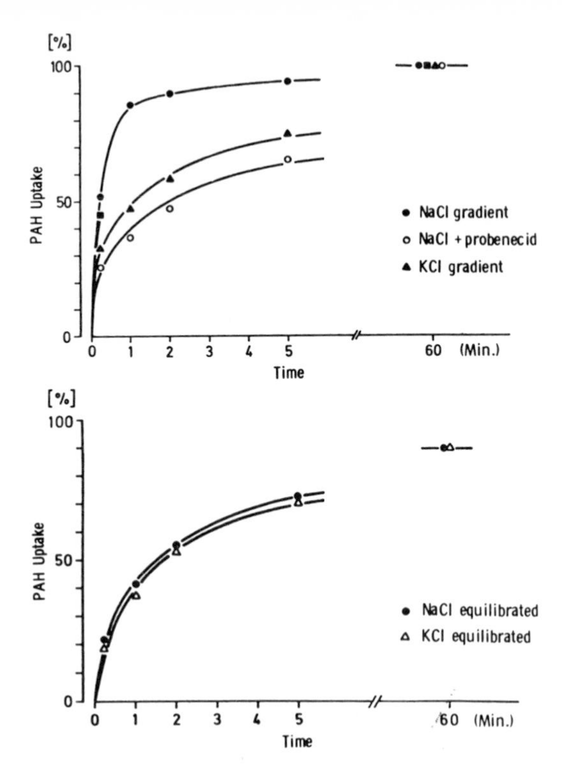

FIG. 7. *Top:* Influence of Na^+ and K^+ gradients on the PAH uptake into basal-lateral plasma membrane vesicles isolated from rat kidney cortex by free-flow electrophoresis. The incubation medium contained 2×10^{-5} M PAH, 0.1 M mannitol, 0.02 M Tris-HEPES (pH 7.4) and (●) 0.1 M NaCl, (○) 0.1 M NaCl + 2×10^{-3} M probenecid, and (▲) 0.1 M KCl. Each point represents the average of 3 experiments. *Bottom:* PAH uptake into basal-lateral plasma membrane vesicles under nonsalt gradient conditions. The vesicles were preincubated in a medium containing 0.1 M mannitol, 0.02 M Tris-HEPES (pH 7.4) and 0.1 M NaCl (●) or 0.1 M KCl (△) for 1 hr and then incubated at 25°C in a medium containing 2.5×10^{-5} M PAH, 0.1 M mannitol, 0.02 M Tris-HEPES, and 0.1 M NaCl or 0.1 M KCl. One typical experiment is given.

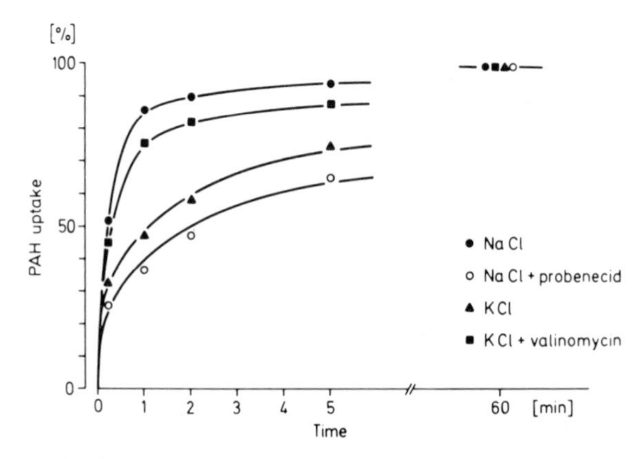

FIG. 8. Influence of Na^+ and K^+ gradients on the PAH uptake into basal-lateral plasma membrane vesicles isolated from rat kidney cortex by free-flow electrophoresis. The incubation medium contained 2×10^{-5} M PAH, 0.1 M mannitol, 0.02 M Tris-HEPES (pH 7.4) and (●) 0.1 M NaCl, (○) 0.1 M NaCl + 2×10^{-3} M probenecid, (▲) 0.1 M KCl, and (■) 0.1 M KCl and valinomycin. Valinomycin was added in an amount of 30 μg/mg protein. The average of 3 experiments are given.

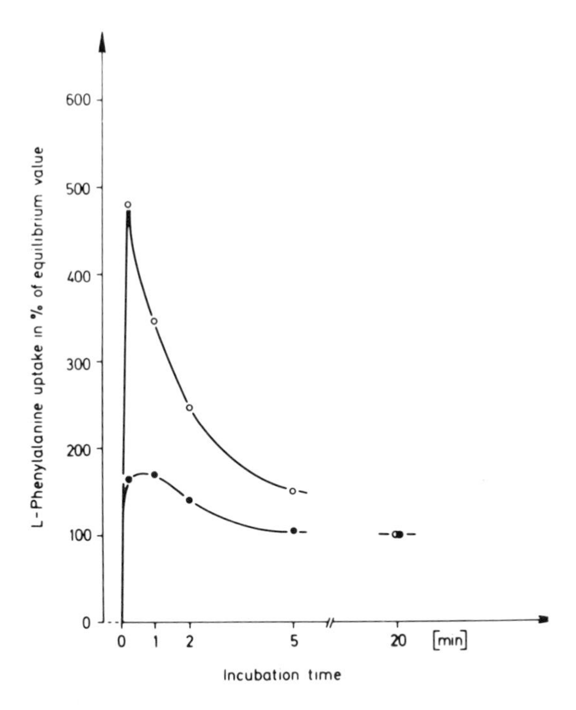

FIG. 9. Influence of potassium diffusion potential on the uptake of L-phenylalanine by isolated renal brush border microvillus vesicles. The membrane vesicles were preloaded with 0.05 M K_2SO_4, 0.1 M mannitol, and 0.02 M Tris-HEPES. The uptake was measured in a medium containing 0.05 M Na_2SO_4, 0.1 M mannitol, 0.005 M K_2SO_4, 0.001 M [^{3}H]L-phenylalanine and 0.02 M Tris-HEPES in the presence (—○—) and absence (—●—) of valinomycin.

exchange across the luminal membrane, studies with valinomycin and carbonyl cyanide *p*-fluoromethoxyphenylhydrazone indicate that this process is electroneutral, i.e., alteration of the transvesicular membrane potential does not change the sodium-driven proton transport from the intravesicular to extravesicular space.

STUDIES OF HORMONE RESPONSIVE SYSTEMS IN THE MAMMALIAN RENAL COLLECTING DUCT

During the past fifteen years it has become generally accepted that hormone action involves a series of sequential steps starting with a recognition event in which the hormone is bound to a discriminatory component (receptor) of the target cell, which functions not only to determine the specificity of hormone action, but also to initiate the chain of propagation events that leads to the final effector process.

Knowledge of the cellular locus and the progression of the sequence of events initiated at the cell membrane by rapidly acting peptide and amine hormones has lagged behind the development of similar information pertaining to the chain of events that constitute the action of the more slowly acting genome-directed hormones (adrenal and gonadal steroids, thyroid hormones, etc.). Recently, however, the separation of purified luminal and contraluminal plasma membranes from the bovine papillary collecting duct has provided some insight into the cellular localization and probable sequence of the known initiating, intermediate, and possibly terminal events in the action of the antidiuretic hormone, arginine vasopressin (3). The findings that the contraluminal membrane, but not the luminal membrane, contains an antidiuretic hormone-sensitive adenylate cyclase, whereas the luminal membrane, but not the contraluminal membrane, contains a cyclic AMP-stimulated membrane-bound protein kinase and its membrane-bound substrate(s) (Fig. 10) suggests that the initial steps in the action of the antidiuretic hormone (ADH) on the kidney takes place at the contraluminal pole of the ADH-target cell, that a subsequent step involves translocation of cyclic AMP through the cytosol or cell matrix, and that the late or terminal steps occur at the luminal pole of the cell where they involve a change in the level of membrane phosphorylation. Similar studies on the rat renal cortex also suggest a role for luminal membrane phosphorylation in the action of parathyroid hormone on the proximal convoluted tubule (15). Our present working hypothesis concerning the mediation of parathyroid hormone action on the proximal tubule and antidiuretic hormone action on the collecting duct is summarized in Fig. 11.

We have hypothesized that the ADH-induced permeability change in mammalian renal collecting duct epithelial cells is a consequence of a cyclic AMP-mediated increase in phosphorylation of the luminal plasma mem-

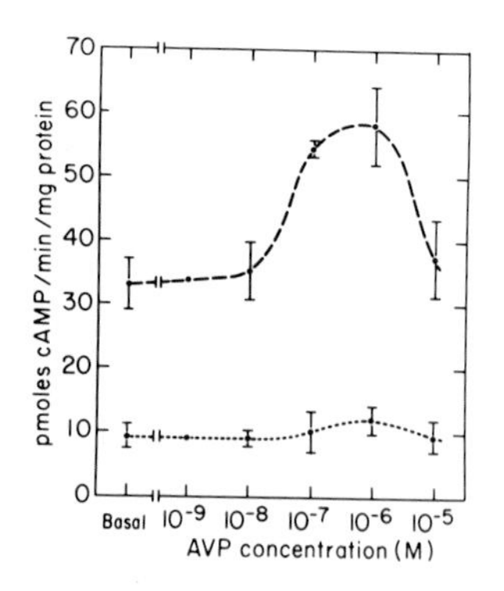

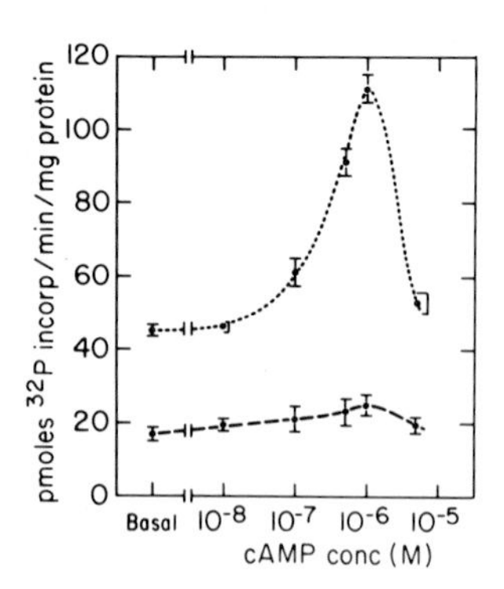

FIG. 10. *Left:* Effect of arginine vasopressin (AVP) on the adenylate cyclase activity of contraluminal membranes (---) and luminal membranes (· · ·) isolated from bovine papillary collecting duct. *Right:* Effect of cAMP on intrinsic protein kinase activity of luminal membranes (· · ·) and contraluminal membranes (---) isolated from bovine papillary collecting duct.

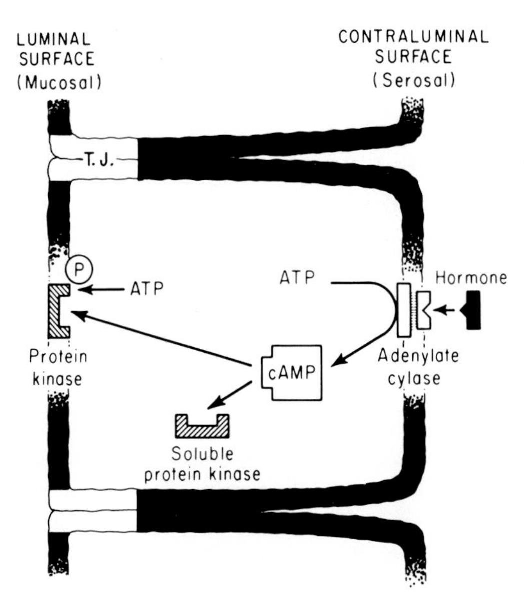

FIG. 11. Hypothetical scheme for the mediation of parathyroid hormone action on the proximal renal tubule and antidiuretic hormone action on the collecting duct of the mammalian kidney.

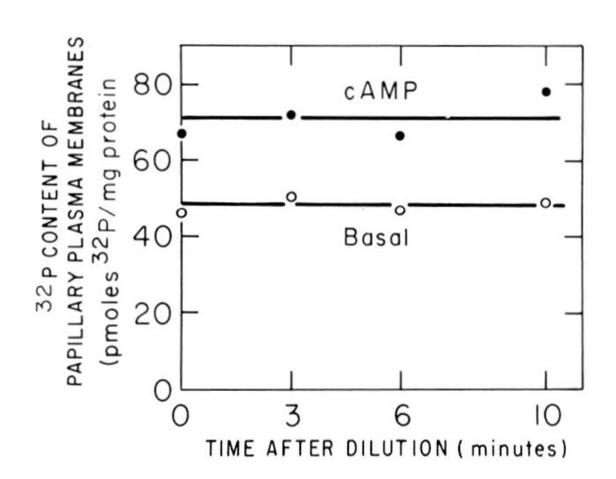

FIG. 12. Lack of dephosphorylation of unfractionated bovine papillary collecting duct plasma membranes in the presence (●) and in the absence (○) of 0.1 μM cyclic AMP. The membranes were phosphorylated by incubation at 30°C for 5 min in a medium containing 50 mM sodium acetate, pH 6.5, 0.025 mM [γ-^{32}P]ATP (0.3 μCi), 10 mM $MgCl_2$, 0.3 mM ethyleneglycol bis(β-aminoethyl ether)-*N,N'*-tetraacetate (EGTA), 20 mM NaF and 2 mM theophylline, and 50–80 mg of membrane protein in a final volume of 0.2 ml. At zero time (5 min after the start of incubation), the samples were diluted 10-fold with buffer to minimize further phosphorylation and the incubation was continued at 30°C for periods of 3, 6, and 10 min.

brane, because we find a cyclic AMP-dependent protein kinase, but not a cyclic AMP-dependent phosphoprotein phosphatase (Fig. 12) operative at this locus. However, others (16,17) have suggested that a dephosphorylation process is involved in the mediation of the action of the antidiuretic hormone, because they find a cyclic AMP-dependent dephosphorylation of a specific protein (protein D) in a membrane fraction from toad bladder. The occurrence of a similar phenomenon in renal membranes cannot be excluded at this time, because it is possible that dephosphorylation of some specific phosphoprotein(s) may be masked by a concurrent phosphorylation of other proteins present in the same membranes. It is also possible that a cyclic AMP-dependent kinase is required to activate the specific phosphoprotein phosphatase for which protein D serves as substrate.

Current studies in our laboratories are directed at the isolation, substrate specificity, and cyclic nucleotide dependency of soluble and membrane-bound kinases and phosphatases of renal cortical and medullary epithelial cells, as well as the distribution of the membrane-bound kinases, phosphatases, and their substrate(s) within the cellular envelope.

ACKNOWLEDGMENTS

The support of the United States Public Health Service (Grant AM-10080 of the National Institute of Arthritis, Metabolism and Digestive Diseases), the Stella and Charles Guttman Foundation, Inc., and the Life Sciences Foundation, Inc. is gratefully acknowledged.

REFERENCES

1. Heidrich, H. G., Kinne, R., Kinne-Saffran, E., and Hannig, K. (1972): *J. Cell Biol.,* 54:232.
2. Kinne, R., Murer, H., Kinne-Saffran, E., Thees, M., and Sachs, G. (1975): *J. Membr. Biol.,* 21:375.
3. Schwartz, I. L., Shlatz, L. J., Kinne-Saffran, E., and Kinne, R. (1974): *Proc. Natl. Acad. Sci., USA,* 71:2595.
4. Murer, H., Hopfer, U., Kinne-Saffran, E., and Kinne, R. (1974): *Biochim. Biophys. Acta,* 345:170.
5. Hannig, K., and Heidrich, H.-G. (1974): *Methods Enzymol.,* 31A:746.
6. Hannig, K., Wirth, H., Meyer, B.-H., and Zeiller, K. (1975): Hoppe-Seylar's *Z. Physiol. Chem.,* 356:1209.
7. Shlatz, L. J., Schwartz, I. L., Kinne-Saffran, E., and Kinne, R. (1975): *J. Membr. Biol.,* 24:131.
8. Kinne-Saffran, E., and Kinne, R. (1974): *Proc. Soc. Exp. Biol. Med.,* 146:751.
9. Kinne-Saffran, E., and Kinne, R. (1974): *J. Membr. Biol.,* 17:263.
10. Frasch, W., Frohnert, P. P., Bode, F., Baumann, K., and Kinne, R. (1970): *Pflügers Arch.,* 320:265.
11. Evers, J., Thees, M., and Kinne, R. (1976): *Biochim. Biophys. Acta* (*in press*).
12. Hoffman, N., Thees, M., and Kinne, R.; *Pflügers Arch.* (*in press*).
13. Murer, H., Hopfer, U., and Kinne, R. (1976): *Biochem. J.* 154:597.
14. Berner, W., and Kinne, R. (1976): *Pflügers Arch.,* 361:269.

15. Kinne, R., Shlatz, L. J., Kinne-Saffran, E., and Schwartz, I. L. (1975): *J. Membr. Biol.,* 24:145.
16. DeLorenzo, R. J., Walton, K. G., Curran, P. F., and Greengard, P. (1973): *Proc. Natl. Acad. Sci. USA,* 70:880.
17. Ferguson, D. R., and Twite, B. R. (1974): *J. Endocrinol.,* 61:501.

Membranes and Disease, edited by L. Bolis, J. F. Hoffman, and A. Leaf. Raven Press, New York, © 1976.

The Chemical Control of Neuronal Activity

Floyd E. Bloom

Laboratory of Neuropharmacology, Division of Special Mental Health Research, IRP, National Institute of Mental Health, Saint Elizabeths Hospital, Washington, D.C. 20032

The electrical activity of nerve cells in the brain is regulated to a major degree by chemical factors which act on or through the membrane. While this is scarcely a novel concept now, the identification of the transmitter chemicals which neurons secrete onto one another has become a long and arduous task for a variety of reasons, described in detail elsewhere (1). Recently, our Laboratory has concentrated upon the types of central synaptic junctions in which the catecholamines, norepinephrine and dopamine are the transmitters. To examine these systems, we have employed a variety of histochemical and electrophysiological indices in order to assess the pharmacological actions of natural substances, and synthetic agonists and antagonists. Our results suggest that the catecholamine systems of the central nervous system may offer some unique insights into one form of interneuronal chemical control system with a profound dependence upon the biological properties of their postsynaptic target cells. As these data have recently been reviewed in depth (2), the present chapter will simply summarize the major pieces of evidence and our present interpretation of these observations.

NEUROTRANSMITTERS AND SECOND MESSENGERS

The now classic studies of Sutherland and Rall (3,4), which evolved into the "second messenger" concept of responses to hormones are also well known. Equally clear is the recognition that the central nervous system is among the richest of tissues in content of adenylate cyclase, and that this enzyme can be influenced *in vitro* by exposure to the catecholamines, both in slices and in cell-free homogenates. In order to develop studies which might test the functional possibility that cyclic AMP formed within a cell is the mediator of a synaptic communication transmitted to that cell by a catecholamine, the following criteria may be considered.

1. The neurotransmitter substance and the activation of the synaptic pathway will regulate the intracellular levels of cyclic nucleotide in the postsynaptic cell population. With few exceptions, however, the methods for detection of cyclic nucleotides cannot resolve them within specific cells,

and therefore a variety of lower resolution methods have been employed to evaluate the effects of electrical stimulation or neurotransmitters in intact ganglia, in intact cells within brain slices, or to determine the effects of transmitter substances in homogenates of brain.

2. The change in intracellular cyclic nucleotide content should precede the biological event triggered by the transmitter or nerve pathway. This criterion is also difficult to satisfy, since the sampling methods of cyclic nucleotide measurement have so far not been sufficiently rapid to detect changes within the period of time in which synaptic potentials are generated. Repetitive activation of a pathway could be used to produce a measurable change, even though it may also result in nonphysiological changes uncharacteristic of the normal mode of a pathway's actions.

3. The effect of the transmitter or nerve pathway in eliciting the physiological event should be altered by drugs which can specifically prevent the hormonal response on the nucleotide cyclase, or which can inhibit the appropriate phosphodiesterase. Thus, potentiation of the effects of a synaptic pathway by phosphodiesterase inhibition or blockade of the effects of a pathway by drugs which block the hormonal activation of the cyclase become critical.

4. Exogenous cyclic nucleotides should be able to elicit the biological event caused by the transmitter or nerve pathway: However, living cells are relatively impermeable to organic phosphates such as cyclic nucleotides. Since nucleotides applied to the exterior of the cell are not only distant from their presumed natural site of production, but exposed to the catalytic action of soluble phosphodiesterases as well, this criterion also carries considerable difficulties for objective satisfaction. Generally, large amounts of exogenous nucleotide with phosphodiesterase inhibitors are required to produce hormone-like effects.

5. To this list of Sutherland's original criteria may now be added the proposal by Kuo and Greengard (5,6), that activation of a phosphotransferase reaction is the major mechanism by which alterations of cyclic nucleotide levels are expressed. Thus, the action of a synthetic cyclic nucleotide ought to emulate the effects of the transmitter or pathway on the parameter of the biological event in direct correlation to the ability of that nucleotide to activate the appropriate protein kinase reaction.

These criteria establish a common yardstick against which to measure the degree of progress in the analysis of a given synaptic connection.

PHYSIOLOGICAL TESTS OF THE SECOND MESSENGER HYPOTHESIS IN THE CENTRAL NERVOUS SYSTEM

In applying the "second messenger" conceptualization of the role of cyclic AMP (see ref. 3), norepinephrine or dopamine would presumably act at certain surface receptors to activate the synthesis of cyclic AMP within the postsynaptic neurons. The intracellular cyclic AMP would then activate subsequent enzymic or molecular events which, among other actions, could result in the changes in cell discharge rate observed when the norepinephrine or dopamine are applied iontophoretically or released by natural synaptic inputs.

The innovative technique of microiontophoresis (see ref. 1) has been quite useful in the assessment and identification of central synaptic transmitter substances, because drugs and other materials with charged groups can be applied from multibarreled micropipettes in minute quantities directly adjacent to single neurons. Iontophoretic administration thus eliminates many of the diffusional and enzymatic barriers which restrict the access of drugs to neuronal receptors when the administration route is parenteral or topical.

When cyclic AMP was first proposed as the mediator of the iontophoretic responses to norepinephrine of cerebellar Purkinje cells (7), emphasis was placed on the actions of iontophoretically applied exogenous cyclic AMP. In the majority of cells tested, but not all, iontophoretic administration of cyclic AMP could produce changes in cerebellar Purkinje cell discharge rate and pattern identical to the effects of norepinephrine (8–11). Cyclic AMP also inhibits hippocampal pyramidal cells (12), cerebral pyramidal tract cells (13,14), and caudate nucleus neurons (15).

At the present time, the most detailed comparison which can be made between the effects of a cyclic nucleotide and the effects of a neurotransmitter or a nerve pathway is with respect to the membrane properties of the test cell. Through the use of an intracellular electrode with intracellular or extracellular iontophoretic drug application the magnitude and direction of the change in membrane potential accompanying the response to the applied substance can be compared to stimulation of the nerve pathway.

For the Purkinje cells of rat cerebellar cortex, the qualitative changes in discharge rate produced by norepinephrine (8) and by the norepinephrine pathway (11) can be mimicked by exogenous cyclic AMP (7) and several of its derivatives (16). The effects of cyclic AMP on transmembrane potential and ionic conductance changes also mimic the effects of norepinephrine and the pathway on those parameters (10,11). In both cases, the membrane responds with hyperpolarization and increased membrane resistance. The effects of cyclic AMP, of norepinephrine, and of the norepinephrine pathway can be potentiated by phosphodiesterase inhibitors (9,10), and the effects of norepinephrine and of the norepinephrine pathway can be antagonized by beta adrenergic blockers (8), which impair the ability of norepinephrine to activate cyclic AMP accumulation in cerebellar cortex and adipocytes. All of these same qualitative comparisons also hold for the hippocampal projection of the central noradrenergic pathway (12,17), and, in preliminary tests, intracellular iontophoresis of cyclic AMP into hippocampal pyramidal cells produces the same hyperpolarization and increase in membrane resistance as does stimulation of the locus coeruleus or iontophoresis of norepinephrine extracellularly (18).

CYTOCHEMICAL TESTS OF THE SECOND MESSENGER HYPOTHESIS

When cyclic AMP or cyclic GMP are coupled to large carrier proteins, the resultant haptene molecule can be employed as an antigen to raise antibodies which will selectively react with either cyclic AMP or cyclic GMP specifically (see 19). In addition to the original application in the radioimmunoassay of cyclic AMP and cyclic GMP, the immunoglobulin fractions of the anti-cyclic AMP antisera can be employed for purposes of immunocytochemistry to localize cyclic AMP or cyclic GMP bound within cells.

Since biochemical assays do not currently permit measurements on single neurons, we sought to determine the staining patterns in cerebellum under conditions which could more closely approximate the electrophysiological studies. Cerebellar biopsies were taken from anesthetized rats prepared as for experiments in electrophysiology. The exposed cerebellar cortex was used to test the actions of topically applied neurohormones and of the norepinephrine synaptic pathway to the cerebellum as revealed by changes in the immunoreactivity to the anti-cyclic AMP immunoglobulin (20). Topical application of norepinephrine in concentrations of 10–100 μM or electrical stimulation of LC resulted in nearly a 5 to 7-fold increase in the number of immunoreactive cerebellar Purkinje cells (from 10% to more than 70% reactivity); much higher concentrations of other inhibitory substances did not increase the immunoreactivity of cerebellar Purkinje cells to AMP immunoglobulin. The increases in immunoreactivity of cerebellar Purkinje cells to LC stimulation could be blocked if the locus coeruleus–cerebellar pathway was first destroyed by treatment with 6-hydroxydopamine. The immunocytochemical approach thus was able to provide some direct indication that the cyclic AMP content of cerebellar Purkinje cells can be increased in response to applied norepinephrine or to activation of the norepinephrine-containing locus coeruleus pathway.

CONCLUSIONS

Despite massive efforts, the role and mechanism by which cyclic nucleotides are involved in synaptic function and other longer term neurobiological phenomena can now only begin to be sketched into perspective. In the CNS, the norepinephrine neurons of the nucleus locus coeruleus inhibit Purkinje cells of the cerebellar cortex and pyramidal cells of the hippocampus in a manner identical to the inhibition produced by exogenous cyclic AMP: the target neurons are hyperpolarized with increased membrane resistance. In these two target areas and in the dopamine-containing inhibitory pathway between substantia nigra and caudate nucleus, the endogenous catecholamine is known to activate adenylate cyclase activity and elevate cyclic AMP content. An increased content of cyclic AMP in cerebellar Purkinje cells during the action of the synaptic pathway can also be demonstrated by immunocytochemical studies. Moreover, the effects of the catecholamines are potentiated by phosphodiesterase inhibitors.

In turn, the effect of the cyclic nucleotide produced at these sites of synaptic transmission is itself mediated through additional intracellular sequences, involving protein kinase (5,6,16) and probably Ca^{2+} as well (1,2,13). Preliminary experiments show that the physiological potency of exogenous synthetic cyclic AMP derivatives varies in close correlation with their ability to activate cyclic AMP-dependent protein kinases. Moreover, the iontophoresis of cyclic AMP intracellularly can produce the same pattern of biophysical responses in hippocampal pyramidal cells, as seen when norepinephrine is applied extracellularly or when the norepinephrine pathway is activated.

In the face of such a powerful biochemical lever for the regulation of neuronal metabolism, it might be wondered whether the electrophysiological effects of postsynaptic inhibition represent the primary message of such synaptic events or whether these electrophysiological effects might not be epiphenomena of a more pervasive, but covert, shift in cellular metabolism which is evoked by these cyclic nucleotide-mediated synaptic stimuli.

Synapses which produce a postsynaptic electrophysiological change mediated by cyclic nucleotides may have been important during differentiation and persisted into adulthood as vestiges to yield intermittent low-level trophic signals (see ref. 21), which are generated only after multiple repetitions such as the intense firing of some norepinephrine neurons during certain stages of sleep (22,23). It may be pertinent to this trophic viewpoint, that the locus coeruleus neurons retain into adulthood the capacity to form collateral axon sprouts (24,26). Alternatively, the cyclic nucleotide–intracellular sequence may represent a form of intercellular chemical communication which evolved early in phylogeny, and which now serves to amplify the effects of relatively small populations of neurons, like the catecholamine neurons, which extend diffusely throughout the nervous system. This chemical amplification could compensate for their limited number (caused perhaps by their schedule of differentiation) in relation to the large number of target cells they must serve. In yet a third scheme, neurons whose synaptic effects are mediated through cyclic nucleotide amplifying steps may "bias" more rapidly transmitting systems which project to the same target neurons, and in which the resultant changes in membrane and intracellular proteins could accentuate or suppress previously established input and output relationships. Under this view, synaptic events which result in intracellular changes in cyclic nucleotides could be considered holistically to transform the postsynaptic cell from one level of metabolic functioning to another by altering one or more interactive enzyme–substrate relationships. For example, in the heart, catecholamines not only increase the force and frequency of the cardiac contractions, they also activate lipolysis and glycogenolysis in order to provide the heart with increased substrate levels for energy metabolism (see ref. 27).

Although a substantial amount of information has been gained concerning the sites and possible mechanisms of action of catecholamine-mediated and cyclic nucleotide-related synaptic events, it is far from clear how prominent or widely used such mechanisms are represented. As a result, we are only now reaching the plateau of knowledge from which such incisive experimental questions can be formulated. Hopefully, at some not-too-distant future session of this continuing body of research into the phenomena of neuronal and other biologic membranes, these answers will be obtained.

REFERENCES

1. Bloom, F. E. (1974): *Life Sci.,* 14:1819.
2. Bloom, F. E. (1975): *Ergeb. Physiolog.* (*in press*).
3. Sutherland, E. W. Oye, I., and Butcher, R. W. (1965): *Recent Prog. Horm. Res.,* 21:623.
4. Rall, T. W., and Sutherland, E. W. (1961): *Cold Spring Harbor Symp.,* 26:347.
5. Kuo, J. F., and Greengard, P. (1969): *Proc. Natl. Acad. Sci. USA,* 64:1349.
6. Greengard, P., and Kebabian, J. W. (1974): *Fed. Proc.,* 33:1059.
7. Siggins, G. R., Hoffer, B. J., and Bloom, F. E. (1969): *Science,* 165:1018.
8. Hoffer, B. J., Siggins, G. R., Oliver, A. P., and Bloom, F. E. (1969): *Ann. NY Acad. Sci.,* 185:531.
9. Hoffer, B. J., Siggins, G. R., Oliver, A. P., and Bloom, F. E. (1972): *Adv. Cyclic Nucleotide Res.,* 1:411.
10. Hoffer, B. J., Siggins, G. R., Oliver, A. P., and Bloom, F. E. (1973): *J. Pharmacol. Exp. Ther.,* 184:553.
11. Siggins, G. R., Hoffer, B. J., Bloom, F. E. (1971): *Brain Res.,* 25:535.
12. Segal, M., and Bloom, F. E. (1974): *Brain Res.,* 72:79.
13. Phillis, J. W., Lake, N., Yarborough, G. (1973): *Brain Res.,* 53:465.
14. Stone, T. W., Taylor, D. A., and Bloom, F. E. (1975): *Science,* 187:845.
15. Siggins, G. R., Hoffer, B. J., and Ungerstedt, U. (1974): *Life Sci.,* 15:779.
16. Siggins, G. R., and Henriksen, S. J. (1975): *Science* (*in press*).
17. Segal, M., and Bloom, F. E. (1974): *Proc. Soc. Neurosci.,* 3:361.
18. Bloom, F. E., Wedner, H. J., and Parker, C. W. (1973): *Pharmacol. Rev.,* 25:343.
19. Siggins, G. R., Battenberg, E. F., Hoffer, B. J., Bloom, F. E., Steiner, A. L. (1973): *Science,* 179:585.
20. Maeda, T., Tohoyama, M., and Shimizu, N. (1974): *Brain Res.,* 70:515.
21. Jouvet, M. (1972): *Eregb. Physiol.,* 64:168.
22. Chu, N-s., and Bloom, F. E. (1974): *J. Neurobiol.,* 5:527.
23. Katzman, R., Bjorklund, A., Owman, C., Stenevi, U., and West, K. (1971): *Brain Res.,* 25:579.
24. Moore, R. Y., Bjorklund, A., Stenevi, U. (1971). *Brain Res.,* 33:13.
25. Pickel, V. M., Segal, M., and Bloom, F. E. (1974): *J. Comp. Neurol.,* 155:43.
26. Mayer, S. E. (1974): *Circ. Res., Suppl. III,* 34:129.

Membranes and Disease, edited by L. Bolis, J. F. Hoffman, and A. Leaf. Raven Press, New York, © 1976.

Sodium Channels and Pumps in Excitable Membranes

J. M. Ritchie

Department of Pharmacology, Yale University School of Medicine, New Haven, Connecticut 06510

Two membrane components have special significance for the function of excitable tissue: the sodium channels through which the sodium ions flow to generate the action potential; and the sodium pumping sites, which use metabolic energy to restore the ionic gradients across the nerve membrane during the recovery process. Other important components of the membrane are also necessary for *normal* function, but they are not absolutely critical for conduction. For example, the potassium channels in myelinated nerve can be blocked completely (by TEA) without much interference with conduction (1): the duration of the action potential merely increases. But no conduction at all is possible when the sodium channels are blocked; and nerve fibers rapidly run down and cease to become excitable when the pumping system for the active extrusion of intracellular sodium is inhibited. Abnormalities in either system, if sufficiently severe, might well be reflected in gross clinical dysfunction. This chapter briefly outlines the present state of knowledge on the character and distribution of these two components in normal tissue. Such experiments, it is hoped, will provide a basis for studying abnormalities present in neurological disease.

PRINCIPLE OF THE METHOD

The principle of the method for determining both the density of the channels and pumps, and their characterization, is simple. Drugs are available that specifically block these two important sites in extremely low concentrations: the sodium channels are blocked by tetrodotoxin (TTX) and by saxitoxin (2,3); and the sodium pumps are inhibited by cardiac glycosides such as ouabain (4). Chemical experiments in which the uptake of each of these drugs by excitable tissue have been studied show that it is possible to identify a saturable component of binding that corresponds with each of the two critical membrane components. Assuming that one molecule of drug combines with a single membrane component, one can thus estimate the density of such components on the axonal membrane by determining the amount of drug taken up at the time physiological function is completely inhibited. For example, Moore, Narahashi, and Shaw (5) found that less than

13 TTX molecules were taken up by lobster nerve at the time that complete block of the action potential occurred; and Keynes, Ritchie, and Röjas (6) found a similar small number for rabbit and crab nerves. However, the drawback in these earlier studies was that no allowance was made for nonspecific binding of the drug, or for binding to some site other than the channels or pumps. Complete binding curves over a large range of concentrations is necessary to resolve this question, and the most convenient way to do this is to use radioactively labeled drugs. Tritium-labeled ouabain has been available for some time (7–11). However, it is only comparatively recently (12,13) that labeled tetrodotoxin and saxitoxin have become available. Most of this chapter will describe, therefore, the more recent work on sodium channels.

SODIUM CHANNELS

The uptake of labeled tetrodotoxin and saxitoxin has now been examined in a variety of tissues, including the nonmyelinated fibers of rabbit, garfish, and lobster nerves (12–14), the squid giant axon (15), frog muscle (16), and rat diaphragm (17). In all these tissues the binding curves are remarkably similar to that shown in Fig. 1 for garfish nonmyelinated fibers. There is a clearly defined saturable component of binding, which is superimposed on a second nonsaturable, linear component. Table 1 shows the maximum binding capacity of the saturable binding components, and the toxin concentration at which saturation is half complete (*K*, equilibrium dissociation constant), for the various tissues studied. In all tissues the maximum binding capacity is less than 400 pmole/g wet dry tissue; and half saturation occurs with concentrations of a few nanamolar. The important question is: does this saturable component represent specific binding to sodium channels?

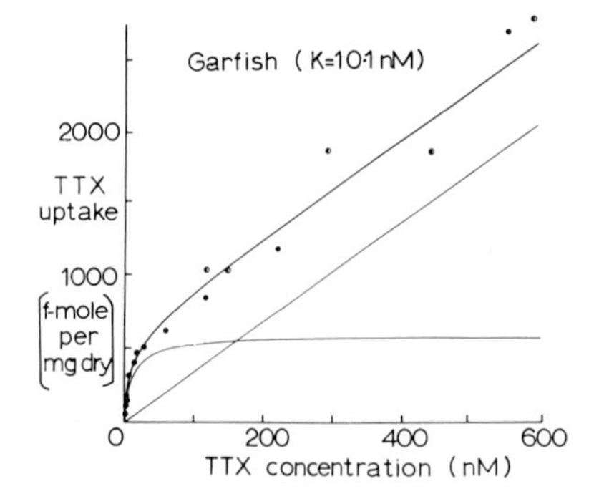

FIG. 1. The uptake of TTX by the olfactory nerve of the garfish at different external concentrations of TTX. The top line is the total binding curve drawn to obey the relation:

$$U = 3.4[\mathrm{TTX}] + 584[\mathrm{TTX}]/(10.1 + [\mathrm{TTX}])$$

where U is the amount of TTX bound in fmole/mg dry tissue and [TTX] is the bathing concentration of TTX in nM. The two components, linear and saturable, are also shown. The equilibrium times were (hr): ◐, 3; ■ (single point), 6; ●, 12. (Taken from Colquhoun et al. (12).

TABLE 1

Tissue	Equilibrium dissociation constant (nM)	Maximum TTX specific binding capacity (fmole/mg wet)	Potassium loss per impulse (temp., 0–10°C) (pmole/mg wet)	Average number of ions per impulse per channel
Nonmyelinated nerve				
Rabbit vagus	3.0	30.7	12.2	0.4×10^3
Garfish olfactory	10.1	60.3	22.5	0.4×10^3
Lobster walking leg	10.2	18.2	–	1.5×10^3
Crab walking leg	–	–	26.6	1.5×10^3
Squid giant axon	7	5.6	0.7	0.1×10^3
Muscle				
Frog sartorius	5	22	3	0.1×10^3
Rat diaphragm	6.1	2.5	–	–

References: Binding experiments: Rabbit, garfish, and lobster nerves (12); crab nerve (6); squid giant axon (Levinson, *personal communication;* 15); frog muscle (16); rat diaphragm muscle (17). Potassium efflux experiments (taken as a measure of the sodium influx): Rabbit nerve (23); crab nerve (22); garfish nerve (11); squid giant axon (from Hodgkin, 1964 (39) assuming 80 cm²/g); frog muscle (see Table 1 in ref. 35).

Identity of the Saturable Binding Component and the Population of Sodium Channels

Voltage-clamp experiments on frog nodes of Ranvier (18) and squid giant axons (19) suggest that a single molecule of tetrodotoxin (T) combines with a single sodium channel (R) to block it, i.e.,

$$T + R \rightleftharpoons TR \tag{1}$$

On this basis the fraction of channels (p_b) blocked at any time by the toxin is given by the hyperbolic, Langmuir, relationship:

$$p_b = \frac{[T]}{[T] + K} \tag{2}$$

where K is the equilibrium dissociation constant of the reaction. In squid giant axon (19) and in frog nodes of Ranvier (18,20,21) the value of K is 1–3 nM.

Most experiments to date have, however, been done on *small* nonmyelinated fibers, to take advantage of the fact that the smaller a cylindrical nerve fiber is, the greater is the proportion of axonal membrane in a given tissue mass. Thus, whereas in squid giant axon the axonal membrane area is only about 80 cm²/g wet nerve, in rabbit and crab nerve it is 6,000–7,000 cm²/g (22,23), and in the olfactory nerve of the pike and garfish it is 41,000–65,000 cm²/g (24,25). Unfortunately, direct electrophysiological determination of K is not possible in these small nonmyelinated nerve fibers because their small size precludes the use of voltage-clamping methods. However, an indirect approach using a computer analysis based on the Hodgkin–Huxley equations (26) has allowed the value of K in these small fibers to be indeed determined solely from electrophysiological experiments, which necessarily study the interaction between the toxin and the sodium channel: the value of K so obtained is about 3 nM. This value is so close to the value obtained in the binding experiments that the saturable component studied in the binding experiments seems clearly identical with the population of sodium channels studied in the electrophysiological experiments. Further evidence for this conclusion comes from relatively recent experiments on the binding of labeled TTX to squid nerve (15). Because of the much smaller area of membrane per gram tissue, the error in the squid experiments is comparatively large. Nevertheless, the value of K determined in the squid binding experiments agrees reasonably well with the corresponding value determined earlier in electrophysiological experiments.

Solubilization of Sodium Channels

An important step in any biochemical study of the sodium channels—their solubilization—has already been achieved by Henderson and Wang (27) and subsequently by Benzer and Raftery (28). Using the detergent Triton X100 these authors have produced a solubilized fraction of the proteins from garfish olfactory nerve that binds TTX (and so presumably are the sodium channels) to the same extent as the original nerve preparation would be expected to do. Unfortunately, all attempts to separate the specific binding fraction from the heterogeneous solution of protein have so far failed.

Normal and Denervated Muscle

Using intact and homogenized muscle Colquhoun et al. (17) have shown that there is a specific component of uptake of toxin by rat diaphragm muscle amounting to 2.5 fmole/mg wet; the equilibrium dissociation constant for this uptake is 6.1 nM. On denervation the specific binding of toxin is about halved, to 1.2 fmole/mg wet. This reduction in density of sodium channels in denervated muscle is consistent with the observation (29,17,38) that the maximum rate of rise of the action potential is reduced 5–10 days after denervation. The decrease in binding of TTX after denervation thus agrees quite well with the impairment of physiological function. Denervation also causes muscle fibres to become sensitive to acetylcholine at regions away from the motor endplate (30), an effect that is well correlated with the appearance of binding sites for α-bungarotoxin at nonjunctional regions (31–34). However, the number of such α-bungarotoxin binding sites in rat diaphragm increases from a normal value of about 4 fmole/mg to about 45 fmole/mg (17). This increase is more than 30 times the amount of TTX-sensitive channels that disappear on denervation, which seems to render implausible the possibility that the acetylcholine receptors that appear after denervation are derived from the very much smaller amount of TTX-resistant sodium channels that disappear.

Characterization of the Sodium Channels in Nerve

The Nature of the Binding Site in the Sodium Channel

Local anaesthetics, veratrine and batrachotoxin, have little effect on the binding of either labeled TTX or saxitoxin (12,13). However, a variety of cations markedly reduce the binding (14). The simplest assumption for such a reduction is that the cation (C) also combines with the channel (R) with an inhibitory equilibrium dissociation constant (K_i) according to the relation

$$C + R \rightleftharpoons CR \tag{3}$$

The fraction of channels blocked at any toxin concentration [Eq. (2)] thus becomes:

$$p_b = \frac{[T]}{\{[T] + K(1 + [C]/K_i)\}} \tag{4}$$

where $[C]$ is the concentration of cation. On this basis Henderson et al. (14) showed that trivalent cations (La^{3+}, Sm^{3+}, Er^{3+}) have a marked affinity for the toxin binding sites ($K_i < 1$ mM), whereas (with the exception of Be^{2+} for which K_i is less than 1 mM) the divalent cations (Mg^{2+}, Ca^{2+}, Sr^{2+}, Ba^{2+}) have apparent dissociation constants in the range 5–50 mM. In general, the monovalent cations (Li^+, K^+, Na^+, Cs^+, TMA^+, and $choline^+$) have a low affinity for the site ($K_i < 100$–1000 mM). Of particular interest are the results with calcium and with the monovalent thallium and hydrogen ions. Compared with the other monovalent cations tested, thallium has a moderately high affinity for the binding site, K_i being approximately 20 mM; the value of K_i for calcium is about 30 mM; and the pH at which half the binding sites are blocked by hydrogen ions is about 5.5. These three values are in excellent agreement with those observed to halve the sodium currents by thallium (36), calcium (37), and hydrogen ions (37) in experiments on voltage clamped frog single myelinated nerve fibers. This agreement provides further strong evidence that the toxin binding component is indeed identical with the population of sodium channels.

The reduction of toxin binding in the presence of a variety of cations has just been attributed to the cations binding to the same site as the toxin and hence competing with the toxin according to Eq. (1) and (3). An alternative possibility is that the cations act on the sodium channel by a general reduction of the local negative surface potential, and hence of the local concentration of the cationic toxin. These two possibilities can be examined experimentally because TTX is a monovalent cation, whereas saxitoxin is divalent. This means that an added cation, such as calcium, would affect both toxins equally if acting by simple competition, whereas saxitoxin being divalent would be much more affected than tetrodotoxin by an effect on the surface potential. Indeed, recent experiments by Hille et al. (21) on the effect of surface charge on the nerve membrane on the physiological action of tetrodotoxin and saxitoxin in frog single myelinated nerve fibres suggest that the result of a 10-fold increase in external calcium concentration (from 2 to 20 mM) is better explained by a change in surface potential (of 7–8 mV) than by simple competition between the toxin and the calcium ion for the same sites.

SODIUM PUMPS

Experiments on the binding of radioactively labeled ouabain to nerve tissue reveal a saturable component of binding that corresponds with the population of sodium pumping sites in the tissue (9–11). For example, as Fig. 2 shows, the curve relating the uptake of ouabain by the nonmyelinated fibers of the garfish olfactory nerve to the extracellular concentration of ouabain can be fitted by the sum of two components: a large saturable component with an equilibrium dissociation constant of 0.5 μM and a much smaller linear component of binding. On the assumption that one molecule of ouabain interacts with each pump site to inhibit it, the size of the saturable component gives the density of sodium pumps in the nerve.

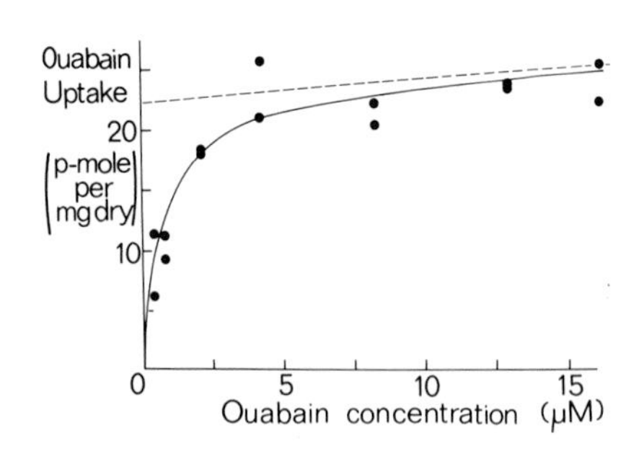

FIG. 2. The uptake of labeled ouabain by garfish olfactory nerve at different external concentrations of ouabain. The nerves were equilibrated for 6 hr with the ouabain. The broken line is the asymptote of the binding curve. The binding curve is the relation:

$$U = 0.91[\mathrm{O}] + 22.3[\mathrm{O}]/(0.51 + [\mathrm{O}])$$

where U is the amount of ouabain bound in pmole/mg dry tissue and [O] is the bathing concentration of ouabain in μM. The total binding curve is a least squares fit to the point that are shown, together with several others not shown at higher concentrations up to 300 μM. (Taken from Ritchie and Straub (11).

DISCUSSION

One important conclusion from these binding studies is that both the sodium pumps and the sodium channels are present in remarkably small amounts on the nerve membrane. For example, the experiments on ouabain binding (Fig. 2) show that the maximum uptake of glycoside by garfish olfactory nerve is 22 pmole/mg dry, which corresponds with only 300 pumping sites/μm^2 membrane, and in rabbit vagus nerve the corresponding values are 4.3 pmole/mg dry and 750 pumping sites per μm^2 (9). For the sodium channels the position is even more extreme: the experiments on TTX binding to garfish nerve (Fig. 1) indicate that in this nerve there are only about six channels per square micrometer. This is a remarkably small number, corresponding with one sodium channel for every 300,000 phospholipid molecules in the nerve membrane. Through each channel a relatively small number of sodium ions (just a few hundred, Table 1) flow to generate the action potential. Similarly, in both rabbit and garfish nerves the pumping sites, though 30–50 times more numerous than the sodium channels, are still few in number. These considerations would seem to imply that a disturbance of function in either of these extremely sparsely distributed critical components of conduction may well lead to disease, even although the gross electromicroscopic picture of the membrane remains relatively normal.

The hypothesis that altered physiological function may be associated with disturbances in the sodium channel system, and be reflected in changes in the TTX binding capacity, is supported by a second finding in these binding studies. For the excitable properties of normal skeletal muscle also depends on TTX-sensitive sodium channels. After denervation, various changes in membrane character occur: for example, the rate of rise of the action potential becomes smaller, and the sodium channels acquire a certain degree of TTX insensitivity. These changes in physiological and pharmacological function are clearly the result (at least in part) of the observed reduction in the density of TTX binding sites in the muscle on denervation.

It seems clear, therefore, that a study of sodium pumps and sodium channels in excitable tissue may be rewarding and might well give clues to the nature of dysfunction in certain neurological diseases.

ACKNOWLEDGMENT

Much of this work was supported by grant NS-08304 and NS-12327 from the USPHS.

REFERENCES

1. Hille, B. (1967): The selective inhibition of delayed potassium currents in nerve by tetraethylammonium ion. *J. Gen. Physiol.,* 50:1287–1302.

2. Kao, C. Y. (1966): Tetrodotoxin, saxitoxin and their significance in the study of excitation phenomena. *Pharmacol. Rev.,* 18:997–1049.
3. Takata, M., Moore, J. W., Kao, C. Y., and Fuhrman, F. A. (1966): Blockage of sodium conductance increase in lobster giant axon by Tarichatoxin (Tetrodotoxin). *J. Gen. Physiol.,* 49:977–988.
4. Skou, J. C. (1971): Sequence of steps in the (Na + K)-activated enzyme system in relation to sodium and potassium transport. *Curr. Top. Bioenerget.,* 4:357–398.
5. Moore, J. W., Narahashi, T., and Shaw, T. I. (1967): An upper limit to the number of sodium channels in nerve membrane? *J. Physiol. (Lond.),* 188:99–105.
6. Keynes, R. D., Ritchie, J. M., and Röjas, E. (1971): The binding of tetrodotoxin to nerve membranes. *J. Physiol. (Lond.),* 213:235–254.
7. Baker, P. F., and Willis, J. S. (1969): On the number of sodium pumping sites in cell membranes. *Biochim. Biophys. Acta,* 183:646–649.
8. Hoffman, J. F. (1969): The interaction between tritiated ouabain and the Na-K pump in red blood cells. *J. Gen. Physiol.,* 54:343–350S.
9. Landowne, D., and Ritchie, J. M. (1970): The binding of tritiated ouabain to mammalian non-myelinated nerve fibres. *J. Physiol. (Lond.),* 207:529–537.
10. Baker, P. F., and Willis, J. S. (1972): Binding of the cardiac glycoside ouabain to intact cells. *J. Physiol. (Lond.),* 224:441–462.
11. Ritchie, J. M., and Straub, R. W. (1975): The movement of potassium ions during the electrical activity and the kinetics of the recovery process in the non-myelinated fibres of the garfish olfactory nerve. *J. Physiol. (Lond.),* 249:327–348.
12. Colquhoun, D., Henderson, R., and Ritchie, J. M. (1972): The binding of labelled tetrodotoxin to non-myelinated nerve fibres. *J. Physiol. (Lond.),* 227:95–126.
13. Henderson, R., Ritchie, J. M., and Strichartz, G. R. (1973): The binding of labelled saxitoxin to the sodium channels in nerve membranes. *J. Physiol. (Lond.),* 235:783–804.
14. Henderson, R., Ritchie, J. M., and Strichartz, G. R. (1974): Evidence that tetrodotoxin and saxitoxin act at a metal cation binding site in the sodium channels of nerve membrane. *Proc. Natl. Acad. Sci. USA,* 71:3936–3940.
15. Levinson, S. R., and Meves, H. (1975): The binding of tritiated tetrodotoxin to squid giant axons. *Phil. Trans. R. Soc. Lond. B,* 270:349–352.
16. Almers, W., and Levinson, S. R. (1975): Tetrodotoxin binding to normal and depolarized frog muscle and the conductance of a sodium channel. *J. Physiol. (Lond.),* 247:483–509.
17. Colquhoun, D., Rang, H. P., and Ritchie, J. M. (1974): The binding of tetrodotoxin and α-bungarotoxin to normal and denervated mammalian muscle. *J. Physiol. (Lond.),* 240: 199–226.
18. Hille, B. (1970): Ionic channels in nerve membranes. *Prog. Biophys. Mol. Biol.,* 21:1–32.
19. Cuervo, L. A., and Adelman, W. J. (1970): Equilibrium and kinetic properties of the interaction between tetrodotoxin and the excitable membrane of the squid giant axon. *J. Gen. Physiol.,* 55:309–335.
20. Schwarz, J. R., Ulbricht, W., and Wagner, H.-H. (1973): The rate of action of tetrodotoxin on myelinated nerve fibres of *Xenopus laevis* and *Rana esculenta*. *J. Physiol. (Lond.),* 233:167–194.
21. Hille, B., Ritchie, J. M., and Strichartz, G. R. (1975): The effect of surface charge on the nerve membrane on the action of tetrodotoxin and saxitoxin in frog myelinated nerve. *J. Physiol. (Lond.),* 250:34–35P.
22. Keynes, R. D., and Ritchie, J. M. (1965): The movements of labelled ions in mammalian non-myelinated nerve fibres. *J. Physiol. (Lond.),* 179:333–367.
23. Howarth, J. V., Keynes, R. D., Ritchie, J. M., and Von Muralt, A. (1975): The heat production associated with the passage of a single impulse in pike olfactory nerve fibers. *J. Physiol. (Lond.),* 249:349–368.
24. von Muralt, A., Weibel, E., and Howarth, J. V. (1975): *Pflügers Arch. Ges. Physiol. (in press).*
25. Easton, D. M. (1971): Garfish olfactory nerve: easily accessible source of numerous long, homogeneous, non-myelinated axons. *Science,* 172:952–955.
26. Colquhoun, D., and Ritchie, J. M. (1972): The interaction at equilibrium between tetrodotoxin and mammalian nonmyelinated nerve fibers. *J. Physiol. (Lond.),* 221:533–553.
27. Henderson, R., and Wang, J. H. (1972): Solubilization of a specific tetrodotoxin-binding component from garfish olfactory nerve membrane. *Biochemistry,* 11:4565–4569.

28. Benzer, T. I., and Raftery, M. A. (1973): Solubilization and partial characterization of the tetrodotoxin binding component from nerve axons. *Biochem. Biophys. Res. Commun.*, 51:939–944.
29. Redfern, P., and Thesleff, S. (1971): Action potential generation in denervated rat skeletal muscle. II. The action of tetrodotoxin. *Acta Physiol. Scand.*, 82:70–78.
30. Miledi, R. (1960): The acetylcholine sensitivity of frog muscle fibers after complete or partial denervation. *J. Physiol. (Lond.)*, 151:1–23.
31. Miledi, R., and Potter, L. T. (1971): Acetylcholine receptors in muscle fibers. *Nature*, 233:599–603.
32. Hartzell, H. C., and Fambrough, D. M. (1972): Acetylcholine receptors. Distribution and extrajunctional density in rat diaphragm after denervation correlated with acetylcholine sensitivity. *J. Gen. Physiol.*, 60:248–262.
33. Berg, D. K., Kelly, R. B., Sargent, P. B., Williamson, P., and Hall, Z. W. (1972): Binding of α-bungarotoxin to acetylcholine receptors in mammalian muscle. *Proc. Natl. Acad. Sci. USA*, 69:147–151.
34. Chang, C. C., Chen, T. F., and Chuang, S-T. (1973): *N,O*-di- and *N,N,O*-tri-(^{3}H)-acetyl α-bungarotoxin as specific labelling agents of cholinergic receptors. *Br. J. Pharmacol.*, 47:147–160.
35. Ritchie, J. M. (1975): Binding of tetrodotoxin and saxitoxin to sodium channels. *Phil. Trans. R. Soc. Lond. B*, 270:319–336.
36. Hille, B. (1975): The receptor for tetrodotoxin and saxitoxin: A structural hypothesis. *Biophys. J.*, 15:615–619.
37. Woodhull, A. M. (1973): Ionic blockage of sodium channels in nerve. *J. Gen. Physiol.*, 61:607–708.
38. Harris, J. B., and Thesleff, S. (1971): Studies on tetrodotoxin resistant action potentials in denervated skeletal muscle. *Acta Physiol. Scand.*, 83:382–388.
39. Hodgkin, A. L. (1964): *The Conduction of the Nervous Impulse.* Charles C Thomas, Springfield, Ill.

Membranes and Disease, edited by L. Bolis, J. F. Hoffman, and A. Leaf. Raven Press, New York, © 1976.

Recent Aspects of the Regulation of ACh Synthesis in Brain Nerve Terminals

P. Lefresne and J. Glowinski

Groupe NB (INSERM U. 114), Collège de France, 75231 Paris Cedex 5, France

A regulatory process for acetylcholine* (ACh) synthesis in brain cholinergic nerve endings was first postulated by Mann, Tennenbaum, and Quastel (1) more than 30 years ago. These authors suggested that end-product inhibition was responsible for the decreasing rate of ACh formation, seen in incubated rat brain slices as the intratissue concentration of the ester was increasing. Recent pharmacological experiments carried out *in vivo* (2) as well as *in vitro* (4,5,39) have confirmed this correlation. In accordance with these findings, the high K^+-stimulated release of ACh from brain slices promotes a reduction in the intracellular content of the ester and an increase in its rate of synthesis (6–9). These data led most authors to propose that the concentration of ACh within nerve endings was responsible for the modulation of the rate of the ester synthesis (10).

The first possible regulation site to be examined was choline acetyltransferase (EC 2.3.1.6.; ChAc). The transesterification reaction catalyzed by this enzyme follows a sequential mechanism: acetyl coenzyme A (AcCoA) binds first, probably to a serine residue of the protein (11) followed by the binding of choline, the second substrate. ACh has been shown to block the second step of the reaction by preventing the binding of choline on the acetylated enzyme (12,13). However, the physiological significance of this inhibition in terms of a regulatory mechanism is doubtful, due to the high concentration of ester required to produce this effect (14).

The identification of two choline uptake processes in brain nerve endings (15,16) suggested another possible regulation site. The so-called low-affinity process (K_m 100 μM) is ubiquitous and therefore of little interest as concerned with the control of ACh synthesis. On the contrary, the "high-affinity" choline permease which appears to be specifically located in the cholinergic nerve endings (17) has been shown to participate in the synthesis of ACh (18). This uptake system is characterized by its high affinity for choline (K_m 1 μM), by its requirement for Na^+ and by its sensitivity to

* Abbreviations used: ACh:acetylcholine; ChAc:choline acetyltransferase (EC 2.3.1.6.); AcCoA:acetyl-Coenzyme A; LDH:lactate dehydrogenase (EC 1.1.1.27.); FH:fumarate hydratase (EC 4.2.1.2.).

some very potent competitive inhibitors (such as hemicholinium-3, HC-3; K_i 2×10^{-8} M).

The importance of the extracellular pool of choline in ACh synthesis was first demonstrated by Birks and MacIntosh (19). They showed that the rate of ACh synthesis was reduced when choline was omitted from the perfusion medium, and restored to normal values when this precursor was subsequently added. Additional evidence for this dependence has been provided by the use of inhibitors such as HC-3 which block both the "high affinity" uptake system and the synthesis process (6,20,21,18). These results led Whittaker and Dowdall (22) to propose that the intracellular concentration of ACh could modulate the activity of the "high-affinity" choline permease. These authors found that the initial rate of choline penetration in squid synaptosomes preloaded with a 10 mM concentration of ACh was decreased by more than 70%, as compared to the control preparation. But the physiological role of this inhibition is questionable, since no decrease in the rate of ACh synthesis could be seen in either rat brain slices (8) or striatal synaptosomes (26) incubated for long periods of time in the presence of high concentrations of the ester.

Another possible site of ACh synthesis regulation could be located in the metabolic pathway leading to the acetyl moiety of the ester. This possibility has never been examined due to the still incomplete knowledge concerning the sequence of the enzymic events leading from pyruvate to ACh. The problem evolves from the metabolic subcellular organization of the cholinergic nerve endings: as in all cells, AcCoA is produced in the mitochondrial matrix and the inner mitochondrial membrane is known to be an impermeable barrier to AcCoA. But the ChAc is known to be located in the cytoplasm (23), and one must postulate the existence of some particular process allowing this enzyme to be supplied with this substrate. The mechanism most-favored to explain the transfer of the acetyl groups from the site of production to the site of utilization is the one involved in fatty acid formation: the intramitochondrially produced citrate crosses the inner membrane through a tricarboxylic acid permease and is then reconverted into oxaloacetate and AcCoA by the cytoplasmic ATP citrate lyase (EC 4.1.3.8.) (24). Some data could be interpreted in terms of AcCoA being the rate-limiting step in the synthesis of ACh: even when rat striatal synaptosomes are overloaded with choline, the maximal rate of the ester synthesis never exceeds 5% of the total ChAc activity of the preparation (21).

This problem of ACh regulation was reexamined on rat striatal synaptosomes using Triton X-100 as a tool for differentiating between the respective importance of the choline and AcCoA supply in the ester synthesis. Low concentrations of this detergent allowed the inactivation of the "high-affinity" choline permease without affecting the mitochondrial activity. The rate of ACh synthesis was then correlated with the changes observed in these two processes.

RESULTS AND DISCUSSION

The experiments were performed by incubating rat brain synaptosomes isolated from rat brain striatum [B fraction of Gray and Whittaker, (25)] during 15 min at 37°C, in a saline medium containing pyruvate, eserine (0.18 mM) and various concentrations of Triton X-100. [^{14}C]ACh synthesized from [2-^{14}C]pyruvate was measured according to the method described by Guyenet et al. (21) and the $^{14}CO_2$ produced in these conditions was estimated as reported previously (26). The activities of the enzymic markers were determined using the following methods: ChAc (27); lactate dehydrogenase (LDH; (28)); fumarate hydratase (FH; (29)).

The permeabilization of the external synaptosomal membranes as a function of increasing concentrations of Triton X-100 was followed by estimating the release in the incubation medium of two cytoplasmic enzymic markers: LDH, being contained in all nerve endings, and ChAc being specifically located in the cholinergic ones. Up to 0.004% of detergent, no large permeability changes could be detected (Fig. 1). A fortiori, using FH as a marker of the mitochondrial matrix, no modification in the membrane properties of these organelles could be seen (Fig. 1). Previous studies have shown that the synthesis of ACh in rat striatal slices (9) as well as in synaptosomes prepared from this brain structure (21) can be a quantitatively estimated by measuring the [^{14}C]ACh produced from [2-^{14}C]pyruvate using the specific radioactivity of this precursor. The rate of ACh formation in nerve endings incubated with Triton X-100 decreases exponentially with increasing detergent concentrations (Fig. 1): a 50% reduction is obtained with a concentration as low as 0.002%.

In order to determine the cause of this early change in the rate of ACh synthesis, the effect of Triton X-100 was examined on (i) the activity of the mitochondria, and (ii) on the "high-affinity" choline uptake system. The activity of the mitochondria was estimated by the amount of $^{14}CO_2$ produced in synaptosomes from [2-^{14}C]pyruvate (26). As shown in Fig. 2A, no significant modification of the activity could be detected up to 0.002% detergent. Since these organelles are very susceptible to the composition of the surrounding medium, it can be assumed that low concentrations of Triton X-100 (0.002%) do not alter the metabolic capacity of the intrasynaptosomal compartments. Therefore, the supply of AcCoA probably remains normal and cannot account for the decrease in the rate of the ACh synthesis. On the contrary, the detergent produces a dramatic reduction in the high affinity choline permease activity which exactly parallels the decrease in the ACh production (Fig. 2B). This correlation between the disappearance of these two synaptosomal functions is similar to that previously described using competitive inhibitors of this uptake process (21).

According to these data the reduction by 50% of the ACh synthesizing

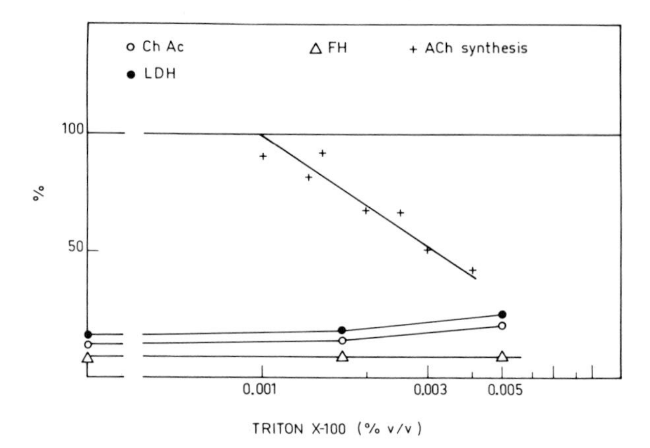

FIG. 1. Effect of the increase of Triton X-100 concentrations on the rate of ACh synthesis, and on the permeabilization of synaptosomal and mitochondrial membranes. *Membrane disruption study:* Rat striatal synaptosomes were resuspended in a saline medium supplemented with 5 mM cold pyruvate. Total ChAc activity was measured in the initial suspension as an index of the number of intact cholinergic nerve endings present in each sample (the value was 0.8 μmoles of ACh synthesized/ml-hr; 0.250 ml/sample). Increasing concentrations of Triton X-100 were then added. After a 10-min incubation at 37°C, samples were centrifuged at 27,000 × g for 10 min. The amounts of enzyme recovered in the supernatants are expressed as percent of the total activities present in the initial suspension. Each value is the mean of data obtained with groups of 4 samples. *Synthesis experiment:* Synaptosomes (ChAc activity 0.9 μmoles of ACh synthesized/ml-hr; 0.35 ml/sample) were incubated for 10 min at 37°C in a saline medium supplemented with [2-^{14}C]pyruvate (5.0 mM, 10 μCi/ml), choline (1 mM) eserine (0.18 mM) and containing increasing concentrations of Triton X-100. The rates of [^{14}C]ACh synthesis are expressed as percent of the control values which were measured in absence of detergent. [^{14}C]ACh was estimated in the hydrochloric phase of the isolation procedure described by Guyenet et al. (21). Each value is the mean of data obtained with groups of 4 samples.

capacity induced by 0.002% Triton X-100 could be attributed to a deficient choline supply induced by the inactivation of the high affinity choline permease. The validity of this assumption was tested by allowing choline to penetrate by passive diffusion into the isolated nerve terminals. Results presented in Table 1 show that an increase in the choline concentration in the incubation medium from 0.02 to 5.0 mM was unable to restore the control rate of ACh synthesis. These results do not favor the interpretation that the decrease in ACh production results from a deficient supply of the nerve endings with choline but would rather indicate that the high affinity system itself participates in the acetylation process. Indeed, only the molecules of choline crossing the membrane through the permease can be used as substrate for ChAc. This could be explained if the translocation of choline were coupled with some biochemical transformation so that these molecules could be distinguished by ChAc from those having penetrated through any other process. As choline itself appears to be the physiological substrate of ChAc, the most obvious conversion that choline could undergo, as cross-

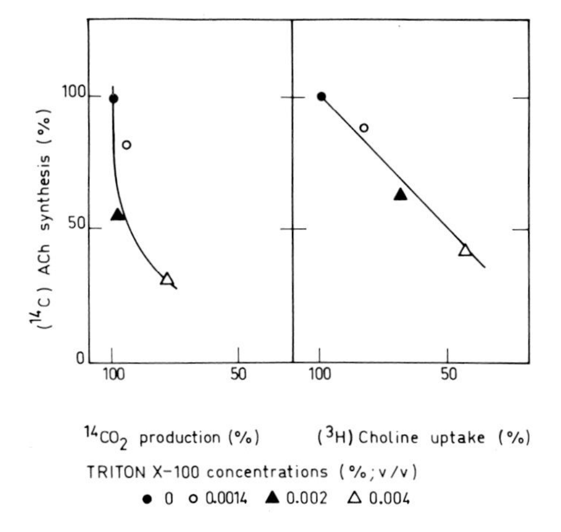

FIG. 2. A: Correlation between the rate of ACh synthesis and the mitochondrial activity as estimated by the amount of $^{14}CO_2$ produced. Synaptosomes (ChAc activity 1.0 μmole of ACh synthesized/ml-hr; 0.35 ml/sample) were incubated in the saline medium supplemented with choline (5 mM), eserine (0.18 mM), [2-^{14}C]pyruvate (1.2 mM, 4.9 μCi/ml), and increasing concentrations of Triton X-100. [^{14}C]ACh synthesized was measured as described in Fig. 1 and $^{14}CO_2$ produced by estimating the radioactivity trapped in 0.2 ml of hyamine (26). Each value is the mean of data obtained with groups of 4 samples. **B:** Correlation between the rate of ACh synthesis and the "high-affinity" choline permease activity. *Synthesis experiment:* Synaptosomes (ChAc activity 0.9 μmoles of ACh synthesized/ml-hr; 0.35 ml/sample) were incubated for 10 min at 37°C in the saline medium supplemented with choline (1 mM), eserine (0.18 mM), [2-^{14}C]pyruvate (5.0 mM, 19 μCi/ml) and increasing concentrations of detergent. Each value is the mean of data obtained with groups of 4 samples. *Uptake experiment:* Synaptosomes were resuspended in the saline medium containing 5 mM pyruvate and increasing concentrations of Triton X-100 (ChAc activity, 1.2 μmoles of ACh synthesized/ml-hr; 0.350 ml). After a 5 min preincubation at 37°C, 20 μl of a solution of [^{3}H]choline were added (1 μM final concentration). The incubation was stopped 30 sec later by adding 6 ml ice-cold saline medium. Passive diffusion controls were similarly treated at 0°C. Each value is the mean of data obtained with groups of 2 samples.

ing the membrane, would be its acetylation. One model to account for the present data would be a reversible enzymic complex located on the presynaptic membrane consisting in the association of one molecule of "high-affinity" choline permease with one molecule of ChAc.

The consequences resulting from this model are consistent with what is known about the subcellular compartmentalization of ACh synthesis. (i) The newly synthesized ACh has been shown to be preferentially released in all the preparations studied so far: superior cervical ganglion (30,31), cortex slices (7,32), diaphragm (33), and electric organ of Torpedo (34). Accurate kinetic studies carried out with this last preparation led Dunant and Israel (35) to propose the existence of a functional link between ACh formation and release. These data then suggest that the site of the ester synthesis is close enough to the presynaptic membrane site of release so that no significant isotopic dilution can occur with the bulk of either the

TABLE 1. *Effect of the concentration of choline on the rate of [^{14}C]ACh synthesis in Triton X-100-treated synaptosomes*

Choline added (mM)	[^{14}C]ACh synthesis (%)	Control (%)
0.02	66	39
0.50	66	40
1.50	53	42
5.00	80	43
Concentration of Triton X-100 (%; v/v)	0.002	0.004

Synaptosomes (ChAc activity: 1.0 μmole of ACh synthesized/ml-hr; 0.350 ml/sample) were resuspended in saline medium containing eserine (0.18 mM), [2-^{14}C]pyruvate (5.8 mM, 14.5 μCi/ml) Triton X-100 and various concentrations of choline. The rates of [^{14}C]ACh synthesis measured in a 10-min incubation period, at 37°C, are expressed as percent of the values obtained in similar conditions but in absence of detergent. Each value is the mean of data obtained with groups of 2 samples.

free or the vesicular preexisting stores of ACh (2). The absolute dependence of the ester formation in synaptosomes from the extracellular pool of choline is now well established (21). It has been shown in the superior cervical ganglion (30) and in the hemidiaphragm (33) that an important, if not the most important, physiological source of choline used in this process originates from the hydrolysis of the released ACh; the precursor has thus to be delivered into the nerve terminals from the presynaptic membrane (3). Yamamura and Snyder (15) reported that 70% of the choline molecules penetrating in rat striatal synaptosomes through the high-affinity permease undergo the acetylation process. This value appears even underestimated as Guyenet, LeFresne, Beaujouan, and Glowinski (36), using the same preparation, observed that the rates of choline uptake and ACh synthesis were equal within the experimental errors (about 140 nmoles/g protein-min) (4). Subcellular studies of ChAc compartmentalization within nerve terminals revealed that, depending on the pH and the ionic strength of the saline medium used, enzyme molecules can bind reversibly to synaptosomal membranes (23,37,38). It is thus conceivable that the ionic properties on the internal side of the presynaptic membrane would allow the reversible binding of part of the ChAc to some specific membrane protein.

ACKNOWLEDGMENTS

The authors are very grateful to Drs. M. Israël and P. G. Guyenet for helpful discussion, and Dr. A. Kato for stimulating criticisms and for revising the manuscript. This work was supported by grants from INSERM (73.5.100.6) and DGRST (73.7.12.27).

REFERENCES

1. Mann, P. J. G., Tennenbaum, M., and Quastel, J. H. (1939): *Biochem. J.*, 33:822–835.
2. Schubert, J. Sparf, B., and Sandwall, A. (1969): *J. Neurochem.*, 16:695–700.
3. Sharkawi, M., and Schulman, M. P. (1969): *Br. J. Pharmacol.*, 36:373–379.
4. Molenaar, P. C., and Polak, R. L. (1970): *Br. J. Pharmacol.*, 46:406–417.
5. Sharkawi, M. (1972): *Br. J. Pharmacol.*, 46:473–479.
6. Browning, E. T., and Schulman, M. P. (1968): *J. Neurochem.*, 15:1391–1405.
7. Richter, J. A., and Marchbanks, R. M. (1971): *J. Neurochem.*, 18:691–703.
8. Greewal, D. S., and Quastel, J. M. (1973): *Biochem. J.*, 132:1–14.
9. Lefresne, P., Guyenet, P., and Glowinski, J. (1973): *J. Neurochem.*, 20:1083–1097.
10. Potter, L. T. (1970): In: *Handbook of Neurochemistry*, pp. 263–284, edited by A. Lajtha. Plenum Press, New York.
11. Roskoski, R. (1973): *Biochemistry*, 12:3709–3714.
12. Kaita, A. A., and Goldberg, A. M. (1969): *J. Neurochem.*, 16:1185–1191.
13. Morris, D., Maneckjee, A. A., and Hebb, C. (1971): *Biochem. J.*, 125:857–863.
14. Glover, V. A. Q., and Potter, L. G. (1971): *J. Neurochem.*, 18:571–580.
15. Yamamura, H. I., and Snyder, S. H. (1972): *Science*, 178:626–628.
16. Haga, T., and Noda, H. (1973): *Biochem. Biophys. Acta*, 291:564–575.
17. Kuhar, M. J., Sethy, V. H., Roth, R. H., and Aghajanian, G. K. (1973): *J. Neurochem.*, 20:581–593.
18. Yamamura, H. I., and Snyder, S. H. (1973): *J. Neurochem.*, 21:1355–1374.
19. Birks, R. K., and Mac Intosh, F. C. (1961): *Can. J. Biochem. Physiol.*, 39:787–827.
20. Marchbanks, R. M. (1968): *Biochem. J.*, 11:533–541.
21. Guyenet, P., Lefresne, P., Rossier, J., Beaujouan, J. C., and Glowinski, J. (1973): *Mol. Pharmacol.*, 9:630–639.
22. Whittaker, V. P., and Dowdall, M. J. (1975): In: *Cholinergic Mechanisms*, pp. 23–42, edited by P. B. Waser, Raven Press, New York.
23. Fonnum, F. (1967): *Biochem. J.*, 103:262–270.
24. Sollenberg, J., and Sorbo, B. (1970): *J. Neurochem.*, 17:201–207.
25. Gray, E. G., and Whittaker, W. P. (1962): *J. Anat.* (*Lond.*), 96:79–88.
26. Lefresne, P. (1974): Thèse de Doctorat d'Etat, Paris.
27. Fonnum, F. (1969): *Biochem. J.*, 115:465–472.
28. Johnson, M. K. (1960): *Biochem. J.*, 77:610–618.
29. Racker, E. (1950): *Biochim. Biophys. Acta*, 4:211–214.
30. Collier, B., and Mac Intosh, F. C. (1969): *Can. J. Physiol. Pharmacol.*, 47:127–135.
31. Collier, B. (1969): *J. Physiol* (*Lond.*), 205:341–352.
32. Molenaar, P. C., Polak, R. L., and Nickolson, V. J. (1973): *J. Neurochem.*, 21:667–678.
33. Potter, L. T. (1970): *J. Physiol.* (*Lond.*), 286:145–166.
34. Dunant, Y., Gautron, J., Israel, M., Lesbat, B., and Manaranche, R. (1972): *J. Neurochem.*, 19:1987–2002.
35. Dunant, Y., and Israel, M. (1975): In: *Cholinergic Mechanisms*, pp. 161–167, edited by P. G. Waser. Raven Press, New York.
36. Guyenet, P., Lefresne, P., Beaujouan, J. C., and Glowinski, J. (1975): In: *Cholinergic Mechanisms*, pp. 137–144, edited by P. G. Waser. Raven Press, New York.
37. Fonnum, F. (1968): *Biochem. J.*, 109:389–398.
38. Tuček, S. (1967): *J. Neurochem.*, 14:531–545.
39. Bertel-Meeuws, M. M., and Polak, R. L., (1968): *Br. J. Pharmacol. Chemother.*, 33:368–380.

Membranes and Disease, edited by L. Bolis, J. F. Hoffman, and A. Leaf. Raven Press, New York, © 1976.

Brain-Specific Cell-Surface Antigens

Kay L. Fields

MRC Neuroimmunology Project, Department of Zoology, University College, London WC1E 6BT, England

The precise role of the cell membrane in the normal development and later function of the nervous system is largely a matter for speculation. Despite our lack of detailed knowledge of the molecular basis of such events as cell migration, cell–cell interactions, or the basis of the specificity of neuronal connections, speculation about events taking place at the cell surface molds the experimental approach and the interpretation of many results in this field. It is clear that nerve cell function depends upon a special, electrically excitable membrane, and also upon connections with other nerve cells forming in a highly specific manner. Synapses are unique specializations of neuronal membranes, where the apparatus for the release of chemical transmitters is closely apposed to the clustered receptors of the postsynaptic cell. Other cell types in the brain also have membrane properties which reflect their function. For example, astrocytes have uptake mechanisms for neurotransmitters, and oligodendroglia produce the closely packed plasma membranes of myelin, which are essential for rapid conduction of the nerve impulse. For these reasons the proteins and properties of the surface membranes of the different cell types of the brain are of considerable interest.

Where high-affinity markers are available for surface structures, such as α-bungarotoxin or cobratoxin, which bind to the acetylcholine receptor, they have been useful for mapping, studying, and purifying the membrane receptor (1,2). However, such convenient tools as specific toxins are unusual. Specific antibody is an alternative which can be exploited in similar ways. Where available through immunization with whole cells, antibody has been used for the assay and purification of membrane proteins such as the histocompatibility (3) and θ (4) antigens. However, for most membrane proteins, neither highly enriched sources of the protein (such as the electric organ for acetylcholine receptors) nor genetically well-defined sources of specific antibody are available. New antigenic systems are needed, especially for functionally defined receptors or other surface antigens.

In immunology, the development of cell surface markers such as θ, Ly, surface immunoglobulin, and immunoglobulin Fc receptors, together with surgical techniques for the depletion of distinct subpopulations of lymphocytes, has led to the identification of the different types of cells, and to the realization that cell interactions are of great importance in the immune response (5). Very briefly, the ways in which specific markers are used by immunologists which might find direct application to the study of cells from the nervous system are (i) the recognition and identification of cell types, (ii) classification of tumors with cell type-specific antisera, (iii) depletion of a mixed population of cells by the use of antisera and complement-dependent killing, and (iv) the separation of cell populations under gentle conditions where viability is retained.

THE USE OF TUMORS FOR IMMUNIZATION—ANTIGENS ON C1300

If viable, separated cell populations from the normal nervous system were readily available, they would be the first choice for developing antisera. However, the separation of different cell types from adult or even embryonic brain is very difficult (6). Subcellular fractions can be obtained, but the analysis of complex sera is very difficult if live, defined cell populations are not readily and reproducibly available.

Some of these problems are circumvented by immunizing with cells derived from nervous system tumors. In the mouse several tumors have long been available, in particular several gliomas and a neuroblastoma, C1300. In the rat many more are easily induced with a carcinogen, ethylnitrosourea, which is highly specific for the nervous system (7). At University College we have induced about 50 such tumors, have transplanted many of them in syngeneic rats, and established cell lines from some of the tumors (8). Several other laboratories have been engaged in similar projects, and several cell lines with properties indicating neuronal differentiation (9) or an origin from glial (10) or Schwann cells (11) are available. With both rat and mouse tumors, some degree of cell type-specific differentiation was retained or could be induced in the tumor cells. An assumption for the immunological work is that these tumor cells also present on their surface some cell-type or brain-specific antigenic determinants.

Another advantage of the tumor system, is that the tumor or cell line serves as a convenient target cell for the assay of the binding of antibodies to cell surface antigens. Complement-dependent cytotoxicity or double layer techniques using radioactive or fluorescein-coupled antiimmunoglobulin have been used with live targets, and are specific for surface antigens.

Two successful antisera against mouse neuroblastoma C1300 have been described (Table 1). Martin (12) raised an antiserum with promising specificity by injecting C1300 ascites tumor cells into rats. The antigen she defined, MBA-1, was on C1300, brain, and to a lesser extent on kidney, whereas other adult tissues were negative. Neither embryonic tissues nor sperm were tested, nor is it known whether this antigen was present on other mouse neural tumors. Attempts to reproduce the antiserum have not been

TABLE 1. *Nervous system cell surface antigens*

Antigen	Antiserum	Tissue Distribution					Neural tumors[a]	Positive cells in brain	Ref.
		Brain	Kidney	Liver spleen	Embryo	Sperm			
1. MBA-1	Rat anti-C1300	++	+	0	?	?	?	?	12
2. C1300 differentiation-specific	Rabbit anti-N18	++	±[b]	0	?	?	0+/1	?	13
3. NS-1	Mouse anti-G26	++	0	0	?	0	2+/4	Glia	15
4. NS-2	Rabbit antiglioblastoma (Sato)	+	0	0	0	0	4+/8	?	17
5. Cerebellar-specific	Rabbit anticerebellar cells	++	?	0[c]	?	?	?	Cerebellar population	18
6. MBA-2	Natural antibody	++	++	0	+	?	4+/11[d]	?	19,20
7. NS-4	Rabbit anticerebellar tissue	++	0	0	+[e]	+	3+/6	Cerebellar, retinal cells	21
8. Rat Common	Mouse anti-33B	++	+	0	++	0	40+/45	?	8
9. Rat Restricted	Mouse anti-33B	0	0	0	0	0	11+/45	?	8
10. F9 teratoma	Syngeneic anti-F9	?	?	0[f]	+	+	0+/2	?	26,27
11. C teratoma stem cell	Syngeneic anti-SIKR	+	+	0	+	+	0+/3	?	28

[a] Immunizing tumor not included.
[b] Kidney had a low binding capacity for 50% of the antibody.
[c] Liver was negative.
[d] All + tumors were neuroblastomas.
[e] Only embryonic neural tissue was positive.
[f] Thymocytes were negative.

successful, perhaps related to the fact that only about 5% of the cytotoxic activity of the original, unadsorbed serum was directed against a brain-specific antigen. The predominant response of the rats was against antigens shared by the tumor with mouse liver or lymphocytes (S. Martin, *personal communication*).

Another successful antiserum against C1300 was raised in rabbits by Akeson and Herschman, who injected differentiated, process-bearing *in vitro* cells (13). They screened their sera and worked only with selected bleeds in which very little antibody was directed against general mouse tissue antigens, more bound to brain, even more to undifferentiated C1300 cells, and the maximum amount bound to differentiated C1300 cells. Most of the immunoglobulin specificity for the differentiated cells was absorbed by brain. It is of great interest to know whether normal process-bearing neurons, or glia, or both express this antigen.

A general problem with the use of tumor cells is the difficulty of determining which cell types in the normal brain have any given surface antigen. Using neuroblastoma one might hope to define neuronal-specific antigens, for example. However, it is equally probable that antigens common to all neuroectodermally derived cells are detected. The use of other neural tumors can give an indication of cell type specificity, but any one tumor may be unrepresentative or atypical, and several tumors of each cell type are not always available. A promising technique for the identification of the normal cell was developed for θ antigen by Mirsky and Thompson (14). Dissociated mouse embryo brain cells were cultured on plastic dishes, and unfixed, live cells were examined *in situ* using immunofluoresence. Cells with the appearance of neurons developed θ antigen after some days in culture, but glial cell types did not. This technique could readily be used for other brain antigens.

OTHER ANTIGENS ON MOUSE BRAIN

NS-1, nervous system antigen-1, is an antigen defined by Schachner (15). The antisera were raised by immunization of syngeneic or genetically related mice with irradiated cells of a transplantable glioma, G26. Cytotoxic antibody was directed against an antigen present in brain tissue of mice and other species. By absorption with normal tissues, Schachner showed the antigen only on central or peripheral nervous tissue, and its absence from other organs. The antigen increased roughly tenfold during the first 3 weeks after birth. Three out of four gliomas had the antigen, but neuroblastoma C1300 and all other nonneural tumors tested were negative. White matter was enriched in the antigen, cerebral cortex was low, and myelin-deficient mice had less antigen. Schachner has concluded that the antigen NS-1 is glial-specific.

Other brain antigens [MBA-1, the C1300 differentiation antigen, and the rat Common antigen (Table 1)] have been found on kidney tissue in low amounts compared to brain, roughly 10% (12,13,8).

In further work on the NS-1 antigen (16), its expression was compared with the levels of two other brain specific proteins, a phosphohydrolase and a cytoplasmic antigen, S-100. These three markers were not coordinately expressed in the cell lines and clonal tumors examined.

Schachner and her colleagues have reported another distinct brain antigen. NS-2 was defined by rabbit antisera against a murine tumor, glioblastoma (Sato) (17). The antigen was on mouse and rat brain, not correlated with myelination, not on retinal neurons, and developed only after birth. The antigen was not on fetal tissue or sperm, nor was it the same as NS-1. It was more abundant on murine glial tumors than on normal brain. The normal cell type(s) bearing the antigen was not clearly defined.

Two reports have appeared of antigens of normal brain cells which may be specific for neurons. Seeds (18) used dissociated, healed cerebellar cells for immunization and for cytotoxicity testing. He found an antigen on a small subpopulation (15%) of his cerebellar cells which was not present on other tissues, or even on other regions of the brain. It is a strong possibility that this is a marker for a neuronal population found only in the cerebellum.

The second example of an antigen which may only be on neurons is MBA-2 (19). This antigen was defined by naturally occuring antibody cytotoxic for a new teratoma-derived neuroblastoma cell line, NB-1. The cytotoxicity was absorbed by brain (of several species), kidney, and embryo, but not by other mouse tissues. The antigen was not on two other mouse neuroblastomas, but was on several human neuroblastoma cell lines. Mouse and human gliomas examined were negative (20).

Schachner, Wortham, Carter, and Chaffee (21) have described an antigen NS-4 which was on developing and adult mouse brain and on sperm. Cerebellar tissues were used for immunization, and cerebellar cells, retinal cells, and several glial tumors were all positive. It was not the same antigen as NS-1 or NS-2. It is of interest to know whether NS-4 and MBA-2 are closely related or even identical. The cell line NB-1 of Martin (19) should be tested for NS-4, and MBA-2 should be assayed on the gliomas which expressed NS-4 in order to confirm the difference in tissue distribution noted in Table 1.

In their discussion of the possible significance of antigens, such as NS-4, common to brain and sperm, Schachner et al. (21) pointed out that two neurologically abnormal mutant mice have defective sperm (22,23). One of the mutants, quaking, is defective in myelination and in sperm maturation (22). The mutant maps near the T/t and H-2 loci. Furthermore, several recessive t alleles affect the production or function of sperm (24) and neurological defects have been noted for other t mutants (22). Thus the proper development of the nervous system and the complex differentiation process resulting in mature sperm both depend upon genes which map near the histocompatibility locus. In nematodes, a high proportion of behaviorally abnormal mutants have defective sperm (S. Ward, *personal communication*).

RESTRICTED AND COMMON ANTIGENS ON RAT NEURAL TUMORS

In the rat one advantage was the large number of available tumors and cell lines, including neuronal, glial and Schwann-cell lines. Immunization of rabbits with *in vivo* tumor cells yielded no specific antisera after absorption with liver and spleen. The first sera which were specific were raised in C3H mice injected with a cell line, 33B (8). This cell line was derived from an ENU-induced central nervous system glial tumor which contained cilia, cell processes, and partial desmosomes, all of which suggest ependymal differentiation.

The anti-33B sera, after absorption with liver and spleen tissues, could be shown to be directed against two antigens. One antigen was not detected on adult or embryonic rat brain, or on any other normal tissue. It was termed the Restricted antigen. Its association with the nervous system is suggested by the finding that it was not limited to the one cell line, but was on six transplanting tumors in our collection. It was also on four cell lines from the Salk Institute, all of which were derived from ENU-induced rat brain tumors (9). The antigen-positive cell lines were not representative of the same cell type by any of several criteria (9). Two cell lines showed neuronal properties and two were nonneuronal, presumptive glial cells. Pfeiffer's cell line RN-22, from a Schwann cell tumor (11) also expressed this antigen. No nonneural cell lines or tumors were positive, including some virally transformed cells.

The Restricted antigen is of unknown significance since neither its normal function nor location has been found. Other tumor-specific antigens have been individually specific to a single tumor, or else possessed by a whole class of similar tumors. The Restricted antigen falls between these two extremes.

Also present in the C3H anti-33B sera were antibodies directed against a brain specific, surface membrane antigen, the Common antigen. It was found on brain, and much less, on kidney. Antigen expression was tenfold higher on the cell line or tumor than on brain. A majority of all our transplanting rat neural tumors were very positive for this antigen, as were the cell lines derived from the tumors. The tumors with the least antigen were the most fibroblastic, and two virus-induced rat lymphomas were negative. With one exception, all cell lines from ENU-induced neural tumors from other laboratories have been highly positive. However, the antigen was not perfectly specific for neural tissue or tumors. It was found on rat embryos of all ages tested (12–20 days), and was on nonneural parts of the embryos. Although normal embryo fibroblasts were negative, other transformed cells of stromal or hepatoma origin (XC and HTC cells) were positive. Quantitative absorption curves have shown that the nonneural transformed cells have $<5\%$ as much antigen as the neural tumor cell lines.

ARE NEUROEMBRYONIC ANTIGENS A SIGNIFICANT OCCURRENCE?

So far, we have discussed three brain antigens, MBA-2, NS-4, and the rat Common antigen, which were found on embryonic tissue. Whereas we have been searching for antigens as useful markers, we are also interested in cell surface molecules which might play a functional role in nervous system development. A most interesting, if still hypothetical, class of such molecules would be those important for cell–cell interactions such as adhesion of similar cells, or recognition events between dissimilar cells. Might not the same pairs of interacting, complementary signal and receptor molecules function repeatedly during development? Reuse of a set of signal/receptor pairs in different contexts would imply that the response of the receptor-bearing cell would be largely determined by its previous developmental history, not at all by the structure of the receptor itself. This is simply a variation of the concept of common signaling mechanisms (cyclic nucleotide, calcium, etc.) now thought to be involved in the triggering of differentiating as well as highly differentiated cells. While the reuse of cell surface signal/receptor pairs would represent an efficient exploitation of such pairs, quite specific mechanisms must be postulated for the control of their synthesis. A relatively modest number of such pairs might be utilized for the majority of all cell–cell interactions. Bodmer (25) has proposed that a major function of the histocompatibility region is to provide "differentiation antigens" or complementary "recognizers" or both, which interact to provide cell to cell recognition during development and morphogenesis. If MBA-2 or the rat Common antigen are "differentiation antigens" in this sense, their presence in embryonic tissues might be due to the reuse in brain of a whole range of cell–cell recognition pairs.

Some evidence of the persistence in the nervous system of a surface antigen present, and possibly essential for early embryonic development has just been found for an antigen in the teratoma system. Artzt and her colleagues (26,27) described an antigen on F9 teratoma cells which is specific for the teratoma stem cells, early embryo cells, and sperm. Sperm from mice heterozygous for the t^{12} allele had only half as much antigen as did wild-type sperm, so that the antigen could be the direct product of the wild-type t^{+} allele. The mutant alleles at the T/t locus cause sharply defined and unique errors in early embryonic and neuroectodermal development. Finding cell surface antigenic changes associated with this morphogenetically important T/t locus supports the hypothesis that cell–cell surface interactions are essential during development (24).

Stern, Martin, and Evans (28) have recently raised an antiserum against

SIKR, another pluripotent, teratoma cell line, and found that it detected an antigen, C, similar to the F9 antigen, in that it was specific for teratoma stem cells, early embryo cells and sperm. However, to their surprise, they found the antigen on both brain and kidney, but not on liver or thymus. The simplest explanation for the persistence of this antigen in brain and kidney is that it plays a similar role there, as it does much earlier in development. Are other, T-locus-associated, antigens found on sperm (24) also to be found on brain or kidney?

Some obvious questions arise from the comparison of the properties of the stem cell antigens F9 and C with MBA-2 and NS-4. (Table 1). NS-4 would seem different from MBA-2 since it was not found on kidney, but could the other three antigens all be very similar or identical? The crucial experiments are obvious: to determine whether the F9 antigen is also in brain, and to ask if Martin's MBA-2-positive neuroblastoma cell line expresses the teratoma stem cell antigens. Structural studies on MBA-2 would also be very interesting, even if it is antigenically distinct from the teratoma stem cell antigens, since the F9 antigen has recently been shown to be structurally similar to the histocompatibility antigen, H-2 (29).

SUMMARY

From this discussion, it is clear that several antigenic systems have been defined for cell surface components of the brain cells of the mouse and rat. Several of these may be cell-type-specific, and should be the starting point for the use of surface markers for brain cell separation and subsequent *in vitro* studies.

The finding of four examples of neuroembryonic antigens among the recently defined brain antigens suggests that this class of cell-surface antigen may be quite prominent and merit more intensive study.

REFERENCES

1. Meunier, J.-C., Sealock, R., and Changeux, J.-P. (1974): *Eur. J. Biochem.*, 45:371.
2. Brockes, J. P., and Hall, Z. W. (1975): *Proc. Natl. Acad. Sci. USA*, 72:1368.
3. Cresswell, P., Turner, M. J., and Strominger, J. L. (1973): *Proc. Natl. Acad. Sci. USA*, 70:1603.
4. Vitetta, E. S., Boyse, E. A., and Uhr, J. W. (1973): *Eur. J. Immunol.*, 3:446.
5. Greaves, M. F., Owen, J. J. T., and Raff, M. C. (1973): *T and B Lymphocytes: Origins, Properties, and Roles in Immune Responses.* Excerpta Medica, Amsterdam.
6. Varon, J., and Raiborn, C. W. (1969): *Brain Res.*, 12:180.
7. Wechsler, W., Kleihues, P., Matsumoto, S., Zulch, K. J., Ivankovic, S., Preussman, R., and Druckrey, H. (1969): *Ann. NY Acad. Sci.*, 159:360.
8. Fields, K. L., Gosling, C., Megson, M., and Stern, P. L. (1975): *Proc. Natl. Acad. Sci. USA*, 72:1296.
9. Schubert, D., Heinemann, S., Carlisle, W., Tarikas, H., Kimes, B., Patrick, J., Steinbach, J. H., Culp, W., and Brandt, B. L. (1974): *Nature*, 249:224.
10. Benda, P., Someda, K., Messer, J., and Sweet, W. H. (1971): *J. Neurosurg.*, 34:310.
11. Pfeiffer, S. E., and Wechsler, W. (1972): *Proc. Natl. Acad. Sci. USA*, 69:2885.

12. Martin, S. E. (1974): *Nature,* 249:71.
13. Akeson, R., and Herschman, H. R. (1974): *Nature,* 249:620.
14. Mirsky, R., and Thompson, E. (1975): *Cell,* 4:95.
15. Schachner, M. (1974): *Proc. Natl. Acad. Sci. USA,* 71:1795.
16. Sundarraj, N., Schachner, M., and Pfeiffer, S. E. (1975): *Proc. Natl. Acad. Sci. USA,* 72:1927.
17. Schachner, M., and Carnow, T. B. (1975): *Brain Res.,* 88:394.
18. Seeds, N. K. (1974): *J. Cell Biol.,* 63:307.
18a. Seeds, N. K. (1975): *Proc. Natl. Acad. Sci. USA,* 72:4110.
19. Martin, S. E., and Martin, W. J. (1975): *Proc. Natl. Acad. Sci. USA,* 72:1036.
20. Martin, S. E., and Martin, W. J. (1975): *Cancer Res.,* 35:2609.
21. Schachner, M., Wortham, K. A., Carter, L. D. and Chaffee, J. K. (1975): *Dev. Biol.,* 44:313.
22. Bennett, W. I., Gall, A. M., Southard, J. L., and Sidman, R. L. (1971): *Biol. Reprod.,* 5:30.
23. Mullen, R., Eicher, E., and Sidman, R. L. (personal communication cited in reference 21) Also Sidman, R. L., this volume.
24. Bennett, D., Boyse, E. A., and Old, L. J. (1972): In: *Cell Interactions* (Proceedings of the 3rd Lepetit Colloquium; edited by L. G. Silvestri), North-Holland, Amsterdam, p. 247.
25. Bodmer, W. F. (1972): *Nature,* 237:139.
26. Artzt, K., Bennett, D., and Jacob, F. (1974): *Proc. Natl. Acad. Sci. USA,* 71:811.
27. Artzt, K., Dubois, P., Bennett, D., Condamine, H., Babinet, C., and Jacob, F. (1973): *Proc. Natl. Acad. Sci. USA,* 70:2988.
28. Stern, P. L., Martin, G. R., and Evans, M. J. (1975): *Cell,* 6:455.
29. Vitetta, E. S., Artzt, K., Bennett, D., Boyse, E. A., and Jacob, F. (1975): *Proc. Natl. Acad. Sci. USA,* 72:3215.

Membranes and Disease, edited by L. Bolis, J. F. Hoffman, and A. Leaf. Raven Press, New York, © 1976.

Cell Surface Properties and the Expression of Inherited Brain Diseases in Mice

Richard L. Sidman

Department of Neuropathology, Harvard Medical School and Department of Neuroscience, Children's Hospital Medical Center, Boston, Massachusetts 02115

The organization of the nervous system seen from the conventional teleologic view seems particularly appropriate to allow neurons to interact with one another as well as with glial cells and with cells of other organs. These interactions are mediated at cell surfaces, so that elucidation of the properties and distribution of surface components is fast becoming the common objective of neuroscientists, whether they use techniques involving microelectrodes, electron microscopy, gel electrophoresis, antibodies, or whatever. In fact, the willingness and competence of its practitioners to use several previously disparate techniques in concert has become the denominating feature of contemporary neuroscience.

Teleology, however, poses traps. We may move further if we reverse the idea that the nervous system is organized to allow surface-mediated cell interactions, in favor of the view that the properties and geometry of surface components determine the very organization itself. From the latter perspective, a central approach to an understanding of how the nervous system works in health and disease is to examine surface-mediated cell behavior and its molecular basis during the development of the nervous system.

This tactic is difficult in practice, for even the immature nervous system is so heterogeneous as to render conventional analytical methods almost useless. Experimental manipulation of the system becomes necessary, but this in turn is far from easy, especially in relatively inaccessible mammalian species, where neither physical insults such as X-irradiation, direct surgical interference, drugs, nor viruses have as yet provided a selective and reproducible general approach.

Mutations provide another alternative, not widely exploited as yet. In mice, more than 500 mutations are known, and almost 300 of these have been mapped on 18 of the 20 chromosome pairs. About 25–33% of the mutations cause major phenotypic abnormalities of behavior, and a number of these have been shown to perturb the development or maintenance of selected classes of nervous system cells (1,2). Whereas the analysis of these mutant phenotypes has not yet led to molecular characterization of any gene product located at cell surfaces with a controlling role in development, the data obtained thus far do indicate clearly that cell–cell interactions mediate crucial aspects of differentiation in the nervous system, and force us at least to begin reformulating problems of neural development in terms of molecular signals, surface receptors, and temporal changes in their distribution.

CELL INTERACTIONS IN THE CONTROL OF CELL POSITION

A given type of neuron occupies a remarkably constant position from individual to individual within a species, and with some exceptions, is even consistent, relative to other types of neurons, between species. Attainment of position appears to be under tight control, presumably, one might think, because the address of a neuron should have something to do with its connections.

A principle that has emerged in the past two decades, mainly based on autoradiographic plotting of temporal changes in the positions of cell nuclei containing labeled DNA after incorporation of ^{3}H-thymidine, is that in general, cells are generated in the developing nervous system at different sites from where they will reside in the mature organ (3). Thus, development involves a remarkable cellular traffic, with cells changing dramatically in shape, rearranging their internal organelles, migrating to initiate or extend the population of new regions, and establishing both relations with new neighbors and new relations with old ones (4). We may ask, is the acquisition of a proper "address" in the developing nervous system an intrinsic property controlled by the cell's own active genome, or is it dependent on cell interactions?

A pertinent test case, described long ago by Ramon y Cajal (5) and confirmed more recently by autoradiography (3), is the genesis of granule cell neurons on the external surface of the cerebellum, followed by translocation of their somas inward past Purkinje cell dendrites and soma, while simultaneously retaining a T-shaped outward-directed process that differentiates into the axon. In electron micrographs of chick embryo cerebellum, Mugnaini and Forstronen (6) noted a close contact of migrating granule cell neurons with Bergmann glial fibers, the radially oriented cytoplasmic processes of the specialized astrocytes of the cerebellar cortex. In parallel Golgi and electron microscopic studies of timed fetal monkey cerebella, Rakic (7) then observed that this contact relationship between young granule cell neuron and Bergmann glial process is invariant and suggested that the surface of the radially oriented glial cell serves to guide the young neuron as it reshapes itself and translocates its soma inward from near the cerebellar surface to the granular layer.

The weaver mutant mouse provides an experimental test of this hypothesis. The mutation is an autosomal recessive, expressed behaviorally in terms of ataxic gait, reduced body size and shortened life span (1). The mature homozygous mutant animal virtually lacks granule cell neurons in its cerebellar cortex, the granule cells having died near the outer surface of the cerebellum at an early postmitotic stage rather than migrating inward (8). The question of whether the granule cells fail to migrate because they are already lethally compromised, or die because they cannot migrate, was answered in favor of the latter alternative when Rezai and Yoon (9), confirmed by Rakic and Sidman (10), showed that in mice heterozygous for the weaver mutation, granule cell migration is significantly slowed, although most granule cells do eventually reach the granular layer and no behavioral abnormality has been recognized in heterozygotes. Our further observations (11) that many Bergmann fibers in heterozygous and homozygous mutant animals display enlarged and irregular profiles, are unusually electronlucent, and contain enlarged mitochondria in the time period when granule cells should be migrating, suggest that we may have concentrated too much on the dying granule cells as direct targets of the weaver genetic locus and should shift some of our attention to the contiguous Bergmann glial processes. Sotelo and Changeux (12) have made essentially similar observations, but prefer the interpretation that the granule cells are the primarily affected elements and that the glial cells are merely delayed in maturation. Irrespective of how this comes to be resolved on the basis of further experimentation, analysis of the mutation to date emphasizes the point that granule cell neurons en route to their final positions do not move in a vacuum, and that interactions at the contiguous surfaces of granule cell neuron and Bergmann glial fiber likely play a role in the complex "addressing" mechanism.

A second test case is the reeler mutation, another autosomal recessive with an unusual phenotypic expression. In several parts of the cerebellar cortex and most regions of the cerebral cortex, the usual laminar arrangement of neuron somas is disturbed, with a tendency for neuronal classes which normally lie deep in the cortex to lie superficially in reeler, and classes which should lie superficially to lie deep (8,13). Since it was recognized that in terms of population statistics, the positioning of neuron somas in cortices bears a systematic relationship to the time of neuron genesis at or near the ventricular surface, with those neurons that form early in embryonic development coming to occupy deep positions in the mature cortex and neurons that form progressively later moving outward past their predecessors and assuming progressively more superficial positions (14), the possibility was considered that the reeler locus might influence the basic schedule of neuron genesis. This was ruled out by ^{3}H-thymidine autoradiographic evidence that each class of neurons in reeler mice is generated at the normal time and in the normal place, and migrates outward at about the normal rate (8,15,16). Furthermore, a detailed anatomical analysis has revealed that each cerebral cortical region lies in the correct position and contains its characteristic types of neurons, from which one may conclude that the vector of migration outward from ventricular surface to cortex is undisturbed (17). A simple inversion of cortical layers was ruled out as the basic expression of the reeler defect by evidence that in cerebral neocortical areas where some classes of neurons do take up reversed laminar positions, others nonetheless find relatively normal positions (15).

These data suggest strongly that some aspect of cell-cell recognition and interaction peculiar to the acquisition of correct position in cortices is at fault in the reeler mutant, and presumably is under control of the wild-type allele at that genetic locus in normal mice. The critical event probably occurs no later than the time a given neuron is just entering the cortex. However, the alternative possibility remains tenable, that the reeler locus controls some intrinsic "addressing" property of the earliest cortical neurons, which in turn dictate the positions of their followers.

A new approach, the synthesis of experimental chimeras between homozygous reeler and normal embryos, gives a conclusive answer in favor of extrinsic control. Mullen (18) has fused 8-cell reeler and wild-type embryos *in vitro* and reimplanted them into "incubator" mothers, a technique of great power in developmental biology (19). One would expect the mouse resulting from this experimental maneuver to contain a mixture of cells of the two genotypes in most or all of its organs, including the brain. The chimeras produced by Mullen to date show either wild-type or reeler behavior, rather than some intermediate, and histologically show almost pure normal or reeler cortices throughout the brain—even when some other organs show a more equal mix as evidenced by a pigmentation marker. Either the nervous system has a very different clonal history from most organs, or there is differential survival of wild type versus reeler cells in different cases, or there is a more typical mix but the cortical cells are not displaying their expected phenotypes. An independent genetic marker for brain cells was needed to resolve this.

In one critical chimera, a glucuronidase histochemical marker provided that independent evidence. This marker, demonstrated by a histochemical method refined by Feder (*personal communication;* see also ref. 2), is suitable for many cell types including Purkinje cells. A reeler, low glucuronidase embryo was fused with an embryo that was wild-type at both loci, i.e., nonreeler and high glucuronidase. The resultant mouse, proved by progeny testing to be a reeler chimera, behaved normally and had a brain of normal structure except for several patches of reeler morphology in the cerebellar cortex, where Purkinje neuron somas were positioned deep to granule cell somas, in reverse of the normal pattern. The essential point is that the "reeler" patches contained Purkinje cells of both high and low glucuronidase types, as did the normal cortical patches. This is clear evidence that neuron position is strongly influenced by factors extrinsic to the neuron (18; Mullen and Sidman, *unpublished observations*).

Whatever these extrinsic factors may be, they are at play very early in development, not later than the time when the earliest Purkinje neurons are migrating outward to form the anlage of the cerebellar cortex, for already at embryonic day 14, Purkinje cells en route to the presumptive cortex lie deep to early cerebellar afferents in reeler, while their counterparts lie superficial to the corresponding afferent fiber plexus in normals (Sidman and Caviness, *unpublished observations*). Cell surface relationships among early embryonic brain cells must now become the focus of our attention.

CELL INTERACTIONS IN SYNAPTOGENESIS

The major principles that have been suggested to account for patterns of connectivity in the nervous system are 1) random formation of connections with subsequent selection based, to choose two among several possibilities, on strengths of adhesion between pre- and postsynaptic elements or on functional activity; 2) initial orderliness based on timing mechanisms; 3) orderliness based on recognition of unique molecular tags that specify the relative position of each cellular unit; and 4) combinations of the above, perhaps different for different parts of the nervous system. This is not the place to review the evidence for or against each of these principles, but simply to point out a few constraints imposed by data obtained in mutant mice.

In the lateral parts of the cerebellum of homozygous mutant weaver mice and throughout the cerebellar cortex in heterozygotes, some granule cell neurons are not only delayed in their rate of migration inward, but remain permanently in ectopic positions in the molecular layer. Do such cells get innervated, and by what? Remarkably, their usual inputs, the terminals of mossy fiber axons, find them by also attaining an abnormal position in the molecular layer where they engage in normal-appearing synapse formation (20). Neither the time nor the place is normal, but a highly specific synaptic association is generated nonetheless. Further, even in areas where virtually every granule cell neuron has disappeared, surprisingly few synapses are seen between cells other than those normally forming synapses with each other (12,20). Despite the myriad alternative choices, a high degree of specificity is retained.

Some of the most dramatic examples of synaptic specificity in the face of abnormal cell addresses are found in the reeler mutant. By electron microscopic criteria, all the classes of normal synaptic relationships and virtually no novel relationships were encountered in the cerebellar cortex even in areas where cell positions were most abnormal (21). By morphological (22,23) and physiological (24) criteria, the major correct connections are made in the primary visual cortex of the reeler mouse, despite the severe abnormalities in neuronal laminar positions. Afferent axons recognize their target neurons, or are recognized by them, despite the apparent handicap of incorrect addresses. Either the recognition is made prior to the acquisition of addresses, i.e., during the stage of cortical neuron migration, or the specificity tags are recognizable virtually independent of absolute and even relative cell position.

A final example will be drawn from the initial study of a new mutant mouse named Purkinje cell degeneration, *pcd* (25). In this autosomal recessive disorder, Purkinje cells degenerate rapidly in the 4th week after birth and more slowly for several weeks thereafter, until no more than a fraction of one percent of the total Purkinje population remains by the 7th week. Loss is selective for Purkinje neurons among all cerebellar cell types, but degeneration of photoreceptor cells in the retina begins at about the same time and procedes slowly over many months, and there is a loss of many mitral neurons in the olfactory bulb at relatively advanced ages. Male *pcd* mice are sterile.

Lee and Shelanski (preliminary data) have solubilized membrane proteins from normal and *pcd* mutant cerebellum and have separated them by slab gel electrophoresis. A band at about 60,000 molecular weight is missing in *pcd* and is markedly reduced in another mutant named nervous, which loses about 85–90% of its Purkinje cells (26), but the band is present in all other normal and mutant cerebella examined. A much higher molecular weight protein also has been described, and likewise is thought to be a Purkinje cell membrane component on the basis of differential studies in normal and mutant mice (27).

The *pcd* cerebellum would seem to be an ideal site for examination of synaptic specificity at combined chemical and morphological levels, since the Purkinje cell axon serves as the sole pathway for transmitting information out of the cerebellar cortex and its synaptic terminals are known to use gamma aminobutyric acid (GABA) as their neurotransmitter. The analysis to date suggests that the specificity mechanisms at synapses may require definition not merely at the level of cell-cell recognition, but at a much higher level of resolution.

Tarlov et al. (28) have found that despite the loss of virtually all Purkinje cells in 60-day-old mutants, the net cerebellar content of GABA, its synthesis *in vitro* from labeled glutamate, and the activity of glutamic acid decarboxylase (GAD, the biosynthetic enzyme) are all normal. Most data are also normal when the deep cerebellar nuclei, the synaptic target area of the Purkinje cells, are dissected out and analyzed separately from the cortex.

These data suggested that some synaptic reorganization had taken place, with other GABA-utilizing axons replacing the lost Purkinje contingent, and prompted an electron microscopic study (Sidman, O'Gorman and Tarlov, unpublished). Purkinje cell axons were found to form more than 80% of the synaptic terminals on the somas of neurons in the lateral nucleus (the mouse's equivalent of the human dentate nucleus, one of the deep cerebellar nuclei). In 60-day-old *pcd* mice these axosomatic terminals had disappeared and, to our surprise, had not been replaced by other synapses, only by a glial coating. This is an impressive example of apparently inviolable synaptic specificity, but it leaves the chemical data unaccounted for. One possibility to be explored is that synaptic reorganization has occurred, with axonal sprouting by a local circuit GABA interneuron, but that is limited to the dendrites, as opposed to the somas, of the deep cerebellar neurons. Sprouting and new synapse formation on dendrites in response to removal of one among several inputs has been described in other systems (e.g., ref. 29). Among the implications are that recognition markers may be differentially distributed along the surface of a given neuron, not merely between neurons, and that when disparate classes of axons lie in proximity, they may compete with each other for synaptic space according to some hierarchy mechanism (2).

To look ahead, one can forsee the development of antisera against surface determinants of nervous system cells (as reviewed by Field, *this volume*), the use of labeled antisera and labeled ligands to localize surface components, and the exploitation of such reagents in biologically relevant tissue culture systems to perturb surface-mediated cell interactions. Some of these approaches have proved fruitful in the study of other cell systems, and are becoming relevant in neuroscience as the questions in this complex field are coming to be more sharply defined.

REFERENCES

1. Sidman, R. L., Green, M. C., and Appel, S. H. (1965): *Catalogue of the Neurological Mutants of the Mouse.* Harvard Univ. Press, Cambridge, Mass.
2. Sidman, R. L. (1974): In: *The Cell Surface in Development,* edited by A. A. Moscona, pp. 221–253. Wiley, New York.
3. Sidman, R. L. (1970): In: *Contemporary Research Methods in Neuroanatomy,* edited by W. J. Nauta and S. O. E. Ebbesson, pp. 252–274. Springer-Verlag, New York.
4. Sidman, R. L., and Rakic, P. (1973): *Brain Res.,* 62:1.
5. Ramon y Cajal, S. (1960): In: *Studies on Vertebrate Neurogenesis,* translated by L. Guth, p. 325. Charles C Thomas, Springfield, Ill.
6. Mugnaini, E., and Forstronen, P. F. (1967): *Z. Zellforsch.,* 77:115.
7. Rakic, P. (1971): *J. Comp. Neurol.,* 141:283.
8. Sidman, R. L. (1968): In: *Physiological and Biochemical Aspects of Nervous Integration,* edited by I. D. Carlson, pp. 163–193. Prentice-Hall, Englewood Cliffs, N.J.
9. Rezai, Z., and Yoon, C. H. (1972): *Dev. Biol.,* 29:17.
10. Rakic, P., and Sidman, R. L. (1973): *Proc. Natl. Acad. Sci. USA,* 70:240.
11. Rakic, P., and Sidman, R. L. (1973): *J. Comp. Neurol.,* 152:103.
12. Sotelo, C., and Changeux, J. P. (1974): *Brain Res.,* 77:484.
13. Sidman, R. L. (1972): In: *3rd Lepetit Colloquium on Cell Interaction,* edited by L. G. Silvestri, pp. 1–13. North-Holland, Amsterdam.
14. Angevine, J. B., and Sidman, R. L. (1961): *Nature,* 192:766.
15. Caviness, V. S., and Sidman, R. L. (1973): *J. Comp. Neurol.,* 148:141.
16. Caviness, V. S. (1973): *J. Comp. Neurol.,* 151:113.
17. Caviness, V. S., and Sidman, R. L. (1973): *J. Comp. Neurol.,* 147:235.
18. Mullen, R. J. (1975): *Genetics,* 80:56.
19. Mintz, B. (1975): *Annu. Rev. Genet.,* 8:56.
20. Rakic, P., and Sidman, R. L. (1973): *J. Comp. Neurol.,* 152:133.

21. Rakic, P., and Sidman, R. L. (1972): *J. Neuropathol. Exp. Neurol.,* 31:192.
22. Caviness, V. S. (1975): *Neurosci. Abstr.,* 1:101.
23. Caviness, V. S. (1976): *Brain Res., in press.*
24. Dräger, U. (1976): *Brain Res., in press.*
25. Mullen, R. J., Sidman, R. L., and Eicher, E. (1976): *Proc. Natl. Acad. Sci.,* 73:208–212.
26. Sidman, R. L., and Green, M. C. (1970): In: *Les mutants pathologiques chez l'animal. Leur interêt pour la recherche biomédicale,* edited by M. Sabourdy, p. 69. CNRS, Paris.
27. Mallet, J., Huchet, M., Pougeois, R., and Changeux, J.-P. *FEBS Lett.,* 52:216.
28. Tarlov, S. R., Beart, P. M., and Sidman, R. L. (1976): *ICN-UCLA Winter Conference on Neurobiology, Squaw Valley, Calif., in press.*
29. Raisman, G., and Field, P. M. (1973): *Brain Res.,* 50:1.

Author Index*

A

B

*Numbers in parentheses are reference numbers.

C

D

E

Mc

M

Subject Index